T5-AFC-731

Soil Pollution

Springer
Berlin
Heidelberg
New York
Barcelona
Budapest
Hong Kong
London
Milan
Paris
Santa Clara
Singapore
Tokyo

Part I
The Interacting Materials

CHAPTER 1

The Soil as a Porous Medium

The soil is the upper layer of the unsaturated zone of the earth. Soils are very diverse in composition and behavior. The solid phase consists of mineral particles of various sizes and shapes and organic matter in various stages of degradation. Plant roots and the living soil population complete the system.

In nature, soils are heterogeneous assemblies of materials, forming porous media. The open boundaries between the solid, liquid and gaseous phases lead to a pattern of continuously changing processes of chemical and biological origin, leading to transient soil properties. The porosity of the soil system is controlled mainly by the association of its mineral and organic parts, soil water having a strong effect. However, the reactivity between the solid and liquid phases could, with time, affect even the stability of the porous medium itself and, consequently, change the open boundaries of the reactive phases.

In most natural soils the solid particles tend to be molded into aggregates or peds, either by a shrink-swell phenomenon under wetting and drying-freezing conditions, or by biologically induced molding, due to soil animals, plant roots, and fungi. This situation again affects the porosity of the soil system, with implications for the transport of water, solutes, nonaqueous liquids, and suspended particles in the unsaturated zone from the land surface to the groundwater.

Since both the soil components and the binding agents govern the porous-aggregation status of the soils, a large spatial heterogeneity is found in nature. Figure 1.1 (Tisdale and Oades 1982) illustrates soil aggregation and soil porosity as affected by the soil components and the binding agents. It might be observed that the ratio between open and closed pores is strongly affected by the agents of formation. A functional classification of the soil pores, given by Greenland (1977), may give a general idea of the effect of soil pore dimensions on the water and solute status in the soil medium (Table 1.1).

In general, soil porosity is the limiting factor in defining the ratio between the solid, aqueous and gaseous phases of the soil medium, the water-air ratio at a given time being controlled by the amount of liquid in the system.

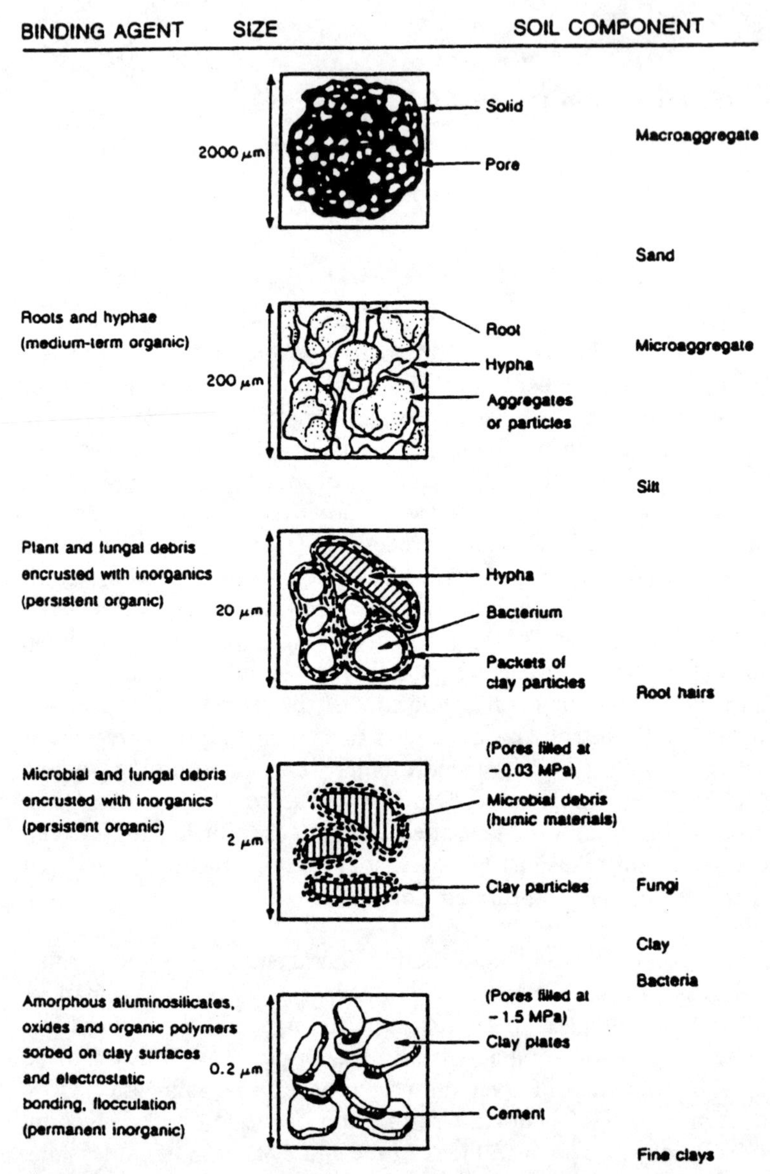

Fig. 1.1. Model of a soil aggregate organization. (Paul and Clark 1989, as adapted from Tisdale and Oades 1982)

1.1 The Solid Phase

From the point of view of potential interactions with various pollutants, the constituents of the soil solid phase should be grouped according to their sur-

Table 1.1. A functional classification of soil pores. (Greenland 1977)

Name	Function	Equivalent cylindral diameter μm
Transmission pores	Air movement and drainage of excess water	> 50
Storage pores	Retention of water against gravity and release to plant roots	0.5–50
Residual pores	Retention and diffusion of ions in solution	< 0.5
Bonding spaces	Support major forces between soil particles	< 0.005

face area. The fate of pollutants is affected by all the components of the soil solid phase. The soil constituents with low surface area could, however, mainly affect the transport of the pollutants as solutes, as immiscible with water liquids, or as vapors. The soil solid phase can also indirectly induce the degradation of the organic pollutants in the soil medium, through its effects on the water/air ratio in the system and, consequently, on the biological activity of the soil. The group of constituents with high surface area controls, besides the transport of pollutants, their retention, and release on and from the soil surface, as well as their surface-induced chemical degradation. It is, therefore, natural that when dealing with the topic of soil pollution, emphasis should be placed on the soil constituents characterized by a large surface area – and these are the clays and clay-organic complexes.

Since our book is addressed not only to soil scientists, but also to a large body of scientists, engineers, and technicians involved in soil pollution problems, we consider it necessary to include a short description of the reactive constituents of the soil.

1.1.1 Clay Materials

The clay materials, which comprise the smallest particles in the soil, are defined as the fraction with particles smaller than a nominal diameter of 2 μm. Many clay minerals have layer structures in which the atoms within a layer are strongly bound to each other, the binding between layers being weaker. Due to this situation, each layer can behave as an independent structural unit. We can find several types of layer arrangements which, however, do not greatly differ in free energy. The most common inorganic structural units to be found in soil clays are the silica tetrahedron SiO_4^{4-} and the octahedral complex MX_6^{m-6b}, formed of a metal unit M^{m+} and six anions X^{b-}. Figure 1.2 shows sheet structures formed by the polymerization of the above two structural units.

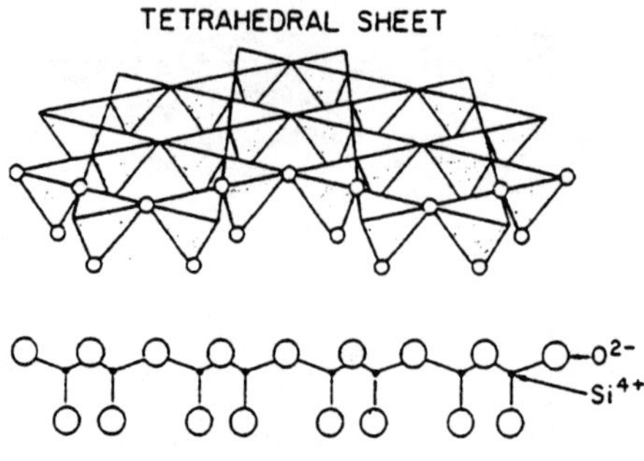

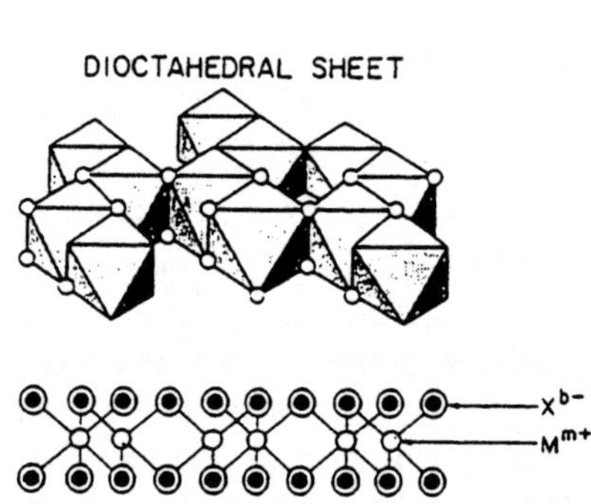

Fig. 1.2. Three layer types of phyllosilicate structure in soil clays. (Sposito 1984)

Brown et al. (1978) emphasized that the architecture of the silicate layer is due simply to the fact that each SiO_4 coordination group shares oxygen atoms with three neighboring SiO_4 groups, to form rings containing six Si and six 0 atoms, each ring being joined to a neighboring ring through shared oxygen atoms. Brown et al. (1978) showed that the other main structural element in layer silicate is an octahedral sheet that contains cations in MO_6 coordination between two planes of oxygen atoms. They summarized the types of octahedral sheets as follows:

"The anion groups coordinated to the M cations are $(OH)_6$, $(OH)_4O_2$ or $(OH)_2O_4$, depending on the structural class of the layer silicate, and as the cation can occupy either four or six of the octahedral interstices, there are six possible types of octahedral sheet, with the following compositions:

Dioctahedral: $Y_2(OH)_6$; $Y_2(OH)_4O_2$; $Y_2(OH)_2O_4$
Trioctahedral: $Y_3(OH)_6$; $Y_3(OH)_4O_2$; $Y_3(OH)_2O_4$

(a) If all the anion groups are OH, the sheet is complete without further coordination of the oxygens; such sheets occur singly, alternating with silicate layers, in the chlorites, and comprise the only structural units in hydroxide minerals such as brucite, $Mg(OH)_2$ and gibbsite, $Al(OH)_3$; they are known as hydroxide sheets.
(b) Octahedral sheets of composition $(OH)_4O_2$ are present in the kaolinite and serpentine group minerals. One plane of groups is entirely (OH), the other has the composition $(OH)O_2$ and the oxygen atoms are shared with tetrahedral Si atoms, forming the apical oxygen atoms of tetrahedral sheet superimposed on the octahedral sheet; this junction represents the ideal structure of the clay mineral kaolinite. Each layer in this group of minerals contains one tetrahedral sheet and one octahedral sheet and the minerals are termed 1:1 layer silicates.

(c) Octahedral sheets of composition $(OH)_2O_4$ occur in the mica, vermiculite, smectite, pyrophyllite, and talc minerals. In these, both planes have the composition $(OH)O_2$ and tetrahedral sheets are superimposed on both sides of the octahedral sheet, one tetrahedral sheet being inverted with respect to the other. The layer unit consists, therefore, of two outer tetrahedral sheets and an inner octahedral sheet, and minerals containing such layers are called 2:1 layer silicates. Because the oxygen atoms in the upper plane of the octahedral sheet are displaced relative to those in the lower plane (by the 'octahedral stagger'), the Si_6O_6 rings of the upper tetrahedral sheet are also displaced with respect to those of the lower tetrahedral sheet.

(d) The chlorites contain two types of layer, a hydroxide sheet alternating with a 2:1 silicate layer, and they have sometimes been referred to as 2:1:1 or 2:2 minerals.

From these structural considerations, it can be seen that the layer silicates can be represented by ions in tetrahedral and octahedral coordination and that the numbers of such ions bear a relatively simple relationship to the oxygen and hydroxyl construction of the mineral."

In extreme cases, a great number of isomorphic substitutions leads to the formation of amorphic compounds classified under the name of "allophane" substances. These materials are mainly 1:1 phyllosilicates with defects containing Al in both tetrahedral and octahedral sheets. Sometimes, they can also exhibit a tubular morphology (e.g., imogolite).

1.1.2 Minerals Other Than Clays

In addition to the clay minerals (i.e., layer silicates), the clay fraction of the soils (<2 μm) could contain a variety of minerals, e.g., oxide minerals, calcium carbonates, or calcium sulfates, etc. Summarizing the character of the soil metal oxides, Gilkes (1990) states that iron oxides (hematite α-Fe_2O_3, maghemite β-Fe_2O_3, goethite α-FeOOH, lepidocrocite β-FeOOH, etc.) are common constituents of soils, with crystals which vary greatly in size, shape, and surface morphology. The surface of iron oxides in soils is often hydroxylated, either structurally or through hydration of the Fe atoms. The crystals of the aluminum oxides that commonly occur in soils [gibbsite $Al(OH)_3$, boehmite β-AlOOH] are also small, but are often larger than the associated iron oxides. Other oxide minerals are generally less abundant than the Fe and Al oxides but, because of their very small crystal sizes and consequently large surface area, they may significantly influence soil properties. For example, the various Mn oxides that are present in some soils can occur as very small ($\sim$10 nm) structurally disordered crystals. Similarly, titanium oxides in soils (rutile, anatase, TiO_2) and even the rare pyrogenic soil mineral corundum (α-Al_2O_3) occur within the clay fraction as approximately 30-nm crystals. The ability of Fe and some other metal ions to undergo redox reactions further increases the role of the metal oxide in the activity of the soil solid phase surface.

Many soils formed from the appropriate parent materials contain significant quantities of relatively high-surface, soluble calcium carbonate ($CaCO_3$) or calcium sulfate ($CaSO_4$). Some agricultural soils may contain more than 50%

Table 1.2. Definitions of soil organic matter components. (Stevenson 1994)

Term	Definition
Litter	Macroorganic matter (e.g., plant residues) that lies on the soil surface
Light fraction	Undecayed plant and animal tissues and their partial decomposition products that occur within the soil proper and that can be recovered by flotation with a liquid of high density
Soil biomass	Organic matter present as live microbial tissue
Humus	Total of the organic compounds in soil exclusive of undecayed plant and animal tissues, their "partial decomposition" products, and the soil biomass
Soil organic matter	Same as humus
Humic substances	A series of relatively high-molecular-weight, yellow to black colored substances formed by secondary synthesis reactions. The term is used as a generic name to describe the colored material or its fractions obtained on the basis of solubility characteristics. These materials are distinctive to the soil (or sediment) environment in that they are dissimilar to the biopolymers of microorganisms and higher plants (including lignin)
Nonhumic substances	Compounds belonging to known classes of biochemistry, such as amino acids, carbohydrates, fats, waxes, resins, organic acids, etc. Humus probably contains most, if not all, of the biochemical compounds synthesized by living organisms
Humin	The alkali-insoluble fraction of soil organic matter or humus
Humic acid	The dark-colored organic material that can be extracted from soil by dilute alkali and other reagents and that is insoluble in dilute acid
Hymatomelanic acid	Alcohol-soluble portion of humic acid
Fulvic acid fraction	Fraction of soil organic matter that is soluble in both alkali and acid
Generic fulvic acid	Pigmented material in the fulvic acid fraction

$CaCO_3$ and almost the same percentage could characterize the sulfatic soils from an arid and semiarid region.

It is hard to estimate the contribution of amorphous materials like allophane or imogolite to the surface activity of the soils. The amorphous materials often coat the crystals present in the soils and, therefore, besides their direct contribution, they could alter the surface properties of the crystalline materials in the soil.

1.1.3 Soil Organic Matter

Soil organic matter is defined as the nonliving portion of the soil organic fraction, and it is a heterogeneous mixture of products resulting from microbial and chemical transformation of organic residues. Although the soil organic matter is, in most cases, only a small part of the total soil solid phase, it is of major importance in defining the physical, chemical, and surface properties of the soil material.

The transformed products of the fresh organic debris have the general name of *humus* but, in reality, they may be composed of humic and nonhumic substances. These substances could be amorphous, polymeric, brown-colored humic substances differentiated on the basis of solubility properties into humic acids, fulvic acids, and humins, and recognizable classes such as polysaccharides, polypeptides, altered lignins, etc., which can be synthesized by microorganisms or arise from modifications of similar compounds. The major components of the soil organic matter and their definitions, summarized by Stevenson (1994), are presented in Table 1.2.

The organic matter extracted from the soils is usually fractionated on the basis of solubility characteristics and the fractions commonly obtained include humic acid (soluble in alkali, insoluble in acid) fulvic acid (soluble in alkali and in acid), hymatomelamic acid (alcohol-soluble part of humic acid), and humin (insoluble in alkali). These dark-colored pigments extracted from the soil are produced as a result of multiple reactions, the major pathway being through condensation reactions involving polyphenols and quinones. According to Stevenson (1994), polyphenols derived from lignin are synthesized by microorganisms and enzymatically converted to quinones, which undergo self-condensation or combine with amino compounds to form N-containing polymers. The number of molecules involved in the process, as well as the number of ways in which they combine, is almost unlimited, which explains the hetero-

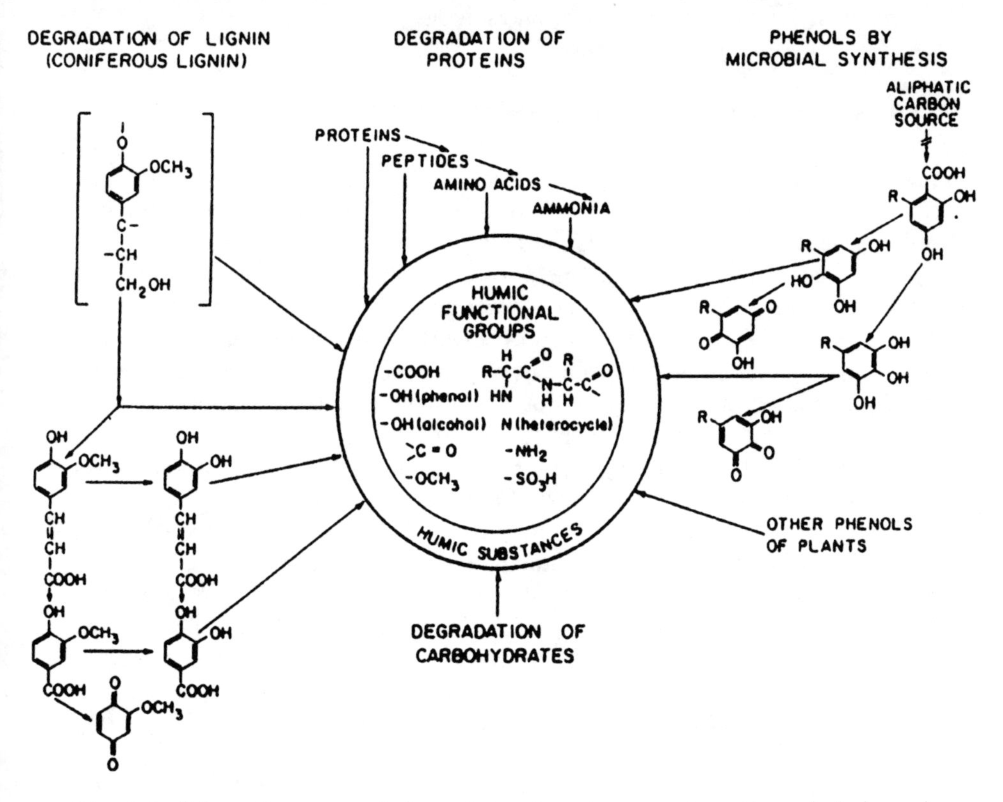

Fig. 1.3. Schematic representation of humic substances in soils and their main components and origins. (Flaig et al. 1975)

geneity of the humic material in any given soil. The structural precursors of humic substances in soils are illustrated in Fig. 1.3.

The major elements in the humic materials are C (50–60%) and O (30–35%). Fulvic acid has lower C but higher O. The percentages of H and N vary between 2 and 6% and that of S from 0 to 2%. The various fractions of the humic substances obtained on a basis of solubility characteristics are part of a heterogeneous mixture of organic molecules which, in different soils and locations, might range in molecular weight from several hundred to several hundred thousands. The average molecular weight range for humic acid is in the order of 10 000–50 000, and a typical fulvic acid will have a molecular weight in the range of 500–7000. The humic fraction of the soil represents a colloidal complex including long-chain molecules or two- or three-dimensional cross-linked molecules whose size and shape in solution are controlled by the pH and the presence of neutral salts. Under neutral or slightly alkaline conditions, the molecules are in an expanded state as a result of the repulsion of the charged acidic groups, whereas at a low pH and high salt concentration, contraction and molecular aggregation occur, due to the charge reduction. These large organic molecules may exhibit hydrophobic properties which govern their interactions with nonionic solutes.

1.1.4 Interactions Between Components of the Solid Phase

The interactions among the various components of the solid phase of the soil strongly affect its surface activity (Wolfe et al. 1990). The surface of the soil solid phase is heterogeneous. It is characterized by multicomponent association among humic substances, clays, metal oxides, $CaCO_3$, and other minerals. In some cases, up to 90% of the soil organic matter was found to be associated with the mineral fraction of the soil (Greenland 1965). On the other hand, there is evidence that much of the surface of clay minerals in soils, specifically in the interlayer spaces of smectites, is not covered with organic matter (Ahlrichs 1972). Metal oxides are also likely to coat the external surfaces of clay minerals, and intercalation of oligomeric hydroxyaluminum species with clays has been reported (e.g., Ahlrichs 1972). Cationic aluminum hydroxyoxides, the charge of which is pH-dependent, may coat clay surfaces, thus reducing the contribution of the clay minerals to the soil cation exchange capacity. This phenomenon is more important in acid soils; as the pH increases, the hydroxyaluminum polymers lose their positive charge, and their effect on the CEC is reduced. The coating of clay surfaces with organic matter and with mineral oxides binds or replaces exchangeable cations that are the sites of many surface-enhanced transformations (e.g., Mingelgrin et al. 1977). Coating may also block access to active sites that are not themselves coated.

The most extended interactions between components of the soil solid phase are those between clay minerals and organic matter. As a function of the properties of the organic compounds and of the clay surface, various me-

Table 1.3. Mechanisms of adsorption for organic compounds in soil solutions. (After Sposito 1984, as adapted from Mortland 1970)

Mechanism	Principal organic functional groups involved
Cation exchange	Amines, ring NH, heterocyclic N
Protonation	Amines, heterocyclic N, carbonyl, carboxylate
Anion exchange	Carboxylate
Water bridging	Amino, carboxylate, carbonyl, alcoholic OH
Cation bridging	Carboxylate, amines, carbonyl, alcoholic OH
Ligand exchange	Carboxylate
Hydrogen bonding	Amines, carbonyl, carboxyl, phenylhydroxyl
Van der Waals interactions	Uncharged, nonpolar organic functional groups

chanisms contribute to the adsorption of organic molecules on the mineral fraction. A summary of the mechanisms involved is presented in Table 1.3. The adsorption mechanisms are expected to operate when dissolved organic matter reacts with the clay surfaces. Theng (1979) emphasized that the quantity of dissolved organic matter adsorbed tends to decrease as the pH increases above pH = 4. Sposito (1984) explained this phenomenon by suggesting that dissolved soil organic matter formed ligand-like surface complexes and, therefore, the predominant adsorption mechanisms are appropriate to anions. It should be emphasized that the hydration status of the mineral phase influences the adsorption of organic molecules, in some cases, by affecting the adsorption mechanism itself, and by decreasing the number of active sites of the mineral solid phase able to interact with the organic molecules.

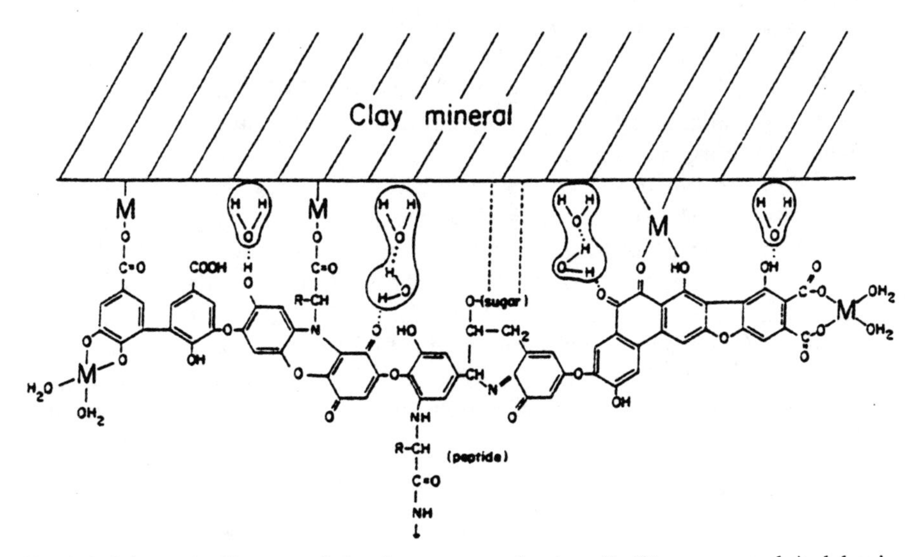

Fig. 1.4. Schematic diagram of clay-humate complex in soil. (Stevenson and Ardakani 1972)

For natural soils, the most important interactions are between clays and humic and fulvic acids. These acids contain a variety of reactive functional groups that are capable of combining with clay minerals. A schematic diagram of a clay-humate complex in soil is presented in Fig. 1.4 (after Stevenson and Ardakani 1972). Since organic anions are normally expelled from negatively charged clays, adsorption of humic and fulvic anions by 2:1 layer type of clay occurs only when polyvalent cations are present on the exchange complex. The main polyvalent cations responsible for the binding of humic and fulvic acids to soil clays are: Ca^{2+}, Fe^{3+}, and Al^{3+}. The divalent Ca^{2+} ions do not, however, form strong coordination complexes with organic molecules, and would be effective only to the extent that a bridge linkage is formed. In contrast, Fe^{3+} and Al^{3+} form coordination complexes with humic substances, and strong bonding of these organic molecules is possible through this mechanism.

When clay minerals are coated with layers of hydrous oxides, their surface reactions are dominated by these oxides rather than by the clays. This is the case for strong bonding through ligand exchange as well as through simple anion exchange.

1.1.5 The Electrified Surface of the Solid Phase

The soil solid surface has a net charge (σ_s) which, in contact with the liquid or gaseous phase, is faced by one or more layers of counter or coions having a net charge separate from the surface charge. Uncharged molecules, such as water molecules, could also be found in this zone, but they do not contribute to the charge density of the surface if they do not dissociate. The adsorption of charge solutes onto the surface of the soil solid phase is subject to both chemical binding forces and the electric field at the interface; it is thus controlled by an electrochemical system (Bolt and van Riemsdijk 1991). There are, however, considerable differences between the surface properties and behavior of soil organic and inorganic colloids.

Some of the functional groups on the clay surface (e.g., M–OH) exhibit electric charges. The size of the charge, as well as its sign, is controlled by the properties of the surface to which the functional groups are bound, and by the composition of the surrounding liquid phase. Sposito (1984) classified the surface charge density of soil clays as follows:

Intrinsic surface charge density (σ_{in}), defined by the number of coulombs per square meter borne by surface functional groups, either because of isomorphic substitutions, or because of dissociation/protonation reactions. The intrinsic charge density could be expressed by the relation:

$$\sigma_{in} = \frac{F(q_+ - q_-)}{S}, \tag{1.1}$$

where F is the Faraday constant, q_+ and q_- are moles of adsorbed ions, and S is the specific surface area (m^2/g).

Structural surface charge density (σ_o), defined as the number of coulombs per square meter, is a result of isomorphic substitutions in soil minerals.
Proton surface charge density (σ_H) is defined as the difference between the numbers of moles of complexed proton charges (q_H) and of complexed hydroxyl charges (q_{OH}) on proton-selective surface functional groups, per unit mass of soil clay; and is expressed by:

$$\sigma_H = F(q_H - q_{OH})S \ . \quad (1.2)$$

The above types of surface charge density could be related by the equation:

$$\sigma_{in} = \sigma_0 + \sigma_H \ . \quad (1.3)$$

Each term from this equation can be measured and may be either positive or negative.

1.2 The Liquid Phase

The soil liquid phase – generally known as the soil solution – is a water solution with a composition and reactivity defined by the properties of the incoming water and affected by fluxes of matter and energy originating from the vicinal soil solid phase, biological system, and atmosphere. This natural, open water system results from a dynamic transformation of the dissolved constituents in various chemical species over a range of reaction time scales. At any particular time, the soil liquid phase is an electrolytic solution, potentially containing a broad spectrum of inorganic and organic ions and unionized molecules.

In a porous medium, two liquid-phase regions can be defined. The first one is that near the solid phase and is considered the most important surface reaction zone of the porous medium system. The near-surface water also controls the diffusion of the mobile fraction of the solute adsorbed on the solid phase. The second region covers the "free" water zone which governs the water flow and solute transport in soils. The composition of the soil liquid phase fluctuates over time due to recharge with rainwater – of variable quality, sometimes approaching that of distilled water – or by irrigation, and as the result of applications of various products (fertilizers, pesticides, etc.) and of waste and effluent disposals. The presently accepted description for soil liquid phase energetic characteristics is based on the concept of matric and osmotic potentials. Matric potential is due to the attraction of water to the solid matrix, and the osmotic potential to the presence of solute in the soil water.

1.2.1 Water Near Soil Solid Phase

In moving through the bulk liquid phase, a water molecule exhibits a series of spatial arrangements which are of great irregularity. In such conditions we are not dealing with a well-defined structure like that of a solid, but with instantaneous structures (I type), comprising molecules in a highly irregular arrangement. With the lengthening of the time scale, two additional types of structure may be defined: vibrationally averaged (V type) and diffusionally averaged (D type), and these highlight the fact that the concept of structure in a liquid water is a dynamic one (Sposito 1984). It is not our aim to extend the discussion on water structure; we are simply reproducing Stillinger's (1980) definition showing that liquid water consists of macroscopically connected random networks of hydrogen bonds with frequent strained and broken bonds that continually undergo topological reformation. The properties of the water arise from the competition among relatively bulky ways of connecting molecules into local patterns characterized by strong bonds and nearly tetrahedral angles and more compact arrangements characterized by more strain and bond leakage. In the presence of an electrolyte, a localized perturbation of the tetrahedral configuration occurs as follows: near the ion, the water molecules are dominated by a dense electromagnetic field resulting in the formation of the primary solvation shell. In the next zone, called the secondary solvation shell, water molecules interact weakly with the ion. Summarizing the properties of an electrolyte solution relevant to their behavior near the clay surfaces, Sposito (1984) shows that the primary solvation shell of a monovalent cation contains between three and six water molecules that exchange relatively rapidly with the surrounding bulk liquid. A secondary solvation shell, if it exists, is very weakly developed. The primary solvation shell of a bivalent cation contains between six and eight water molecules that move with the cation as a unit. A secondary solvation shell containing about 15 water molecules develops as the cation concentration decreases, and it also moves with the cation as a unit.

The configuration of soil water could be altered near the phyllosilicate surfaces. The siloxane surface influences the character of the water due to the nature of their charge distribution and the nature of the complexes formed between the cation and the surface functional groups. Both the type of charge and the degree of charge localization, as well as the valence and size of the complexed cations, control the features of the water molecules near the surface. Each mineral group (kaolinite, vermiculite, or smectite) with its own surface properties affects the near-surface water properties in a different way. The adsorbed water in the case of kaolinite is a bulk liquid water, whereas the water adsorbed on a vermiculite-smectite-type mineral is an aqueous solution, because of the presence of exchangeable cations on the 2:1 layer silicates. Sposito (1984) showed that a simplified consensus is that the spatial extent of adsorbed water on a phyllosilicate surface is about 1.0 nm (two to three layers of water molecules) from the basal plane of the clay mineral.

The interlayer water of clay minerals being structurally different from bulk liquid water or water in aqueous solution (Mamy 1968; Sposito and Prost 1982), it is to be supposed that the chemical reaction in this region is affected by a perturbed water structure. Another ability of the interlayer water is to be replaced by a much less polar solvent. When the layer charge of the clay is low and widely diffused over surface oxygens (e.g., smectites), a strong hydrogen bond to the layer is not necessary. The same pattern is also followed when the cation is large and univalent. Under these conditions, a much less polar liquid (e.g., nitrobenzene) can completely replace the interlayer water (Farmer 1978). These properties of the near-surface water are of great importance in dealing with the problem of pollution by both organic and inorganic molecules.

The water retention on oxides and hydrooxides of aluminum and iron is through the hydroxyl groups, but it can also involve oxide bridges and water coordinated with structural cations.

Water is retained on organic surfaces, at a molecular level also. Farmer (1978) stated that the principal polar sites where water adsorption occurs are likely to include carboxylic groups, phenolic and alcoholic groups, oxides, amines, pectones, aldehydes, and esters. Ionized carboxylic groups and their associated cations are likely to have the greatest affinity for aqueous solutes. It should not be forgotten that, in addition to polar sites, the organic surface exhibits important hydrophobic regions which are largely involved in the retention of organic micropollutants.

1.2.2 The Aqueous Solution

The natural processes creating an equilibrium between the solid, gaseous, and liquid phases control the composition of the soil aqueous phase. The chemical composition of the soil aqueous solution at a given time is the end product of all the reactions to which the liquid water has been exposed in the soil environment.

The thermodynamic properties of the soil aqueous solution are expressed in terms of a single-species solution activity coefficient for each molecular constituent. The composition of the soil aqueous solution should, however, be considered on the basis of molecular speciation in the soil solution, which, in turn, is related to biological uptake exchange reactions and transport through soils.

The amount of soil aqueous solution to be found in a soil under unsaturated conditions varies with the physical properties of the medium. Table 1.4 gives, as an example, some data about the amount of free water found in various soils. The composition of the soil aqueous solution fluctuates as a result of evapotranspiration or addition of (e.g., rain or irrigation) water to the system. The changes in the solution concentration, as well as the rate of change, are higher in soils with a lower retention capacity than in those with high retention capacity. These are referred to as the buffer properties of the soil.

Because of the "diversity" of the soil solid phase, as well as of changes in the amount of water in the soil as result of natural and human influences, it is difficult to make generally valid statements concerning the chemical composition of the soil aqueous solution. For the upper layers of the soil – the root zone – the composition of the soil aqueous phase is characterized by a rather

Table 1.4. Water-holding capacity of selected soils (A horizon) from Israel

Type of soil	Water rentention capacity, % w/w			
	Saturated paste	At 1/3 atm	At 15 atm	Air-dried (hygroscopic water)
Typic Torrispsamment	26.9	4.74	2.42	1.05
Calcicxerollic Xerochrept	28.2	6.28	3.44	1.21
Calcic Haploxeralf	34.9	16.0	6.2	1.77
Typic Haplargid	40.6	25.6	1.27	3.1
Lithic Xerorthent	45.5	26.4	9.0	2.20
Typic Haploxeralf	47.7	25.2	11.7	4.13
Calcic Palexeralf	50.1	20.0	8.7	3.4
Vertic Palexeralf	58.8	31.4	15.8	5.4
Lithic Haploxeroll	71.2	36.5	20.5	6.62
Typic Chromoxerert	95.6	49.0	34.3	9.6

low total salt concentration and is, in general, close to that of the rain or irrigation water.

The chemical characteristics of the soil aqueous solution could be summarized as follows:

Acidity-alkalinity of the soil aqueous solution, measured as pH, is affected by the quality of the applied water (acid for rain – neutral or alkaline for irrigation) and buffered by the soil buffer system.

Salinity or total salt concentration is usually expressed as total dissolved solids (TDS) or as the electrical conductivity of the solution (EC). The major fractions of the anions comprise Cl^-, SO_4^{2-}, and NO_3^- and the common cations are Ca^{2+}, Mg^{2+}, Na^+, and K^+. It may be assumed that composition of the soil solution varies between the composition of the water entering the system and that of the solution in equilibrium with the solid phase and the products applied to the soil. The fate of some major anions (e.g., nitrates) in the system is characterized, however, by a fluctuating concentration. The major cationic species are strongly influenced by interactions with the soil solid phase, and the phosphate behaves in a similar manner.

Inorganic trace elements such as alkali and alkaline earth cationic materials, transition metals, nonmetals, and heavy metals may be found in the soil

Table 1.5. Diffusion constants of CO_2, O_2, and N_2 in air and water, and their solubility in water at 20 °C. (Paul and Clark 1989)

	Diffusion constant ($cm^2 s^{-1}$)		Solubility in H_2O (cm^3 L^{-1})
	Air	Water	
CO_2	0.161	0.177×10^{-4}	8.878
O_2	0.205	0.180×10^{-4}	0.031
N_2	0.205	0.164×10^{-4}	0.015

aqueous phase. The trace elements are of natural or anthropogenic origins. The adsorption reaction is a most significant mechanism in distributing trace elements between the soil solid and liquid phases. The complexation of inorganic trace elements by organic ligands found in solution influences the equilibrium status between the two phases and affects their concentration in the soil aqueous solution.

Organic trace compounds – of natural or anthropogenic origin – can also be found in the soil aqueous phase. The phase distribution of the organic trace compounds in the soil medium is controlled by the nature and properties of the soil colloids, the chemical and physicochemical characteristics of the organic molecules, and the nature of the soil environment.

1.3 The Soil Gaseous Phase

Soil porosity on the one hand and the soil moisture content on the other are the factors governing the volume of the soil gaseous phase. From the point of view of soil pollution, the soil gaseous phase, referred to as the soil atmosphere, can assist the movement of organic molecules in the vapor phase and of water. The gaseous transport phase through pores which are not available for the liquid phase makes the soil atmosphere an important channel for soil pollution by volatile toxic chemicals. On the other hand, the transport of water as vapor into the pores might lead to the formation of a layer of water which covers the potentially available sites for nonpolar gaseous pollutants, thus reducing pollutant fixation on the solid phase.

Soil atmosphere composition CO_2, N_2, and O_2, which are the major gases in the atmosphere, are also those found in the soil atmosphere. Gases arising from biological activity, such as nitrogen oxides, may be present at any time, but, because of their high reactivity with soil components and their susceptibility to biological activity, they are usually transitory (Paul and Clark 1989). In general, in well-aerated soils the O_2 content is around 20% and that of CO_2 between 1 and 2%. In heavy soils with high moisture content, however, the CO_2 content of the soil atmosphere may reach values as high as 10%. The concentration of CO_2 in the soil atmosphere could be related to both cropping techniques and soil properties (Henin 1976). The composition of the soil atmosphere changes with depth and with time under cultivation.

The composition of the soil atmosphere might change as a result of gas dissolution into the liquid phase. The solubility of gases in water depends on type of gas, temperature, salt concentration, and the partial pressures of the gases in the atmosphere. The most soluble gases are those that become ionized in water (CO_2, NH_3, H_2S), but O_2 and N_2 are much less soluble (Table 1.5).

The CO_2 concentration in the soil atmosphere might be different in small and large pores and might vary as a function of the aerobic or anaerobic activity of the soil microbial population. Paul and Clark (1989) showed that a change from aerobic to anaerobic metabolism occurs at an O_2 concentration of less than 1%. Thus, the overall aeration of a soil is not as important as that of individual crumbs and aggregates. Calculations show that water-saturated soil crumbs larger than 3 mm in radius have no O_2 in their center (Harris 1981). This means that aerobic and anaerobic zones may coexist in a soil even under unsaturated conditions.

1.4 The Soil as Biological System

Soil organisms are an integral part of the soil medium and promote a continuous interaction between the living and nonliving soil populations. Both physical and chemical properties of the soil solid phase are affected by organism activity. The soil populations also affect the properties of the soil liquid and gaseous phases. The free-living organisms of the soil biota include bacteria, fungi, algae, and fauna.

1.4.1 Soil Biota: Components and Distribution

Viruses grow only within the living cells of other organisms. Bacteria are the most numerous of the microorganisms in soil. Paul and Clark (1989) show that both energy source and carbon source are useful for describing basic physiological differences among bacteria, as among organisms generally. Those using light as their energy source are termed phototrophic, and those deriving their energy from a chemical source, chemotrophic. If CO_2 is used as the cell carbon source, the organism is termed lithotrophic. If cell carbon is derived principally from an organic substrate, the organism is organotrophic. Essentially the same differentiation expressed in the terms lithotrophic and organotrophic is provided by the terms autotrophic and heterotrophic, respectively. The majority of known bacterial species are chemoorganotrophic and are commonly referred to as heterotrophs. Photolithotrophs include the higher plants, most algae, cyanobacteria, and green sulfur bacteria. Chemolithotrophs use divers energy sources, e.g., NH_4^+, NO_2^-, Fe^{2+}, S^{2-}, and $S_2O_3^{2-}$. The obligate chemolithotrophs use the same basic physiological pathway (the Calvin cycle) found in most other organisms in metabolizing their own cell constituents. Their apparent inability to use any known external source of organic carbon is possibly linked to their lack of permeases to move organic molecules across cell membranes. Organic molecules must be synthesized within the cell.

The fungi embrace eukaryotic organisms (molds, mildews, rusts, smuts, yeasts, mushrooms, and puffballs) and of the soil organisms, these organotrophs are primarily responsible for the decomposition of organic residues, despite the fact that they are outnumbered by the bacteria. Typically, the fungi form slender filaments or hyphae, which collectively form the mycelium soma or thallus, reaching several decimeters in diameter. Algae are the most widely distributed of all green plants. Terrestrial forms are founded on, and in, the soil and soil parent material. Lichens are symbiotic associations of fungi and algae, forming a single thallus or undifferentiated body. Protozoa and metazoa are, respectively, unicellular and multicellular organisms, developing in the soil and forming the soil mesofauna. Both soil physicochemical properties and the structure of the soil microbial communities could be affected by the activity of the above groups of organisms.

The population density of microbial organisms in the soil profile, which generally follows the distribution pattern of organic matter, decreases with depth. However, the population density does not continue to decrease to extinction with increasing depth, but reaches a very low and constantly declining density. Distribution of microorganisms within the soil fabric might also be observed. Bacteria show attachment to sand grains by fine fibrillae extending from their cell walls and by extracellular mucilaginous polysaccharides (mucigels). For clustering of sand grains, a combination of microbial and plant mucigels, fungal hyphae, and small rootlets is required. Most bacteria are larger than clay particles, and usually carry a negative charge. Mucigels are macromolecules that do not move around in soil and, therefore, it is probable that very fine clay particles migrate to mucigels and organisms to initiate aggregate formation. Pore diameters in microaggregates are 0.2–2.5 μm, and in macroaggregates, mostly 25–100 μm. Small pore neck sizes prevent entry of organisms; therefore, only a few of the small pores can be invaded by bacteria.

Chemical analysis of the soil organic matter in microaggregates shows that the contained sugars are mostly of microbial origin. Macroaggregates permit aeration, water entry and drainage, diffusion of soluble compounds, and occupancy by organisms. Their pores are colonized by microflora and microfauna, and some pore necks may permit entry of members of the mesofauna. Large macroaggregates (>1 mm) may show penetration by roots or close association with them. Macroaggregates are more intensively colonized by organisms and are metabolically more active sites than the soil taken as a whole. Chemical analysis of macroaggregates shows greater contents of nutrients (carbon, nitrogen, sulfur, phosphorus) and a greater proportion of sugars of microbial origin than is found in the soil generally (Paul and Clark 1989).

1.4.2 Microorganism Bioactivity

Life being compared with a series of enzyme reactions, it is natural that the activity of soil enzymes is of primary importance in dealing with microorganism bioactivity. Table 1.6 lists the common soil enzymes and the reactions they catalyze. Some enzymes are constitutive (e.g., urease) and are routinely produced by cells, others are adoptive or induced and are formed only in the presence of a susceptible substrate. Soil enzymes, being proteins, are

Table 1.6. Some soil enzymes found in soil and reactions they catalyze. (Paul and Clark 1989)

Enzyme	Reaction catalyzed
Oxidoreductases	
Catalase	$2H_2O_2 \rightarrow 2H_2O + O_2$
Catechol oxidase (tyrosinase)	O-Diphenol + 1/2 $O_2 \rightarrow$ O-quinone + H_2O
Dehydrogenase	$XH_2 + A \rightarrow X + AH_2$
Diphenol oxidase	P-Diphenol + 1/2 $O_2 \rightarrow$ P-quinone + H_2O
Glucose oxidase	Glucose + $O_2 \rightarrow$ gluconic acid + H_2O_2
Peroxidase and polyphenol oxidase	A + $H_2O_2 \rightarrow$ oxidized A + H_2O
Transferases	
Transaminase	R_1R_2-CH-N^+ H_1 + R_1R_4 CO $\rightarrow$ R_1R_4-CH-N^+ H_1 + R_1R_2 CO
Hydrolases	
Acetylesterase	Acetic ester + $H_2O \rightarrow$ alcohol + acetic acid
α- and β-Amylase	Hydrolysis of β (1 → 4) glucosidic bonds
Asparaginase	Asparagine + $H_2O \rightarrow$ aspartate + NH_1
Cellulase	Hydrolysis of β (1 → 4) glucan bonds
Deamidase	Carboxylic acid amide + $H_2O \rightarrow$ carboxylic acid + NH_3
α- and β-Galactosidase	Galactoside + $H_2O \rightarrow$ ROH + galactose
α- and β-Glucosidase	Glucoside + $H_2O \rightarrow$ ROH + glucose
Lipase	Triglyceride + 3 $H_2O \rightarrow$ glycerol + 3 fatty acids acids
Metaphosphatase	Metaphosphate → orthophosphate
Nucleotidase	Dephosphorylation of nucleotides
Phosphatase	Phosphate ester + $H_2O \rightarrow$ ROH + phosphate
Phytase	Inositol hexaphosphate + 6 $H_2O \rightarrow$ inositol + 6-phosphate
Protease	Proteins → peptides + amino acids
Pyrophosphatase	Pyrophosphate + $H_2O \rightarrow$ 2-orthophosphate
Urease	Urea → 2 NH_1 + CO_2

often entrapped in soil organic and inorganic colloids and, therefore, a large background of extracellular enzymes, not directly associated with the microbial biomass, can be found in the soil (Paul and Clark 1989). The microbially induced transformations control the availability and cycling in the soil medium of nutrients such as carbon, nitrogen, sulfur, and phosphorus. The microbial activity, including decomposition, can be described by the kinetics of corresponding transformations. The reaction rates are generally expressed as functions of the concentration of one or more of the substrates being degraded. Several types of reaction kinetics cover the microbially mediated decay in the soil medium: (1) zero-order kinetics, in which the rate of transformation of the substrate is not affected by changes in its concentration; (2) first-order kinetics, in which the rate of transformation of the substrate is proportional to its concentration; and (3) hyperbolic reaction kinetics, in which the rate of transformation approaches some maximum with time. These reaction kinetics can be mathematically expressed, and microbially induced transformations in the soil medium can be predicted accordingly.

1.5 Soil Heterogeneity

Soil properties on the field scale are spatially variable, with a difference of sometimes more than an order of magnitude occurring within relatively small fields (Bresler et al. 1984; Yaron et al. 1985). Although few published data are available, it is assumed that the soil mineral and organic matter contents, and thus its adsorption capacity, vary as well. Beckett and Webster (1971) calculated that the coefficient of variation for organic matter content for a single field will be 25–30%. This variability is apparently inherent to some degree in all field soils. Soil heterogeneity may result, for example, from microrelief effects, natural cracks, inclusion of different materials, variations imparted during formation processes, and from human activity. Soil heterogeneity can affect pollutant behavior in relation to adsorption, transport, and persistence.

The interpretation of the variability of soil constituents and physical properties has been an object of interest for pedologists and soil physicists (e.g., Biggar and Nielsen 1976; Dagan and Bresler 1979; Burgess and Webster 1980; Bouma 1981, 1989; Nielsen et al. 1983; Bresler et al. 1984; Webster 1985). The concept of representative elementary volume (REV), developed and used for porous media (Hubert 1955; Bear 1972), was transferred to soil science. In this concept the soil continuum is considered as a summation of representative elementary volumes (REV), and the definition and measurement of the soil properties are made on volumes of soils greater than this volume. Volume scales selected for the various soil properties differ and the REV would be different for physical, chemical, or microbiological parameters. The whole soil is considered to be divided into domains of microscopic, macroscopic, and megascopic effects, and the variations within each domain increase from the microscopic to the megascopic.

When a pollutant reaches a soil, its concentration in the liquid phase is dependent not only on the soil constituents but on their domainal distribution. For example, in the microscopic domain, the concentration of the solution will fluctuate depending upon the size and distribution of the pores and of the soil aggregates. Solute diffusion into aggregates will affect the adsorption-desorption process, the rate of chemical conversion, and the ratio between anaerobic and aerobic microbial activity. The partitioning between the air, water, and solid phases in the microscopic domain may also affect the volatilization of pollutants. In the macroscopic or megascopic domain, where large pores or soil cracks exist, convective transport, together with diffusion into and out of the soil aggregates, will define chemical behavior in the soil. As a consequence, the spatial distribution of soil properties affects and governs pollutant behavior in the soil system.

The spatial variability of a soil is not confined to field conditions; it occurs even on the laboratory scale generally used for defining soil parameters relevant to soil pollution estimation and prediction. There are two reasons. The first is related to the variability among the samples used in laboratory measurements, which must be taken into account in the sampling procedure. The second reason lies in the intrinsic variability of the sample. In a recent work of Wise (1993), the effect of laboratory-scale variability on the nonlinear sorptive behavior in a porous medium was proved. The author considered a porous medium to be composed of spatially distributed particles, individually characterized by randomly distributed sorptive capacities and selectivity coefficients. Comparing the validity of Langmuir's and Freundlich's isotherms for individual particles and soil aggregates, Wise (1993) observed a discrepancy between the results obtained with an individual particle, and those with an ensemble of particles, in composing the media. As the variability in underlying sorptive property increases, the Langmuir isotherm ceases to describe the behavior of the aggregates of individual particles well, under either static or dynamic conditions. The estimation of pollution hazard from the laboratory-obtained parameters should, therefore, consider the variability among the individual particles forming the analyzed soil sample, in order to validate the results obtained.

References

Ahlrichs JL (1972) The soil environment. In: Goring CAI, Hamaker JW (eds) Organic chemicals in the soil environment, vol 1. Marcel Dekker, New York
Bear J (1972) Dynamics of fluids in porous media. Elsevier, New York, 323 pp
Becket PTH, Webster R (1971) Soil variability: a review. Soils Fertil 34: 1–15
Biggar JW, Nielsen DR (1976) Spatial variability of leaching characteristics of a field soil. Water Resour Res 12: 78–84

Bolt GH, van Riemsdijk WH (1991) The electrified interface of soil solid phase. In: Bolt GH et al. (eds) Interactions of soil-colloid-soil solution interface. Kluwer, Dordrecht, pp 37–115
Bouma J (1981) Soil morphology and preferential flow along macropores. Agric Water Manage 3: 235–250
Bouma J (1989) Using soil survey data for quantitative land evaluation. Adv Soil Sci 9: 177–215
Bresler E, Dagan G, Wagenet RJ, Laufer A (1984) Statistical analysis of salinity and texture effects on spatial variability of soil hydraulic conductivity. Soil Sci Soc Am J 48: 16–25
Brown G, Newman ACD, Rayner JH, Weir AH (1978) The structure and chemistry of soil clay material. In: Greenland DJ, Hayes MBM (eds) Chemistry of soil constituents. John Wiley, Chichester, pp 29–179
Burgess TM, Webster R (1980) Optimal interpolation isarithmic mapping of soil properties. J Soil Sci 31: 315–331
Dagan G, Bresler E (1979) Solute dispersion in unsaturated soil at field scale. Soil Sci Soc Am J 43: 461–466
Farmer VC (1978) Water on partial surfaces. In: Greenland DJ, Hayes MHB (eds) The chemistry of soil constituents. Wiley, Chichester, NY pp 405–449
Flaig W, Beutelspacher H, Rietz E (1975) Chemical composition and physical properties of humic substances. In: Gieseking JE (ed) Soil components 1: Organic compounds. Springer, Berlin Heidelberg New York
Gilkes RY (1990) Mineralogical insights into soil productivity: an anatomical perspective. Proc 14th Int Congress of soil science, Kyoto, Japan. Transaction Plenary Papers, pp 63–75
Greenland DJ (1965) Interactions between clays and organic compounds in soils. Part 1. Mechanisms of interaction between clays and defined organic compounds. Soils Fertil 28: 415–420
Greenland DJ (1977) Soil damage by intensive arable cultivation: temporary or permanent? Philos Trans R Soc Lond B 281: 193–208
Harris RF (1981) Effect of water potential on microbial growth and activity. In: Par J et al. (eds) Soil Science Society of America, Madison
Henin S (1976) Cours de physique du sol. Orstom, Paris, 156 pp
Hubert MK (1955) Darcy's law and the field equation of the flow of underground fluids. Trans Ann Inst Mineral Meteorol Engin 207: 222–239
Mamy J (1968) Recherche sur l'hydration de la montmorillonite: propriétés diélectriques et structure du film d'eau. Thèse Fac Sci Paris, Ann Agron 19: 175–220
Mingelgrin U, Saltzman S, Yaron B (1977) A possible model for the surface-induced hydrolysis of organophosphorus pesticides on kaolinite clays. Soil Sci Soc Am J 41: 519–529
Mortland MM (1970) Clay organic complexes and interactions. Adv Agron 22: 75–120
Nielsen DR, Wierenga PJ, Bigger JW (1983) Spatial soil variability and mass transfer from agricultural soils. In: Chemical mobility and reactivity in soil systems. Soil Sci Soc Am Spec Publ 11. Soil Science Society of America, Madison, pp 65–78
Paul FA, Clark FE (1989) Soil microbiology and biochemistry. Academic Press, New York, 273 pp
Sposito G (1984) The surface chemistry of soils. Oxford University Press, New York, 234 pp
Sposito G (1986) Thermodynamics of the soil solution. In: Sparks DL (ed) Soil physical chemistry. CRC Press, Boca Raton, 147–179
Sposito G, Prost R (1982) Structure of water adsorption on smectites. Chem Rev 82: 53–573
Stevenson FJ (1994) Humus chemistry, 2nd edn. John Wiley, New York, 460 pp
Stevenson FJ, Ardakani MS (1972) Organic matter reactions involving micronutrients

in soils. In: Mortvedt JJ, Giordano PM, Lindsay WL (eds) Micronutrients in agriculture. Soil Science Society of America, Madison, pp 79–114

Stillinger FH (1980) Water revisited. Science 209: 451–453

Theng BKG (1979) Formation and properties of clay-polymer complexes. Hilger, London

Tisdale JM, Oades JM (1982) Organic matter and water stable aggregates in soil. J Soil Sci 32: 141–163

Webster R (1985) Quantitative spatial analysis in the field. Adv Soil Sci 3: 1–70

Wise WR (1993) Effects of laboratory scale variability upon batch and column determination of nonlinear sorptive behaviour in porous media. Water Resour Res 29: 2983–2992

Wolfe NL, Mingelgrin U, Miller GC (1990) Abiotic transformations in water, sediments and soil. In: Cheng HH (ed) Pesticides in the soil environment. Soil Sci Soc Am Book Ser 2. Soil Science Society of America, Madison, pp 103–167

Yaron B, Gerstl Z, Spencer WF (1985) Behaviour of herbicides in irrigated soils. Adv Soil Sci 3: 121–211

CHAPTER 2

The Soil Pollutants

We consider as a pollutant any chemical of natural or anthropogenic origin which accumulates in the soil medium and changes the natural soil equilibrium, as a result of human activity. A modern society, however, cannot be developed without the use of natural and synthetic compounds which may be added to the various ecosystems. The art of safeguarding the environment consists of minimizing the amount of out-system compounds reaching the land surface and restraining as much as possible the pollution-generating sites by controlling their disposal habits.

This chapter will deal with the pollutants reaching the land surface, with emphasis on their chemical structure and properties, which in the final state control their behavior in the soil. For didactic reasons, we have grouped the soil pollutants into inorganic, inorganic-organically bound compounds, and organic toxic compounds. This chapter will not include tabular data of the properties of the soil pollutants, but only a blend of illustrative characteristics of the major pollutants reaching the soil porous media.

2.1 Inorganic and Inorganic-Organically Combined Pollutants

In this group of pollutants we will include nitrates, phosphates, salts, and trace elements. The main characteristics of these compounds are that many of them are found in the soil in inorganic and organic forms, their behavior changing according to their speciative status.

2.1.1 Nitrogen Forms

The main inorganic N compounds detected in soils include NO_3^-, NO_2^-, exchangeable NH_4^+, mineral-fixed NH_4^+, dinitrogen gas (N_2), and nitrous oxide (N_2O_2). The nonexchangeable (fixed NH_4^+) may constitute over 50% of total N in some soils, but in most of the agricultural land it constitutes no more than 10%. Fixation and release of NH_4^+ are discussed in Chapter 6.

NO_3^-, NO_2^-, and exchangeable NH_4^+ are considered the most important inorganic N compounds, and under natural conditions their concentration generally ranges from a few to a few tens of ppm (Allison 1973). The N concentration in the soil upper layer fluctuates during the year and is affected by the climatic changes (e.g., from about 10 ppm during the winter to about 50 ppm during the spring in a temperate climate). In contrast to fixed NH_4^+, the distribution of which is correlated with depth in the presence of clay material, the distribution of NO_3^- is not correlated with a specific soil characteristic, but usually follows the water movement through the soil. Exceptions are soils with significant anion exchange properties, or other sorption sites, where the mobility of NO_3^- is less appreciable (e.g., soils with high Fe-oxides-hydroxides content).

Stevenson (1982) stated that 90% of the N in the surface layer of most soils is organically combined. He defined the following group of forms of organic N in terms of acid hydrolysis fractionation: acid-insoluble N, amino acid N, amino sugar N, and hydrolyzable unknown N (HUN) fraction. The N distribution in soils exhibits a climate effect: a higher percentage of the N in soils occurred as amino acid N and as amino sugar N in the warm climates; lower values were observed at cool temperatures and in the arctic zone. The high resistance of organic N complexes to soil microbial attack is known; it can be explained by the nature of the N-organic complexes (e.g., NH_3^- or NO_2^--lignin or humic substance complexes) by their adsorption on clay mineral surfaces and by the fact of some of the organic N being entrapped in small pores, inaccessible to microorganisms (Stevenson 1982).

Jenny (1941) showed that in uncultivated soils the organic matter and N content attains a steady-state level that is governed by the soil-forming factors (climate, relief, vegetation and organisms, parent material, and time).

The N cycle in soils is an integral part of the overall cycle of N in nature (Fig. 2.1); it is maintained in soils, not only through fixation of molecular N_2 by microorganisms, but also through the return of NH_4^+ and NO_3^- in rainwater. The N balance in soils is controlled by biological N_2 fixation, mineralization, and conversion of organic forms of N to NH_4^+ and NO_3^- (ammonification-nitrification), or by utilization of NH_4^+ and NO_3^- by plants and microorganisms (assimilation-immobilization). An important source of N is fertilization of agricultural lands and from land disposal of wastes and sewage effluents. An additional source of nitrogen is that which originates from acid rains. The natural and man-made contents of nitrogen oxides (NO_x) in the atmosphere originate from lighting, volcanic eruptions, and, mainly, from biological processes. Most man-made emissions derive from point sources characterized by spatial variations of NO_x level, peaks being observed over the cities. These emissions induce the acidification of rain and the NO_x forms a solute which reaches the land surface via the rainwater.

Nitrogen losses from soil occur mainly through crop removal, volatilization, and leaching. Under suitable conditions, NH_3 can be lost from soils by volatilization (see Chap. 4). An additional cause of losses results from bacterial denitrification to N_2O and various intermediate reduction products. All bacteria involved in denitrification are facultative anaerobes and their denitrifying

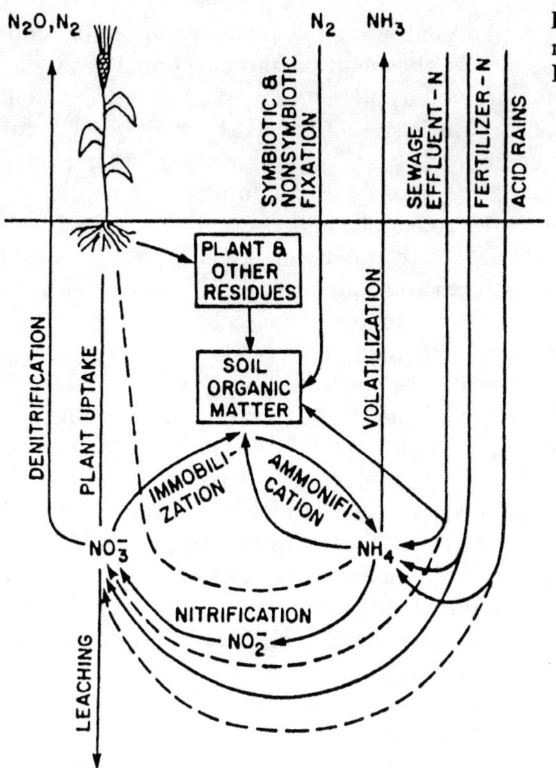

Fig. 2.1. Major components of nitrogen cycles in soils. (After Feigin et al. 1991)

activity occurs under a soil with a poor drainage, in a temperature range of 5–25 °C, and at soil pH around 7.5, in the presence of a good supply of decomposable organic matter. Leaching encompasses NO_3^- and NH_4^+ to a lesser degree; it follows the water transport pattern through soils.

Adverse effects of nitrogen encompass both human health and the terrestrial environment, and these occur when excess of NO_3^- and its reduction product NO_2^- (N-nitroso compounds) are found.

2.1.2 Phosphorus Forms

Soil phosphorus forms include organic and inorganic fractions, consisting of compounds characterized by different solubility and availability for various reactions, and with specific potential contributions to the pool of labile or available P in the soil. The percentage of each P fraction varies greatly. The organic fraction is 5–90% of the total P, with the higher percentages being typical of organic soils.

The principal inorganic P forms in soils are Ca orthophosphates, adsorbed orthophosphates, and occluded phosphates (Lindsay 1979; Mengel 1985). The formation of Ca phosphates is promoted under alkaline conditions in the presence of a high Ca concentration. Since the pH and the level of Ca in calcareous soils are high, Ca phosphates are the predominant stable compounds in these soils. The various Ca phosphate compounds found in soils are characterized by specific solubility and availability levels. The order of solubility of the common phosphate ions is: $H_2PO_4^- > HPO_4^{2-} > PO_4^{3-}$. The formation of stable phosphate compounds (tricalcium phosphate and apatite) is favored at high pH, while Al and Fe phosphates are formed at low pH.

The level of total P in soils varies greatly, lying between 0.1 and 0.8 g P kg^{-1} dry soil and 500 and 2000 kg P ha^{-1} in the top 0.2-m layer. Typical levels of P in the subsoil solution of unfertilized soil range from 0.001 to 0.1 mg P l^{-1} (Hook 1983). The level of P available to plants, as determined by various chemical and biological methods (Olsen and Sommers 1982), comprises only a small fraction of the total P present. The concentration of water-soluble P is 0.03–3 mg kg^{-1} dry soil (Ryden and Pratt 1980), with most of the published values below 1 mg P kg^{-1}. Application of P to the soil, as fertilizer, manure, or in effluent, results in an immediate rise in the level of water-soluble P in the soil. However, due to adsorption and precipitation reactions taking place in the soil, the level of soluble P declines rapidly with time. Phosphorus uptake by plants also reduces the level of soluble P in the soil, but usually at a much slower rate. The actual level of water-soluble P in soil is usually very low, and the movement of P through the soil is restricted (Feigin et al. 1991). Phosphorus contained in substances such as fertilizer, organic amendments, and sewage effluents, when added to the soil, soon becomes an integral part of the soil P.

Summarizing the transformation of P in the soil medium, Feigin et al. (1991) emphasized that P in fertilizers, consisting of soluble P, becomes part of the labile P in soils. Effluents, depending on the chemical properties of the P they contain, may also contribute to the pool of fresh organic P. Manures and other materials containing organic P may behave in a similar way. The labile P is available to plants and is therefore immobilized in plant materials (grain, stover, or root) and through microbial activity as fresh, and, later, also as stable organic P. Some of the plant P, such as grain P, is removed from the field, while the rest is reincorporated into the soil, finding its way into the pool of fresh organic P. Decomposition of soil organic matter (fresh and stable pools) releases P back into the pool of labile P. The rate of this process depends on the nature of the soil organic matter, and on soil conditions such as pH, moisture, temperature and aeration.

The peak concentration of labile P, which occurs immediately after the addition of soluble P to the soil, declines within a few hours or, at most, within a day or two, as P is transformed into other forms, mainly by binding to the pools of active and stable inorganic P. The relative proportions of the labile, active and stable inorganic-P forms are constant, providing a constant supply of P as it is taken up by the crop.

The predominant processes involved in the transformation of labile P to less soluble forms are sorption, precipitation, and occlusion of inorganic P. The formation of insoluble Ca-phosphate compounds is the main process responsible for the reduction in the level of labile P in calcareous soils. The high concentration of Ca and the high pH characteristic of these soils promote the process. Sorption processes are often the major explanation for the removal of labile P by noncalcareous soils mainly on Fe- and Al-oxides (Ryden and Pratt 1980).

Phosphorus loss from soil occurs mainly through crop removal runoff and leaching. The uptake of P by agricultural crops, for example, varies greatly (10–100 kg $ha^{-1}yr^{-1}$) with most values being between 20 and 50 kg $ha^{-1}yr^{-1}$. Leaching of phosphate also varies as a function of P form, of the soil properties, and of the amount of water (rain – irrigation) passed through the system. An important retention factor is the soil adsorption capacity for P, the amount in excess of that adsorbed should be leached into depth. Ryden and Pratt (1980) showed, however, that the actual capacity of soil to adsorp P is often much greater than that predicted by laboratory tests, and that alternate drying and wetting restores the adsorption capacity of the soils. On the other hand, cases of faster-than-predicted movement of P have been recorded and explained in terms of the preferential flow of the solute. An additional factor which should be considered in P retention by soils is its precipitation. Runoff from agricultural lands is a major route for phosphate transport to surface water. Adverse effects of phosphorus encompass both human health, as well as the soil and water environment and those occurring when the chemical is in excess. Man's activities affect the global phosphorus cycle to an ever-increasing degree. The phosphate in sewer discharge and from farm erosion is already affecting freshwater and marine ecology in parts of North America, Europe, and Asia.

2.1.3 Salts

The soil liquid phase, as well as all water reaching the soil, is characterized by a specific salt content. The distribution of the ions between the soil liquid and solid phases is governed by the exchange properties of the solid phase and controlled by the ion exchange process. As an example, chemical examination of solutions extracted from soils from arid and semiarid climatic areas reveals Na^+, Ca^{2+}, Mg^{2+}, K^+, Cl^{3-}, SO_4^{2-}, HCO_3^- and CO_3^{2-} as major ionic components.

Under special conditions — mainly as a result of human activity — the salt presence in the soil, as reflected by the quantity and its ionic ratio, may become harmful to the environment. Among the anions with a potential to become pollutants, sulfur and chloride should be considered, as well as magnesium and sodium, among the cations.

2.1.3.1 The Anions

Both anions, sulfur and chloride, could reach harmful levels for the soil environment as a result of human activity and consequently become pollutants.

Sulfur forms to be considered as potential pollutants originate from both earth and atmospheric sources. Sulfur in soil occurs as sulfate minerals, sulfur-bearing minerals, adsorbed on mineral surfaces, as components of the soil organic matter, and as sulfides in the soil liquid phase. The atmospheric sulfur consists primarily of SO_2, but other gaseous sulfur compounds could be found.

Inorganic sulfur might be found in soils as mineral sulfur, adsorbed sulfate, or as solutes. In mineral form, sulfur may occur as precipitating calcium or magnesium sulfates, or as metal sulfides such as FeS_2 or ZnS_2. The sulfide minerals occur primarily under anaerobic conditions and oxidize to sulfuric acid when exposed to air. Inorganic sulfur represents only 5–10% of total sulfur in soils (Neptune et al. 1975).

Sulfur-containing compounds in the soil organic matter include the amino acids cystine and methionine and their related compounds. Organic sulfur has been fractionated (Anderson 1975) into the following fractions: reducible sulfur, ester sulfate, sulfur, carbon-bonded sulfur, etc. Release of organic sulfur is related to the rate of organic matter decomposition. Sulfur released from organic matter is oxidized to sulfate by sulfur-oxidizing microorganisms.

The gaseous sulfur dioxide (SO_2), resulting from human activity as an emission from power stations or industrial and urban complexes, is dissolved in the water, inducing its acidification, and in this form reaches the land surface. Since these man-made emissions are, in part, of point source origin, their effect on the soil sulfur balance will be only a local one.

An additional deposition of sulfur on the soil surface resulting also from an anthropogenic source is the sulfate. It can be found in irrigation or drainage water and reaches the soil as a result of irrigation with saline water or due to the rise in the drainage water sulfate level as a result of a malpractice in controlling the sulfur balance in irrigation systems. In a soil, it may also be changed as a result of the use of $CaSO_4$ for soil reclamation.

Chloride is an additional component of the earth water which could become a pollutant when added to the soil system under uncontrolled management. Under natural origin, the chloride can reach the soil by dissolution of crystalline material and deposition reactions which may be either reversible or irreversible. Mainly, the ion Cl^- is found as a solute in the soil liquid phase, as a free salt, or adsorbed on positively charged soil surfaces. The Cl^- balance in soil is disturbed as a result of irrigation with saline water rich in Cl^- or by a chlorotic groundwater which rises into the upper part of the soil as a result of a poor drainage system.

2.1.3.2 The Cations

Sodium and magnesium are *earth cations*, which in the soil solution are balanced by soluble anions. Under special anthropogenic circumstances, these cations could become harmful for the environment, and consequently become potential pollutants.

Magnesium is the fourth most abundant metal in living organisms. In the soil, it is a constituent of many soil minerals and is present as exchangeable magnesium on the cation exchange complex and in the soil liquid phase as solute. Small amounts of magnesium may also be found bound with soil organic ligands.

The magnesium minerals are, in general, constituents of soil clays (e.g., smectite, vermiculite, illite); the more soluble magnesium minerals (e.g., dolomite, magnesium sulfate, etc.) become reduced in concentration by weathering or leaching. The data from the literature indicate a range of 0.015–1.02% for total magnesium in soils. Total magnesium in soils increases with increasing clay content.

Irrigation with saline water characterized by a high magnesium content can be noted as an anthropogenic source of magnesium in soils, which may affect the soil physical properties and fertility. Barber (1984) showed that the exchangeable magnesium in soils usually varies from 0.5 to 14 cmol kg^{-1} with magnesium saturation of the exchange capacity, usually 10–20%. Exchangeable magnesium is usually 20–60% of total magnesium. Soil solution magnesium is in equilibrium with exchangeable magnesium, but aridic soils have soil solutions with much higher solute concentrations due to the lack of natural leaching.

Magnesium fixation and release from soils depend on soil pH. Chan et al. (1979) showed that in soils with pH-dependent or variable charge, magnesium was exchangeable at pH values below 6.0 and became nonexchangeable at a pH about 6.5. The release of magnesium occurs from both nonexchangeable and exchangeable sources. An additional source of magnesium could be found in alkaline water where such water sources are used for irrigation. Applied under this condition, magnesium could remain in the soil liquid phase, mainly when the soil exchangeable complex is saturated with calcium cations.

Sodium The only important halide mineral in soil is NaCl. It is a very soluble mineral, quickly removed from soils where rain or irrigation are high enough to cause deep leaching. The presence of this very soluble mineral in soils requires that its rate of accumulation should be greater than its leaching rate. However, due to the exchange processes, Na^{+} from the aforementioned soluble source can exchange Ca^{2+} and Mg^{2+} to the soil exchangeable complex and accumulate in soils according to their exchange capacity. An accumulation of sodium will occur in soils when irrigation with saline water is practiced.

Sodium is a soil chemical concern when it occurs in excess, generally >15% of the exchangeable cations. When found in the soil liquid phase, sodium can exchange the calcium cation from the soil complex, leading to the deterioration of the soil quality by affecting its hydrophysical properties. High sodium concentration in the liquid phase can become toxic for many agricultural crops.

Sodium carbonate forms which are associated with alkali soils are: nahcolite ($NaHCO_3$), soda ($Na_2CO_3.10H_2O$), and trona ($Na_2CO_3.NaHCO_3$–$2H_2O$). The synthesis of Na_2CO_3 minerals is a result of a physicochemical or a biological process. The physicochemical process consists of leaching of a soil with highly sodic water, followed by leaching with water of low salinity, or by weathering of igneous rocks to produce bicarbonates of Na^+, Ca^{2+}, and Mg^{2+}. Upon the evapotranspiration of the resultant solution, $CaCO_3$ and $MgCO_3$ minerals will be precipitated and the remaining solution will contain Na^+ and HCO_3^-. Further drying of the soil will induce an increase in the concentration of Na^+ and the formation of Na_2CO_3. The biological formation of Na_2CO_3 in soils is a result of a reduction process.

2.1.4 Potentially Toxic Trace Elements

Potentially toxic trace elements form a major group of compounds involved in soil pollution. Originating either from soil parent material or from anthropogenic sources, the trace elements could become toxic for the immediately surrounding biota when their concentrations exceed crop and animal tolerance limits for these elements. In the meantime, a concentration of trace elements in soil could form a major source of pollutants for the ground and surface water.

Adriano (1986) defined "trace elements" as those elements that occur in natural and perturbed systems in small amounts and that, when present in sufficient concentrations, are toxic to living organisms. We can still consider valid the Nriagu (1984) estimation of the amount of trace elements mined and used by our advanced society: 0.5, 20, 240, 250, and 310 million tonnes of Cd, Ni, Pb, Zn, and Cu, respectively. Adriano (1986) quoted other estimations concerning anthropogenic As, Cd, Pb, Cu, and Zn which are currently being disseminated via the atmosphere to distant ecosystems at about 22 000, 7000, 400 000, 56 000, and 214 000 tonnes annually, respectively. It is generally admitted that the inputs from anthropogenic sources exceed the contributions from natural sources by severalfold.

While the trace elements originating from soil parent material, are, on a short time scale, almost in a steady state, those introduced into the soil as a result of human activity are, in general, continuously increasing. Among the anthropogenic sources we can include: commercial fertilizers (as defined micronutrients or as impurities in the basic fertilizers, mainly phosphates), the impurities accompanying industrial soil reclamation products (gypsum and lime), the trace elements used directly bound onto organic biocides, mining residues, and automobile emission products. A major source of trace element pollution is land disposal of municipal and animal refuse and of sewage water and sludges.

In the present chapter, only the trace elements having an adverse environmental effect will be discussed. When dealing, however, with the nature of the potentially toxic trace elements in soils, we will also consider their specific forms. Their behavior is defined by their specific form rather than by their total concentration. In this regard, Adriano's (1986) table summarizing both plant and animal response is reproduced (Table 2.1). For further details on properties and behavior of potentially toxic trace elements, the reader is addressed

Table 2.1. Effects of trace elements on plant and animal nutrition. (Adriano 1986)

Element	Essential or beneficial to		Potential toxicity to		Comments
	Plants	Animals	Plants	Animals	
Ag	No	No		Yes	Interacts with Cu and Se
As	No	Yes	Yes	Yes	Phytotoxic before animal toxicity; may be carcinogenic
B	Yes	No	Yes		Narrow margin, especially in plants
Ba	No	Possible			Insoluble; relatively nontoxic
Be	No	No	Yes	Yes	Speciation important; carcinogenic
Bi	No	No	Yes	Yes	Relatively nontoxic
Cd	No	No	Yes	Yes	Narrow margin; enriched in food chain; carcinogenic; itai-itai disease
Co	Yes	Yes	Yes	Yes	Relatively nontoxic; high enrichment factor; carcinogenic
Cr	No	Yes	Yes		Speciation important; Cr^{4+} very toxic; otherwise relatively nontoxic; carcinogenic
Cu	Yes	Yes	Yes		Easily complexed in soils; narrow margin for plants
F	No	Yes	Yes		Accumulative toxicity for plants and animals
Hg	No	No		Yes	Enriched in food chain; aquatic accumulation; minamata disease
Mn	Yes	Yes	<pH 5		Wide margin; toxic in acid soils; among the least toxic
Mo	Yes	Yes		5–20 ppm	High enrichment in plants; narrow margin for animals
Ni	No	Yes	Yes	Yes	Very mobile in plants; relatively nontoxic; carcinogenic
Pb	No	No	Yes	Yes	Aerial dispersion and primarily surface deposited; cumulative poison
Sb	No	No		Yes	Insoluble; relatively nontoxic
Sc	Yes	Yes	Yes	4 ppm	Narrow margin for animals; interacts with other trace metals
Sn	No	Yes		Yes	Relatively nontoxic; very low uptake by plants
Ti	No	Possible			Insoluble; relatively nontoxic; possibly carcinogenic
Tl	No	No		Yes	Very mobile in plants
V	Yes	Yes	Yes	Yes	Narrow margin and highly toxic in animals; high enrichment factor; carcinogenic
W	No	No			Very mobile in plants; very rare and insoluble
Zn	Yes	Yes			Wide margin; easily complexed in soils; may be lacking in some diets; relatively nontoxic

to Adriano's monograph, published in 1986, which was used by us in describing the properties of the potentially toxic trace elements.

Adopting the Adriano (1986) nomenclature, among the potentially toxic trace elements we will include group micronutrients such as: Zn, Mn, Cu, Fe, Mo, and B. Additional potentially toxic trace elements to be discussed are As, Cd, and Cr. A particular case is that of radioactive trace elements which may reach the soil by accident following an unsafe disposal. Since the behavior of radioactive compounds in soils is similar to that of nonradioactive trace elements, they will not be discussed separately.

The anthropogenic origin of trace elements introduced into the soil environment covers the whole spectrum of human activity and cannot be artificially grouped and discussed according to their provenance. We will select only, as an example, a limited number of trace elements having an environmental impact, which will be discussed in alphabetical order, as follows:

Arsenic (*As*) belongs to group V-A and has an atomic weight of 74.992. The more common oxidation states are: 0, III, and V, as bound covalently with most metals and stable organic compounds. As sources we mention soil parent material, airborne emissions. landfill wastes, etc.

In uncontaminated soils, the total As in general does not exceed 20 ppm, in the form of insoluble compounds with Al, Fe, and Ca. In particular cases, the water-soluble fraction can reach about 6% of the total As in soils (Akins and Lewis 1976). This explains the movement of As (as sodium arsenite) in the soil profile. The reduced form of arsenite is four to ten times more soluble in water than the oxidized state (V). In a strongly reducing soil environment, elemental As and arsine (III) can exist: in an aerobic environment, arsenite (V) is stable, and arsenite (III) can be the predominant form under moderately reducing conditions created by flooding (Deuel and Swoboda 1972). Both represent oxidative and reductive transformation of meta-arsenates.

Boron (*B*) belongs to group III-A of the Periodic Table and is a nonmetallic element. At room temperature B is inert, but strong oxidizing agents can affect it, forming borates. Boric oxide (B_2O_3) is acidic, and soluble in water. The soluble form — originally from irrigation water — can be a source of pollution for the irrigated soil. As B sources can be mentioned the parent material, water resources, fertilizers, sewage sludges, and coal combustion effluents.

Total B in normal soils reaches 2–100 ppm, with an average of 30 ppm, and it depends largely on the soil parent material. While in a clay-organic soil the average B content is 40–500 ppm, in sandy soils it reaches only about 5 ppm. In the heavy organic soils, B is mainly distributed in the upper layer of the soil profile, while in the sandy soils it could be distributed almost equally along the profile. In many cases, the B presence is associated with saline-alkaline soils.

Boron can be found in four well-defined forms: water-soluble, adsorbed on clay surfaces, fixed in the clay mineral lattices, and organically bound. The nonionized species $B(OH)_3$ is the predominant one in the soil solution and occurs under a pH up to 9. Polymeric forms of B are unstable in soils.

Cadmium (*Cd*) is a soft metal with an atomic weight of 112.40. It has eight stable isotopes. It is a transition metal in group II B and is almost always divalent in stable compounds, the most common one being CdS. It can form hydroxides and complex ions, for example, with ammonia or cyanide [e.g., $Cd(NH_3)_6^{4-}$ and $Cd(CN)_4^{2-}$]. A series of complex organic amines, sulfur complexes, and chelates can also be mentioned. The Cd ions can form insoluble compounds, usually hydrated with carbonate, arsenate, phosphate, or oxalate ions. As Cd sources can be mentioned, besides the soil parent material, sludges and effluents (mainly industrial), fertilizers, atmospheric fallout, and radioactive waste disposal.

The total Cd in noncontaminated soils originates from the soil parent material and ranges from < 0.1 to 10 ppm. Cd is fairly mobile in the soil profile and is almost uniformly distributed. Cd contamination induces an enrichment in Cd content of the upper layers of the soil. In soils, Cd is found in various forms and speciations which were summarized by Adriano (1986) as follows: *exchangeable* phase — adsorption of Cd by electrostatic attraction to negatively charged exchange sites on clays, organic particulates, and hydrous oxides; *reducible* (hydrous-oxide) phase — adsorption or coprecipitation with oxides, hydroxides, and hydrous oxides of Fe, Mn and, possibly, Al, present as coatings on clay minerals or as discrete particles; *carbonate* phase — carbonate precipitation in soils high in free $CaCO_3$ and bicarbonate, and alkaline in reaction (precipitation is also likely to occur with phosphate); *organic* phase — complexation with the organic fraction by chelation or organic binding. The complexes may vary in stability from easily decomposable to moderately resistant to decomposition; *lattice* phase — fixation within the crystalline lattices of mineral particles, sometimes known as *residual* fraction; *sulfide* phase — very insoluble and stable compounds of Cd sulfides occurring in poorly aerated soils, such as rice paddy soils; and *solution* phase — exists in soil solutions in either the ionic or complexed form.

The properties of the soil solid phase and the composition of the soil solution, together with the initial Cd concentration, as well as the environmental conditions (soil moisture status and temperature) define the Cd forms or species.

Chromium (*Cr*) is a member of group VI-B of the Periodic Table and has an atomic weight of 51.996. It has four stable isotopes, and five radioactive isotopes, the one mostly used as a tracer being ^{51}Cr. It is commonly found in oxidation states 0, III, and VI, the III state being the most stable. The trivalent form has a great tendency to coordination with oxygen- and nitrogen-containing ligands. The hexavalent form is the most toxic of all the oxidation states of Cr. As Cr sources can be mentioned, besides the parent material, industrial waste disposal (e.g., paper mills, petrochemicals, steel, plating, etc.), fertilizers, and sewage sludges, as well as municipal effluents. Deposition of atmospheric particulates is also a source of Cr accumulation in soils. Total Cr concentration in soils ranges, in general, between 5 and 1000 ppm, but there are

soils characterized by a lower concentration. Berrow and Reaves (1984) reported a mean value of 50 ppm for world soils. As a function of its oxidation state and complexing agents, Cr can be found at various depths in the soil profile. In general, however, the reports on the pattern of profile distribution of Cr are inconsistent. Cr added to a soil can be oxidized or reduced, dissolved in the soil solution, adsorbed on the mineral and organic solid phase, chelated by an organic ligand, or precipitated as insoluble compounds. Reduction of Cr (VI) to Cr (III), for example, is slower in an alkaline soil than in a neutral or acid one. Factors affecting the fate of Cr in the soil environment are: pH — by increasing its solubility in the aqueous phase and affecting its reduction rate; oxidation state — increasing the toxicity of the compound in the hexavalent form and its mobility, due to the anionic nature of the species; electron donors or acceptors — by affecting the reduction process in the case of electron donors [e.g., organic matter, Fe (II), etc.] or enhancing oxidation in the case of electron acceptors (e.g., Mn). Some organic acids to be found in the soil medium (e.g, citric, gallic, acetic, etc.) may serve as chelators for a species like Cr (III) or electron donors for Cr (VI). In an anaerobic environment, with a low redox potential and pH, Cr would exist in the Cr(III) state.
Copper (*Cu*) belongs to group 1-B of the Periodic Table and has an atomic weight of 63.546. It is characterized by two natural and two radioactive isotopes; ^{64}Cu with a half-life of 128 h being the most commonly used as tracer. In nature, Cu occurs in the I and II oxidation states as native metals and as other forms such as sulfate, sulfosalts, carbonates, etc. Cu reaches the soil as industrial wastes, food additives, disease and protection agents for both animal and vegetable populations, fertilizers, manures, sludges, and effluents. Also, Cu can reach a soil surface as an atmospheric particulate, high deposition being observed over a long distance from the emitting urban-industrial source. The primary route of Cu into the soil is from the parent material itself.

In normal agricultural soil, the total Cu content ranges between 1 and 50 ppm, but values up to 250 ppm have been registered. In general, it is strongly adsorbed and is one of the least mobile elements, mostly remaining uniformly distributed along the soil profiles. In soils, Cu can occur in various forms: in soil solution, ionic and complexed, in normal exchange sites, on specific sorption sites, occluded in the soil oxide materials, in organic residues and, finally, in the lattice structure of primary and secondary minerals. McLaren and Crawford (1973) and McBridge and Bouldin (1984) found that the bulk of available soil Cu reserve residues is in the organically bound fraction. Organic-bound forms of Cu dominate most arable soils, and the enrichment of the soil with Cu (organic or inorganic) is reflected in the increase in most fractions. Mattigod and Sposito (1977) suggested that the major inorganic form of complexed Cu^{2+} in neutral and alkaline soil solutions would be $CuCO_3$, since in highly alkaline soils $Cu(OH)_4^{2-}$ anions become the predominant soluble species. Among the anions in soil solutions, sulfate and chloride may form complexes with Cu^{2+} in a saline environment (McBride 1981; Doner et al. 1982). The mobility and availability of Cu is largely dependent on soil pH, organic matter content, the presence of iron, manganese, and aluminum oxides, and clay soil mineralogy.

Lead (*Pb*) belongs to group IV-A of the Periodic Table and has an atomic weight of 207.19. It has oxidation states of II and IV, four stable isotopes and one radioactive form (^{212}Pb with a half-life of 10 h). The Pb salts may be slightly soluble in water (chloride and bromide) or almost insoluble (carbonate and hydroxide).

Total Pb concentration in uncontaminated natural soils ranges from <1.0 ppm to about 20 ppm (Nriagu 1978). The average content of organic soils is, however, about three times greater than that of mineral soils (Reaves and Berrow 1984). An increase in soil Pb content is due to human activities as follows: use as an additive in gasoline constituents with all the consequent global contamination by particulate emissions, accumulation near roadways, and industrial use in urban areas; smelting; and, finally, soil disposal of sewage sludges and effluents. Commercial fertilizers, as well as some crop-protection chemicals containing traces of Pb, could increase the Pb content of agricultural soils. The distribution of Pb in the soil profile decreases with depth and this declining profile is explained by its retention on the organic matter generally found in the upper layer of the soil profile.

In soil, Pb is found in various forms and speciation states. In arid soils, exchangeable, sorbed, organic carbonate and sulfide fractions have been found (Sposito et al. 1982) and in a humid area hydrous-oxide forms have been detected (Alloway et al. 1979). Under certain conditions, Pb can be transformed by microorganisms, and a volatile compound — tetramethyl Pb — has been detected. The forms of Pb in soils are controlled by the soil mechanical composition, mineralogy, organic matter content, and the solution pH.

Manganese (*Mn*) is a member of group VII-A of the Periodic Table and has an atomic weight of 54.938. It can exist in compounds in a series of oxidation states; its stable salts are those in oxidation states II, IV, VI, and VII; the lower oxides are basic and the higher ones acidic. Besides its natural origin from metamorphic, sedimentary, and igneous rocks, Mn can reach the land surface as the result of human activity. Mn is used as fertilizer in inorganic (MnO) or organic (Mn EDTA) forms. It has been identified in sewage sludges and effluents dispersed on lands. Particulate emission from the iron and steel industry and from fossil fuel combustion are additional aeolian sources of the Mn present in soils.

There are many Mn forms in soils due to the fact that it forms compounds in several oxidation states: it forms nonstoichiometric oxides in mixed oxidation states, coprecipitates with iron oxides and, in the solid phase, exhibits amphoteric behavior interacting specifically with both anions and cations. The forms of soil Mn are classified as water-soluble, exchangeable, adsorbed, chelated, or complexed, secondary clay minerals, primary minerals, and insoluble metal oxides. In the presence of organic matter, Mn complexes particularly with humic acid; changes with pH in the availability of Mn are also due to complexation with organic matter. It has been found, however, that

Mn chelates in soils are unstable, either Fe or Ca being able to substitute for Mn.

One of the most important factors determining Mn toxicity in soils is the soil pH. It has, however, been proved that soil Mn availability-toxicity is the result not only of soil factors but of a combined effect of soil properties and root exudates. Mn interaction with other ions may also affect its availability-toxicity. As an example, addition of Fe to the soil solution may also reduce the Mn toxicity. The redox status of soils influences the solubility of Mn. In the III and IV oxidation states Mn occurs as precipitate in an oxidizing environment, whereas Mn in oxidation state II is dominant in solution and in the solid state under reducing conditions. As previously stated, the presence of organic matter in soil, as well as its application in a chelating form as fertilizer, can also affect the existing status of Mn in the soil environment.

Mercury (*Hg*) belongs to group II-B of the Periodic Table, and has an atomic weight of 200.59. It has seven stable isotopes. Hg can be found in three stable oxidation states: 0, I, and II. It may be found as stable mercuric sulfides and sulfosalts or in the form of inorganic or organic complexes. In nature, Hg occurs in all kinds of rocks, mainly as sulfide. Hg mineral consists essentially of cinnabar and metocinnabar. Anthropogenic Hg sources include agricultural and industrial uses. Sources of airborne Hg include combustion of fossil fuels, chlorine manufacturing, Hg mining and smelting, etc.

Total Hg in normal soils has an average level in the order of 0.1 ppm, but in contaminated soils it can reach 15 ppm. For agricultural soils the Hg average level has been found not to exceed those found in virgin soils, indicating that soil enrichment by Hg due to agricultural practices has not occurred except for some particular cases, where Hg compounds were used as pesticides. Two contradictory patterns describe the Hg distribution in the soil profiles: it is accumulated in the upper layers of the soil, and is fairly mobile with depth. The accumulation of Hg in the upper layers is due to retention by the clay materials and organic matter, and its presence in the soil C horizon occurs under excessive leaching by rains or irrigation.

Hg is very unstable in the soil environment. Many mercurial compounds — organic and inorganic — decompose to elemental Hg, which may volatilize, be converted to HgS, or complex with inorganic ligands. This transformation is a result of a biological process. In the soil environment Hg could be strongly chelated by soil organic matter, with humic substances containing S keeping Hg in a soluble form. The inorganic complexes of Hg II are mainly formed with chloride and hydroxide, and become fairly stable in water. Hahne and Kroontje (1973) showed that the precipitation of Hg occurred only if the concentration of Hg (II) as $Hg(OH)_2$ exceeded 207 ppm. Chlorides, one of the most active complexing agents for Hg, will affect the metal solubility. As in the case of other metals discussed above, the soil pH, organic matter, texture, and mineralogy are factors affecting the mobility and availability of Hg. Besides these factors, the Hg carriers can influence its stability and mobility in the soil. Redox potential influences the stability of the solid phase and, for example, the insoluble HgS produced under reducing conditions can be converted into $HgSO_4$ under an oxidizing environment.

Molybdenum (*Mo*) belongs to group VI-B and occurs as five isotopes. Its atomic weight is 95.94. Mo has five possible oxidation states (II, III, IV, V, VI) with a predominance for the states IV and VI. In dilute solution the predominant form of soluble Mo is molybdate anion.

In natural soils, the Mo content is dependent on the soil parent material and levels of 0.5–5 ppm are considered normal. Besides the parent material origin, Mo reaches the soil as a result of soil fertilization practices in oxidation state VI, as MoO_4^{2-} ion. It is also frequently found in sewage sludges dispersed on lands, and may result from coal combustion and mining processes. In general, Mo accumulates in the soil upper layer due to its preference for organic matter. In an alkaline environment, Mo leaching is enhanced, but leaching of Mo is observed also under other conditions.

Mo can be found in various forms in soils, as follows: fixed within the crystal lattice, adsorbed on soil material as an anion, bound with organic matter, exchangeable, and water-soluble.

The species of Mo comprising soluble Mo are anionic with MoO_4^{2-} as the predominant ion, and this occurs above pH = 4 (Lindsay 1972). The pH of the soil solution determines the speciation of Mo and, consequently, its behavior in the soil medium. It has been established that the dependence of the Mo solubility on the hydroxy ion concentration and, consequently, the concentration of Mo in soil solution are inversely proportional to the square of H^+ ion concentration. The forms of Mo depend also on the properties of soil materials and on the presence of iron and aluminum oxides in the soil solid phase. Soils high in organic matter and under poorly drained conditions (neutral to alkaline) are those with high soluble Mo. Reactions of Mo species with phosphorus, sulfur, nitrate, or other nutrients could affect the equilibrium between the solid and soluble compounds in the soil medium.

In general, very few toxic effects of Mo on plants have been observed, but in isolated cases where unfavorable environmental conditions are created in the soil system, an excess of Mo could occur which becomes toxic to plants. In general, Mo originating from soils does not represent a hazard for animals or humans.

Nickel (*Ni*) belongs to group VIII of the Periodic Table and has an atomic weight of 58.71. It is characterized by five stable isotopes. Ni forms stable complexes with many organic ligands. Under anaerobic conditions, sulfide may control the solubility of Ni. Normally, Ni occurs in the 0 and II oxidation states.

The literature reports 20–40 ppm as the content of Ni for normal soils, but considerable variations may occur due to the parent material quality as well as anthropogenic sources, including metal-ore smelting, sewage sludge disposal, phosphate rock use as fertilizers, etc. In soils, nickel occurs in several chemical forms including: exchange sites, specific adsorption sites, fixed within the clay lattice, adsorbed on sesquioxides, and, finally, fixed in soil organic matter. Ni occurs in the ionic form and is complexed with either organic or inorganic ligands. No well-defined pattern of Ni distribution in the soil profile has been observed in natural soils; however, in some cases, a higher concentration has been observed in the soil layer rich in organic matter or having a relatively high clay content. In soils in general, Ni (II) is stable over a wide range of pH and redox conditions (Cotton and Wilkinson 1980). Ni halides and salts of oxo-acids are generally soluble in water, while Ni carbonate is almost insoluble. The solubility of Ni is inversely related to pH. Organic matter, by complexing the metal, contributes also to increasing its solubility. As in the case of other metals, Ni in excess can have toxic effects on plant, animal, and human development.

Selenium (*Se*) is a member of group VI, has six stable isotopes, and can be easily oxidized from Se (0) to Se (VI). The stable form in nature is selenite (SeO_3^{2-}), which is stable under alkaline and mildly acid conditions. Se can also form a large number of organic compounds. Se is inconsistently dispersed in soil parent materials and frequently enriches carbonate debris in sandstorms. A primary source of the Se in nature is volcanic eruptions, and metallic sulfides are associated with igneous activity. As secondary sources, the biological pools in which Se has accumulated may be mentioned. This particular trace element has become a pollutant mainly in the areas characterized by a rich Se content of the parent material. Industrial sources, as well as by-products of phosphatic fertilizers, also contribute to the soil enrichment in Se. Total selenium in natural soil has a level of 0.1–2 ppm; it reaches an average content of 6 ppm in seleniferous soils. Se becomes a pollution problem mainly in the arid and semiarid areas where, under an uncontrolled agricultural management, Se originating from cretaceous shales has become an environmental hazard. In soils, Se is mainly accumulated in the layers characterized by a high clay content, the role of soil organic matter being less defined. The form of Se largely affects its distribution in the soil profile. The organic form originates from decayed seleniferous vegetation, while inorganic Se occurs as metal selenides, selenites, or selenates. The predominant mobile inorganic form of Se in soils is selenite. In soils, Se could be biomethylated, producing volatile organic metabolites. This process is the result of soil microbial activity.

Zinc (*Zn*) belongs to group IIB, has an atomic weight of 65.37, and six radioactive isotopes (with half-lives between 55 min and 245 days). The oxidative state of Zn is exclusively II. The total Zn in soils depends on the nature of the parent materials, organic matter, texture, and pH. A normal soil, on the average, contains 10–300 ppm Zn, but a wider range (1–900 ppm) has also been mentioned (Adriano 1986). Zinc fertilizers, sewage sludges, and atmospheric dust of industrial origin are anthropogenic sources of Zn accumulation in soils. As in the case of other metals, Zn in soils can be found in various forms as follows: water-soluble, ionic, or complexed with organic ligands, exchangeable, extractable, occluded in soil hydrous oxides, precipitated, and as a constituent of the lattice structure of the clay minerals. In the soil medium, Zn complexes with humidified materials; the less stable complex is associated with the phenolic hydroxyl and weakly acidic carboxyl groups, while the more stable complex is associated with strongly acidic carboxyls. Complexation increases with the humification of the organic matter. Chelating agents, either natural or synthetic, are important in defining Zn mobility in soil, due to the fact that chelated Zn is generally more soluble in water than elemental Zn.

The behavior of Zn in soil is affected by the soil pH. When the pH is above 7.0, the availability of Zn becomes very low since its compounds are very soluble only in an acidic environment. Redox conditions also have a pronounced effect on the solubility of Zn. Under flooding conditions, relatively insoluble zinc sulfide is formed due to strong reducing conditions. Organic matter has an indirect effect on Zn solubility, the compound being temporarily immobilized by bicarbonates, hydroxides, and organic ligands originating from the decomposition of organic residues. The Zn-fertilizer interaction is critical for the trace element solubility, since Zn also forms complexes with N and P [e.g., $ZnNO_3^+$, $Zn(NO_3)_2^0$, $ZnH_2PO_4^+$, $ZnHPO_4^0$, etc].

Other potential toxic trace elements could be described, but it is not the aim of the authors to cover the whole spectrum of compounds. Elements such as antimony, barium, cobalt, fluorine, helium, tin, titanium, and vanadium can be found in the soil medium as a result of the origin (parent materials) or as industrial residues from disposals as a result of human activity.

2.2 Toxic Organic Residues

The molecular properties of the toxic organic chemicals, in particular the electronic structure, water solubility, and the ability to volatilize, are of great importance in defining their behavior in the soil medium. The ability of molecules to ionize, for example, is the primary reason why many toxic organic chemical interactions with soils are pH-dependent. For nonionic compounds, aqueous solubility (hydrophilic property) is the most important molecular parameter correlated with sorption, since, in general, highly soluble nonionic chemicals exhibit low affinity for the soil surface, while the opposite holds true for chemicals of low solubility. There are also other molecular properties relevant to chemical behavior in soil, such as shape, size, configuration, polarity and polarizability, acidity, or volatility.

It is not our aim to tabulate the properties of the toxic organic chemicals which have the potential to become soil pollutants. There are hundreds or thousands of synthetic products which, as a result of a chemical and biological degradation, may be transformed into new compounds, and it is almost impossible to summarize their properties. For didactic reasons we will group the potential organic soil pollutants as small and macromolecules, and for each group we will discuss those physicochemical properties relevant to their behavior in the soil medium. Selected examples of toxic organic chemicals will be used to illustrate our presentation.

2.2.1 Small Organic Molecules

Organic chemicals characterized by small molecules reach the soil surface mainly as a result of crop-protection practices or during the land disposal of industrial wastes. Crop-protection chemicals are applied to the soil directly or reach the soil as spray application to crops. In this case, we are dealing with a nonpoint-source pollution, while in the case of industrial residues, the pollution site is a well-defined point source. When these substances reach the soil, various degradation and transfer processes occur and the chemical properties of the compound, such as molecular structure, ionic charge and ionizability, polarizibility, volatility, and water solubility will determine which process will predominate.

Some general considerations and properties of the major groups of organic chemicals with small molecules will be discussed, using selected compounds as typical examples. More details on their properties can be found in the *Pesticide Manual* (1991) and various reviews (e.g., Saltzman and Yaron 1986; Calvet 1989; Senesi and Chen 1989; Hayes and Mingelgrin 1991).

2.2.1.1 Pesticides

One of the most representative compounds from the group of the small organic molecules pollutants are the pesticides.

Cationic organic molecules. Bipyridylium herbicides such as diquat and paraquat are the main compounds from this group. Those herbicides that belong to the bipyridylium quaternary ammonium class are characterized by a positive charge and are heterocyclic compounds. Diquat and paraquat (Scheme 1), the two herbicides of this group, are mainly contact herbicides, but may also be used as aquatic biocides. They are available commercially as dibromide and

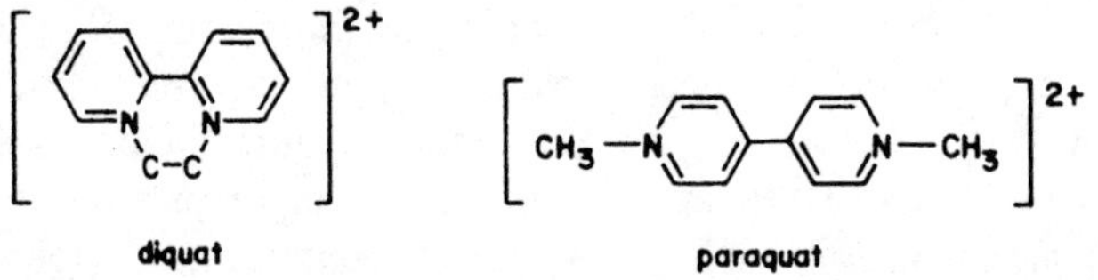

diquat paraquat (Scheme 1)

dichloride salts, respectively, and are characterized by high solubility in aqueous solutions, where they dissociate readily to form divalent cations. Diquat and paraquat are nonvolatile and do not escape as vapor from aquatic or soil systems. They are readily photodecomposed on exposure to sun or UV light, but are not photodecomposed when adsorbed onto particulate matter. These compounds are able to form well-defined charge-transfer complexes with phenols and many other electron donor molecules.

Basic organic molecules. Several members of the s-triazine family were selected as typical examples of basic compounds. The triazines are heterocyclic nitrogen derivatives, the ring structure being composed of nitrogen and carbon atoms. Most triazines are symmetrical (Scheme 2). The substituent at the R_1 positions determines the ending of the common name. With chlorine atom, the common name ends in *-azine*; with methylthio groups, *-tryn*; and with methoxy groups ($-OHC_3$) *-ton*. The solubility of the compound is determined by the R_1 substituent, with the $-OCH_3$ substitution resulting in the highest solubility. The presence of electron-rich nitrogen atoms confers on s-triazines the well-known electron donor ability, i.e., weak basicity, and the capacity to interact with electron acceptor molecules, giving rise to electron donor-acceptor (charge-transfer) complexes.

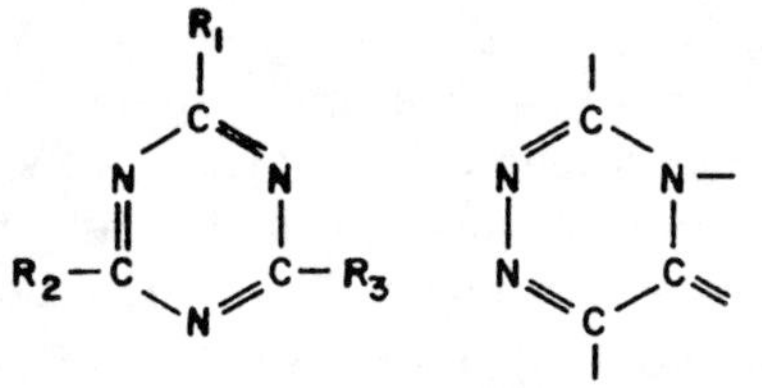

symmetrical triazine asymmetrical triazine

(Scheme 2)

Symmetrical triazines have low solubilities in water, the 2-chloro-s-triazines being less soluble than the 2-methylthio and 2-methoxy analogs. Water solubility increases at pH values where strong protonation occurs, i.e., between pH = 5.0 and 3.0 for 2-methoxy- and 2-methylthio-s-triazine, and at pH = 2.0 or lower for 2-chloro-s-triazines. Structural modifications of the substituents significantly affect solubility at all pH levels. The s-triazines, and especially the chloro-s-triazines are partly photodecomposed in aqueous systems by UV and IR radiations, including sunlight, while methoxy-substituted compounds are not photodegradable. Most s-triazines are relatively volatile, so that they can be lost from aquatic and soil systems by volatilization processes (Weber 1972).

Acidic organic molecules. This group of compounds comprises various families of chemicals, including substituted phenols, chlorinated aliphatic acids, chlorophane-oxyalkanoic acids, and substituted benzoic acids, which possess carboxyl or phenolic functional groups capable of ionizing in aqueous media to yield anionic species. These materials range in acid strength from strong acid trichloroacetic acid (TCA) to relatively weak acids such as 4-(4-chloro-o-tolyloxy) butyric acid (MCPB). For example, the benzoic herbicides are derivatives of benzoic acid (Scheme 3) which contain chlorine atoms, methoxy, or amino groups. In general, the benzoic herbicides are applied to both plants and soil. Dicamba and 2,3,6-TBA are two benzoic herbicides with similar chemical structures. 2,3,6-TBA is the common name for 2,3,6-trichlorobenzoic acid. In dicamba, the chlorine atom at the number 2 position is replaced by a methoxy ($-OCH_3$) group. Chlorinated aliphatic acids show the highest water solubility and the strongest acidity among this group of chemicals, owing to the strong electronegative inductive effect of the chlorine atoms which replace the hydrogens in the aliphatic chain of these acids.

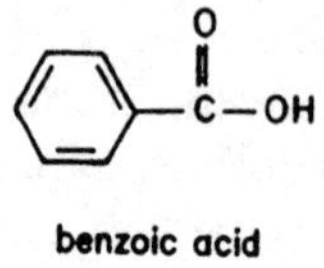

$C_6H_5-C(=O)-OH$

benzoic acid

(Scheme 3)

The water solubilities of unionized phenoxyalkanoic acids are low, as they have a considerable lipophilic component. Most commercial formulations of these herbicides, however, contain the compound in soluble salt form; thus, the anionic species predominate in neutral aqueous systems, while at low pH levels they are present in the molecular rather than the anionic form. These herbicides may undergo reactions of alkylcarboxylic acids, aromatic compounds and ethers (Yaron et al. 1985).

Acid pesticides (such as 2,4-D, 2,4,5-T, picloram and dinoseb) are characterized by their ability to ionize in aqueous solutions, forming anionic species. At pH values lower than their dissociation constant, the molecular form will be present, and increasing the pH above pKa will favor dissociation. Both 2,4-D and 2,4,5-T are among the most widely known and used phenoxy alkanoic acid pesticides. In general, the small organic molecules characterized as "phenoxys" have the following structural formula (Scheme 4):

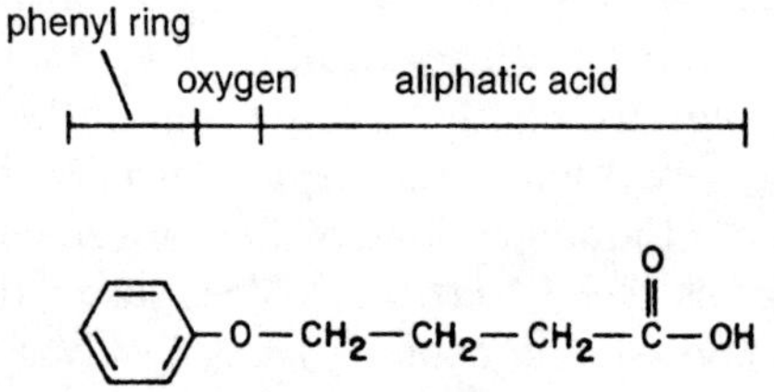

phenyl ring — oxygen — aliphatic acid

$C_6H_5-O-CH_2-CH_2-CH_2-C(=O)-OH$

(Scheme 4)

The phenyl ring is attached to an oxygen, which is attached to an aliphatic acid. The length of the carbon chain of the aliphatic acid determines the name of the herbicide. Among the small acid molecules we may also mention halogenated benzoic acids, used as herbicides, and chlorinated aliphatic acids, used as insecticides.

Nonionic organic molecules. The small organic molecules included in this group do not ionize significantly in aqueous systems and vary widely in their chemical compositions and properties. In this class of compounds are hundreds of pesticides belonging to such chemically diverse groups as chlorinated hydrocarbons, organophosphates, carbamates, ureas, anilines, anilides, amides, uracils, and benzonitriles. The great differences among the properties of these groups, and even among compounds within a group, are reflected in the variability of their adsorption behavior. The only common feature of the adsorption mechanism that has been clearly demonstrated for all the nonionic pesticides is the predominant role of the organic colloids.

The following examples of nonionic organic molecules were arbitrarily selected, and they are used only to illustrate some properties of the compounds included in this group.

The organophosphorus pesticides are part of the phosphoric acid ester group with a general formula of the type (Scheme 5):

$$
\begin{array}{l}
\quad\quad O(S) \\
RO \diagdown \;\| \\
\quad\quad P-O-X \\
\quad\; \diagup \\
RO \quad\quad (S)
\end{array}
\qquad \text{(Scheme 5)}
$$

where R is an alkyl group.

These esters are stable at neutral or acidic pH but susceptible to hydrolysis in the presence of alkalis, where the P-O-X ester bond breaks down. The rate of the process is related to the nature of the constituent X, to the pressure of catalytic agents, and to pH and temperature.

The family of chlorinated hydrocarbons is among the most widely known insecticide group and includes: DDT, toxaphene, lindane, chlordane, aldrin, dieldrin, etc. Since the compounds in this family exhibit long persistence in soils, their use has decreased dramatically during the past few years. With the exception of lindane, all chlorinated hydrocarbons exhibit very low solubility in water. DDT is about ten times less soluble than the other compounds of this family, and is, therefore, considered to be immobile as a solute in soil systems. Endrin, dieldrin, and aldrin show higher water solubility and are, therefore, slightly mobile in soils. The vapor pressure of chlorinated hydrocarbons varies widely: from low (DDT, endrin, and dieldrin) to moderate (toxaphene and aldrin), high (chlordane and lindane), and very high (heptachlor). Volatilization of DDT from soils and other surfaces is, therefore, almost insignificant.

Organophosphates are more toxic than chlorinated hydrocarbons, particularly to humans, but they exhibit lower persistence in soils. This behavior is also related to an enhanced water solubility and higher vapor pressure of organophosphates. Malathion and parathion insecticides are known to be chemically hydrolyzed and biodegraded by microorganisms in soil systems. The most-used organophosphate herbicide is glyphosate (Senesi and Chen 1989).

Phenyl carbamates and substituted ureas are examples of nonionic herbicides. Substituting three of the hydrogen atoms of urea with other chemical groups such as phenyl, methyl, and/or methoxy groups produces effective herbicides. The chemical structure of substituted urea is (Scheme 6):

$$
\begin{array}{ccccc}
H & & O & & R_2 \\
\backslash & & \| & & / \\
N & - & C & - & N \\
/ & & & & \backslash \\
R_1 & & & & R_3
\end{array}
\qquad \text{(Scheme 6)}
$$

Most urea herbicides are nonselective and are usually applied to soil, where they are absorbed by roots and translocated apoplastically to the upper part of the plant. The amount of herbicide translocated differs among the various substituted urea compounds. Diuron is widely used as a preemergence herbicide for control of a large number of weeds. Monuron has a much greater solubility in water than diuron and was the first urea-type herbicide produced.

Carbamate and thiocarbamate herbicides derive their base structural formula from carbamic acid (Scheme 7):

Carbamate: $R_1{-}N(H){-}C({=}O){-}O{-}R_2$

Thiocarbamate: $(R_1)(R_2)N{-}C({=}O){-}S{-}R_3$

(Scheme 7)

In the thiocarbamates, one of the oxygen atoms is replaced by a sulfur atom. Other substitutions also occur.

The carbamates are soil-(e.g., propham and chlorpropham) or foliar-applied (e.g., asulam, phenmedipham) herbicides. They are characterized by a high specific selectivity on particular crops. Propham applied to the soil is readily translocated to the plants when absorbed by roots. Chlorpropham is more persistent in the soil than propham itself. Other carbamates such as barban and phenmedipham are mainly applied to the foliage, and their interaction with the soil is of limited significance.

Thiocarbamates are usually applied to the soil. Their high volatility necessitates quick incorporation into the soil or application via irrigation water. Thiocarbamate pesticides are typified by butylate, which is a soil-incorporated compound translocated upward to shoots via the apoplast system, following root uptake. EPTC (ethyl dipropylethiocarbamate) is a volatile soil-incorporated herbicide that is absorbed by seeds, roots, and emerging shoots in contact with treated soil. Metham (vapam) is formulated as a water-soluble compound. It is a temporary soil fumigant, able to control germinating weed seeds and perennial weeds, and is applied and incorporated into the soil.

2.2.1.2 Hydrocarbons

Organic solvents and petroleum products are an additional group of potential organic pollutants. In general, these products exhibit a low solubility in water, and most of them are apolar. Regarding their origin, the polar material content (wt.%) in crude oil ranges from 2.9 (Prudhoe) to 17.9 (Kuwait). The solubility of petroleum hydrocarbons in water increases with the increase of hydrocarbon polarity. Figure 2.2 (after Clark and Brown 1977) shows, for example, the fractional distillation distribution of crude and refined oils. The hydrocarbon groups are in themselves mixtures of components with various carbon structures (e.g., *n*-paraffins ranging from C_{11} to C_{32}; iso-paraffins, comprising 1-ring to 6-ring cycloparaffins; aromatics; benzene; toluene; C_8 to C_{11} aromatics, etc.) The proportions of the various hydrocarbon groups and the composition of each group are determined by both the origin of the crude oil and the distillation procedure. Since the petroleum product is not a homogeneous com-

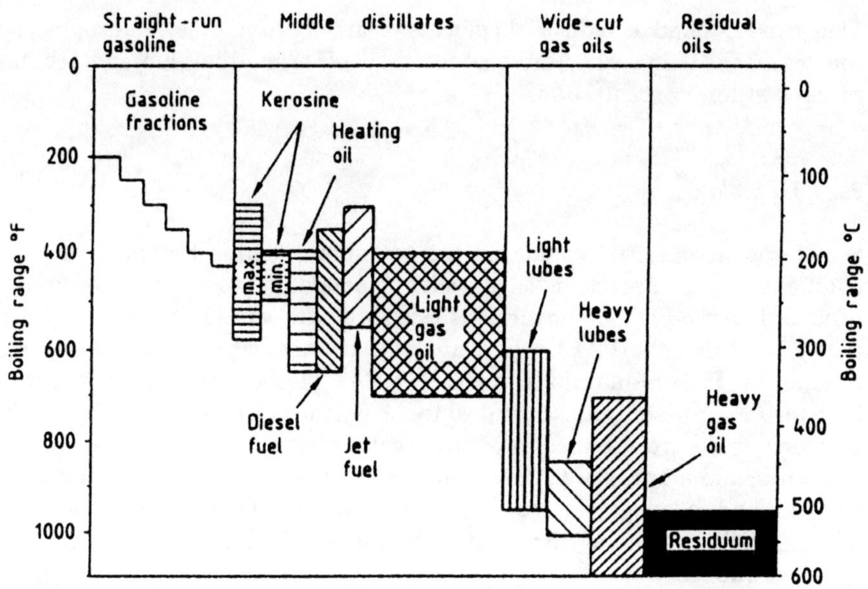

Fig. 2.2. Fractional distillation distribution of crude and refined oils. (Clark and Brown 1977)

pound, but a mixture of hydrocarbons with various physical and chemical properties, its behavior in porous media (e.g., volatilization, dissolution, adsorption) will be controlled not by the ideal behavior of each hydrocarbon component of the product, but by the properties of the mixture. It may be observed that the gasoline fraction is characterized by the lowest boiling point and is, therefore, the most volatile of the refined products. Oils, with the highest boiling points, are the least volatile petroleum fraction. The evaporation rate of a specific hydrocarbon is a function of its vapor pressure, which is inversely related to the carbon number and molecular weight. Compounds with molecular weights greater than n-C_{15} will continue to evaporate over long periods of time, although these rates become insignificant after 100 h. As mentioned previously, each petroleum product is a mixture of hydrocarbons with differing vapor pressures. Consequently, the loss of volatile hydrocarbons will lead to increases in the density and kinematic viscosity of the remaining petroleum products. Rostad (1976) showed, for example, that the kinematic viscosity of a Norwegian crude oil increased — due to evaporation of highly volatile compounds — in 48 h from 18.78 to 56.02 (mm^2s^{-1}), and its density from 0.862 to 0.894 g ml^{-1}. More details on petroleum hydrocarbon properties and their soil pollution effects can be found in the reviews of Clark and Brown (1977) and Yaron (1989).

Among the polar small organic molecules we can also find the polynuclear aromatic hydrocarbons, consisting of two or more benzene rings arranged in various configurations, with hydrophobicity increasing with molecular weight.

They can be found as industrial effluents or arising from other anthropogenic sources (Cole 1975). They are characterized by a long half-life and are hazardous at low concentrations.

2.2.1.3 Phtalates

Senesi and Chen (1989) emphasize the potential pollution hazard created by additions of small organic molecules such as phtalates. Most of the industrially produced *ortho*-phtalic anhydride is esterified with various alcohols to form phtalic acid diesters (PAE), which are used mainly as plasticizers. For PVC plastics, the PAE-resin ratio is usually 1:2. PAE plasticizers generally occur in monomeric forms, only loosely linked to the polymers and, therefore, are easily extruded. PAEs are also used as pesticide carriers and insect repellents, in dyes, cosmetics, and lubricants. PAEs are lipophilic or lyophobic liquids of medium viscosity and low vapor pressure. Di-2-ethylphtalate (DEHP) and di-octylphtalate (DOP) are among the most widely used plasticizers for plastics and synthetic rubbers.

Although DEHP is a moderately viscous liquid, practically immiscible with water, and has a very low vapor pressure, it has been detected in various aqueous solutions together with other PAEs. These chemicals are mostly regarded as pollutants from industrial sources, as their levels in the environment are high enough to threaten human health, directly or indirectly, through disturbance of the ecological balance.

2.2.1.4 Selected Relevant Properties

The small organic compounds were grouped and described above on the basis of their molecular properties. In the discussion to follow we will try to focus our discussion on selected relevant properties of the small organic molecules, with regard to their behavior in the soil environment and under fluctuating environmental conditions. As previously mentioned, we will not enlist the whole range of small organomolecules, but only one will be chosen to illustrate the relevant properties to be discussed. Details on the above topic may be found, for example, in the comprehensive review of Hayes and Mingelgrin (1991) on the interaction between small organochemicals and soil colloidal constituents.

Charge separation and dipole moment should be considered important factors controlling the interaction of the small molecules with a surface. The charge separation expresses the difference in electronegativity between various moieties of the molecule. The dipole moment is dependent on both the size of the separated charge and the distance of separation between the centers of charge in the molecule. Recently, it was demonstrated (Tsvetkov et al. 1990; Vimond-Laboudigue 1994) that, typical organic molecules being too large to be approximated by point-charge dipoles, the charge density in the interacting moiety in the molecule may correlate better than the dipole moment with the strength of interaction with a surface.

Molecular size and steric effects also affect the organic molecule-surface interaction. Experimental results suggest that large molecules are more strongly held than small molecules with similar shapes, charge characteristics and polarities. The steric factor can affect both the molecule retention on the surface and its surface enhanced transformation. The importance of the steric factor decreases, however, when the molecule is in a more free state, such as the solute in an aqueous solution.

Electron density distribution affects the binding of ionized and ionizable molecules to charged surfaces. Charges are delocalized on aromatic and conjugated aliphatic structures. The distribution of charge can be calculated by a semiempirical quantum mechanical procedure, as suggested by Burdon et al. (1977) for a series of soil organic molecules.

Solubility in water or in organic solvents of the small organic molecules controls the molecule retention on surfaces, due to the competition for the active site between the solvent and the solute.

Hayes and Mingelgrin (1991) presented the forces acting between solvent molecules in a variety of organic solvents and in water, and the reader will find aspects of that review relevant to the topic of this chapter.

Nonpolar organic compounds are thought to penetrate into voids in the liquid water structure as the result of thermal motions. Such filler molecules are considered to be structure makers. The mechanism of dissolution of polar molecules in water is different. Polar sorptives disrupt the hydrogen bonding, which gives a degree of structuring to liquid water, and new hydrogen bonds are formed between the sorptives and the water (Hayes and Mingelgrin 1991).

Volatility of the molecule should also be considered as a major relevant property governing the adsorption, transport, and persistence of the chemicals in the soil medium. Volatilization is defined as the loss of chemicals from surfaces in the vapor phase. Vaporization of chemicals, followed by movement into the atmosphere or into the unsaturated zone is related to their vapor pressures or vapor densities, and is affected by temperature and by the surrounding air movement (see Chap. 4).

2.2.2 Organic Macromolecules

The organic macromolecules are from two origins: natural and synthetic.

Naturally occurring macromolecules arising from proliferation of microorganisms and through the chemical transformation of organic materials in the soil medium include, on the one hand, humic and fulvic substances and, on the other, positively and negatively charged polysaccharides. The naturally occurring macromolecules could not be listed as soil pollutants, but they can play an important role in the soil pollution process as ligands of various inorganic and organic toxic chemicals (Clapp et al. 1991). A polymer is characterized by a flexible chain of statistical elements (segments) which affect its adsorption on soil solid phase.

Synthetic organic macromolecules could reach the land surface as compounds used in soil hydrophysical property reclamation, or as pollutants following human activities.

Amongst the soil reclamation products, the following may be included: (1) poly(vinyl) alcohol, (2) poly(acrylamide), (3) poly(ethylene glycol) or poly (ethylene oxide) and (4) poly(vinyl pyrolidine). All these polymers are water-soluble and nonionic, and their chemical structures are shown in Fig. 2.3.

The above products are, in general, not toxic, but they can become a hazard by accidental disposal, in high concentrations, on the land surface. An example of this polymerization of organic molecules is observed by studying the fate of phenate under land-fill sites. The phenols could be polymerized by the mineral surface reactions of phenol with clays, leading to the formation of colored products which, in general, are of a higher molecular weight than the parent compounds. It was found that the phenolic compounds from the Ca-clay-2, 6-dimethyl phenol complex can reach values as high as 3000 m/e (Sawhney 1989). Figure 2.4 shows the configuration of phenolic compounds identified in methanol extracts of clay-organic complexes.

The surfactants are an additional group of organic molecules, which reach the soil surface mainly as a result of sludge or effluent disposal. In the last 20 years, the biodegradable detergent has replaced the persistent one and anionic

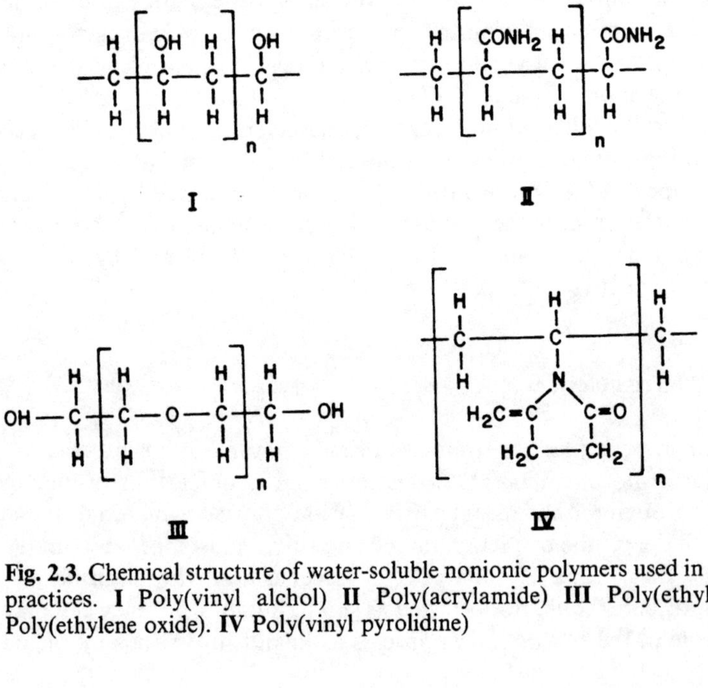

Fig. 2.3. Chemical structure of water-soluble nonionic polymers used in soil reclamation practices. **I** Poly(vinyl alchol) **II** Poly(acrylamide) **III** Poly(ethylene glycol) or Poly(ethylene oxide). **IV** Poly(vinyl pyrolidine)

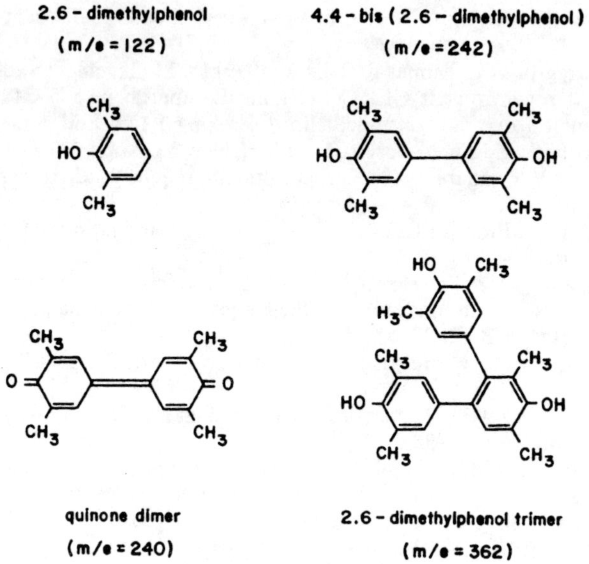

2.6 - dimethylphenol (m/e = 122)

4.4 - bis (2.6 - dimethylphenol) (m/e = 242)

quinone dimer (m/e = 240)

2.6 - dimethylphenol trimer (m/e = 362)

Fig. 2.4. Phenolic compounds identified in methanol extract of clay organic complexes. (After Sawhney et al. 1984)

surfactants such as the alkyl-aryl-sulfonate (LAS) have become the most popular detergents. The structure of this compound consists of a straight-chain alkyl group, a benzene ring, and a sulfonate group. The structure and the geometrical shape of the molecules can be different in different members of the anionic surfactant group, and this affects their behavior in soil. As an example, we can give the structures of three anionic surfactants: LAS — linear alkyl sulfonate, DDBS — dodecyl benzene sulfuric acid, and NaLS — sodium lakuryl sulfate. DDBS and LAS surfactants contain relatively bulky substituent groups, while NaLS has a filiform geometrical structure. The structure defines the interaction with the soil mineral phase (Acher and Yaron 1977). The anionic surfactants in water and aqueous solution are detected by a methylene blue method which depends on the formation of blue-colored salt when methylene blue reacts with the anionic surfactant. This group of surfactants is known as methylene blue-active substances (MBAS).

References

Acher AJ, Yaron B (1977) Behaviour of anionic surfactants in a soil-sewage effluent system. J Environ Qual 6: 418–420

Adriano DC (1986) Trace elements in the terrestrial environment. Springer, Berlin Heidelberg New York, 533 pp

Akins MB, Lewis RJ (1976) Chemical distribution and gaseous evolution of arsenic-74 added to soils as DSMA^{+4}As. Soil Sci Soc Am J 40: 655–658

Allison FE (1973) Soil organic matter and its role in the crop production. Elsevier, Amsterdam, 63 pp

Alloway BHJ, Gregson M, Gregson SK, Tanner R, Tills A (1979) In: Management and control of heavy metals in the environment. CEP Consultants, Edinburgh, pp 545–548

Anderson G (1975) Sulfur in soil organic substances. In: Gieseking J (ed) Soil components 1. Organic Compounds. Springer, Berlin Heidelberg New York, pp 333–341

Barber SA (1984) Soil nutrient bioavailability – a mechanistic approach. Wiley-Interscience, New York, 398 pp

Berrow ML, Reaves GA (1984) In: Proc Int Conf Environmental Contamination. CES Consultants, Edinburgh, pp 334–340

Burdon J, Hayes MHB, Pick ME (1977) The electron density distribution in pyridinium and bipyridinium compounds and its relevance to their adsorption by expanding lattice clay. J Environ Sci Health B 12: 37–51

Calvet R (1989) Adsorption of organic chemicals in soils. Environ Health Perspect 83: 145–177

Chan KY, Davey BG, Geering HR (1979) Adsorption of magnesium and calcium by a soil with variable charge. Soil Sci Soc Am J 43: 301–304

Clapp CE, Harrison R, Hayes MHB (1991) Interactions between organic macromolecules and soil inorganic colloids and soils. In: Bolt CH et al. (eds) Interaction at the soil colloid-soil solution interface. NATO ASI Ser E 190: 409–469

Clark RC, Brown DW (1977) Petroleum-properties and analysis in biotic and abiotic systems. In: Mailins DC (ed) Effects of petroleum in arctic and sub-arctic marine environments and organisms. Academic Press, London, pp 1–89

Cole JA (1975) Ground water pollution in Europe. Water Inf Center, Port Washington, New York

Cotton FA, Wilkinson G (1980) Advanced inorganic chemistry. Wiley, New York, 314 pp

Deuel LE, Swoboda AR (1972) Arsenic solubility in reduced environment. J Soil Sci 36: 276–278

Doner HE, Pukite A, Yange E (1982) Mobility through soils of certain heavy metals in geothermal brine water. J Environ Qual 11: 389–394

Feigin A, Ravina I, Shalhevet J (1991) Irrigation with treated sewage effluent. Springer, Berlin Heidelberg New York, 224 pp

Hahne HCH, Kroontje W (1973) Significance of pH and chloride concentration on behaviour of heavy metal pollutants: mercury (II), cadmium (II), zinc (II) and lead (II). J Environ Qual 2: 444–450

Hayes MHB (1987) Sorption and chemical transformation processes of small organic chemicals in soil. Trans XIII Congr Int Soil Sci, Hamburg, 1986, 6: 584–595

Hayes MHB, Mingelgrin U (1991) Interactions between small organic chemicals and soil colloidal constituents. In: Bolt GH et al. (eds) Interactions at the soil colloid-soil solution interface. NATO ASI Ser E 190: 323–408

Hook JE (1983) Movement of phosphates and nitrogen in soils following application of municipal wastewater. In: Nielson DW, Elrik DE, Tanji KK (eds) Chemical mobility and reactivity in soil system. Soil Sci Soc Am Spec Publ 11: 241–255

Jenny H (1941) Factors of soil formation. McGraw-Hill, New York, 281 pp

Lindsay WL (1972) In: Nortvedt JJ, Giordano PM, Lindsay WL (eds) Micronutrients in agriculture. Soil Science Society of America, Madison, pp 41–57

Lindsay WL (1979) Chemical equilibria in soils. John Wiley, New York

Mattigod SV, Sposito G (1977) Estimated association constants for some complexes of trace metals with inorganic ligands. Soil Sci Soc Am J 41: 1092–1097

McBride MB (1981) Long-term reactions of copper (II) in a contaminated calcareous soil. In: Lonegran JF, Robson AD, Graham RD (eds) Copper in soils and plants. Academic Press, New York, pp 25–45

McBride MB, Bouldin DR (1984) Long-term reactions of copper (II) in a contaminated calcareous soil. Soil Sci Soc Am J 48: 56–59

McLaren RG, Crawford BV (1973) Studies on soil copper I. The formation of copper in Soils. J Soil Sci 24: 172–181

Mengel K (1985) Dynamics and availability of major nutrients in soils. Adv Soil Sci 2: 67–134

Neptune AML, Tabatabai MA, Hanway JJ (1975) Sulfur fractions and carbon-nitrogen-phosphorus-sulfur relationship in some Brazilian and Iowa soils. Soil Sci Soc Am Proc 39: 51–55

Nriagu JD (ed) (1978) The biogeochemistry of lead in the environment. Elsevier, Amsterdam, 247 pp

Nriagu JD (ed) (1984) Changing metal cycles in human health. Springer, Berlin Heidelburg New York, 445 pp

Olsen SR, Sommers LT (1982) Phosphorus. In: Page AL, Miller RH, Keney DR (eds) Methods of soil analysis 2: Chemical and microbiological properties. Agronomy 9: 404–430

Reaves GA, Berrow ML (1984) Total lead concentration in Scottish soils. Geoderma 32: 1–8

Rice MA, Kamparth EJ (1968) Availability of exchangeable and nonexchangeable Mg in sandy Coastal Plain Soils. Soil Sci Soc Am Proc 32: 386–388

Rostad H (1976) Behaviour of oil spills with emphasis on the North Sea. Report of the Continental Shelf Inst, Trondheim

Ryden JC, Pratt PF (1980) Phosphorus removal from wastewater applied to land. Hilgardia 48: 1–36

Saltzman S, Yaron B (eds) (1986) Pesticides in soils. Van Nostrand Reinhold, New York, 379 pp

Sawhney BL (1989) Movement of organic chemicals through landfills and hazardous waste disposal sites. In: Sawhney BL, Brown K (eds) Reaction and movement of organic chemicals in soils. SSSA Spec Publ 22. Soil Science Society of America, Madison, pp 447–474

Sawhney BL, Isaacson PJ, Gent MPN (1984) Polymerization of 2,6-dimethylphenol on smectite surfaces. Clays and Clay Miner 33: 123–127

Senesi N, Chen Y (1989) Interactions of toxic organic chemicals with humic substances. In: Gerstl Z, Chen Y, Mingelgrin U, Yaron B (eds) Toxic organic chemicals in porous media. Springer, Berlin Heidelberg New York, pp 37–90

Sposito G, Lund JL, Chang AC (1982) Trace metal chemistry in arid zone soils amended with sewage. Study I. Fractionation of Ni, Cu, Zn, Cd and Pb in solid phases. Soil Sci Soc Am J 46: 260–264

Stevenson FJ (ed) (1982) Nitrogen in agricultural soils. Am Soc Agron, Madison, Wisconsin, 484 pp

Tsvetkov F, Mingelgrin U, Lahav N (1990) Cross-linked hydroxy-Al-montmorillonite as a stationary phase in liquid chromatography. Clays Clay Minerals 38: 380–390

Vimond-Laboudigue A (1994) Etat et localisation d'un pesticide (le dinoseb) fixé par l'hectorite et la vermiculite decylamonium. DSc Thèse No 1340. Université des Sciences, Lille, 122 pp

Weber JB (1972) Interaction of organic pesticides with particulate matter in aquatic and soil system. In: Faust SD (ed) Fate of organic pesticides in the aquatic environments. Adv Chem Ser III. American Chemical Society, Washington DC, pp 55–120

Worthing CR, Hance JR (eds) (1991) The pesticide manual – a world compendium. British Crop Protection Council Suffolk, 1141 pp

Yaron B (1989) On the behaviour of petroleum hydrocarbon in the unsaturated zone: abiotic aspects. In: Gerstl Z, Chen Y, Mingelgrin U, Yaron B (eds) Toxic organic chemicals in porous media. Springer, Berlin Heidelberg New York, pp 211–230

Yaron B, Gerstl Z, Spencer WF (1985) Behaviour of herbicides in irrigated soils. Adv Soil Sci 3: 121–211

Part II
Pollutants Partitioning Among Soil Phases

CHAPTER 3

Pollutants-Soil Solution Interactions

Solubility equilibrium is the final state to be reached by a toxic chemical and the soil water phase under well-defined environmental conditions. Equilibrium provides a valuable reference point that is useful for ascertaining the chemical reaction.

Equilibrium for a reaction

$$A + B \rightleftharpoons AB \tag{3.1}$$

can be expressed in terms of the formation constant (K_f) where

$$K_f = \frac{AB}{A.B} \ . \tag{3.2}$$

When the reaction is inverse

$$AB \rightleftharpoons A + B \ , \tag{3.3}$$

we are dealing with a dissociation process where the constant (K_d) was

$$K_d = \frac{A.B}{AB} \quad \text{with } K_f = \frac{1}{K_d} \ . \tag{3.4}$$

Equilibrium constants can be expressed either on concentration basis (K^c) either on activity basis (K°) or as mixed constants (K^m) in which all terms are given in concentration except for H^+, OH^-, and e^- (electron) which are given in activities.

Soil systems comprise a dynamic and complex array of inorganic and organic constituents and at any moment the concentration of pollutants in the soil liquid phase is governed by a series of reactions such as acid-base equilibria, oxidation-reduction equilibria, complexation with organic and inorganic ligands, precipitation and dissolution of solids and, finally, ion exchange and adsorption. The rates at which these reactions occur, together with the rates of degradation and/or biological uptake, control the concentrations of the pollutants in the soil liquid phase. Figure 3.1 (adapted from Mattigod et al. 1981) shows the interrelation between the above reactions in the soil system and the soil solution.

A short description of the reactions involving the transfer of protons and/or electrons and affecting the solubility equilibria follows.

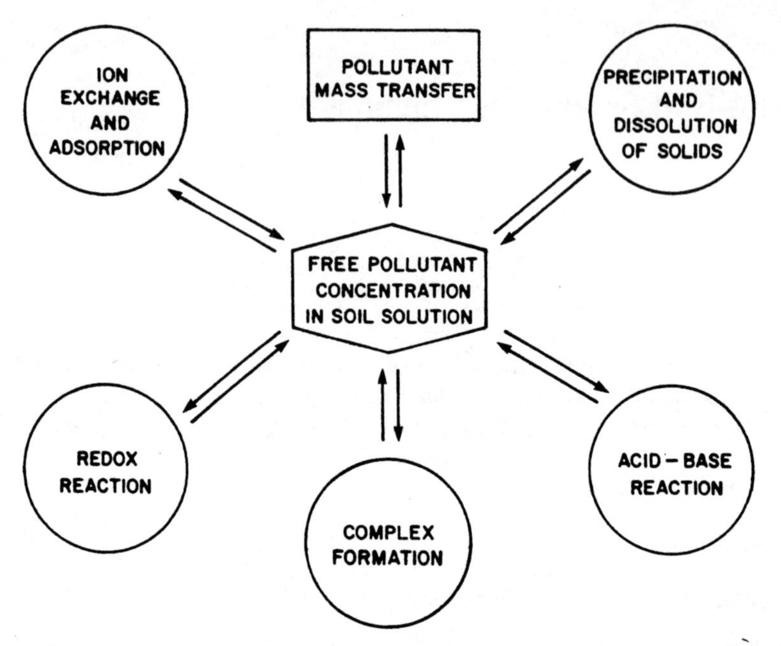

Fig. 3.1. Principal processes controlling the concentrations of free pollutant solutions. (After Mattigod et al. 1981)

3.1 Acid-Base Equilibria

The group of reactions covering the transfer of protons governs the acid-base equilibrium relation. In this case the proton donor is an acid and the acceptor a base.

$$\text{Acid} \rightleftharpoons \text{base} + \text{proton} \tag{3.5}$$

with the following equilibrium constant:

$$K_A^\circ = \frac{[H^+][\text{base}]}{[\text{acid}]}\ , \tag{3.6}$$

where K_A° is an acidity constant, [ ] are activities.

This equation stipulates that for a given ratio of the activities of a particular acid and its conjugate base, the proton activity has a fixed value. In a system where we have a complex of two bases and two acids, the relationship becomes

$$\text{acid}_1 + \text{base}_2 \rightleftharpoons \text{base}_1 + \text{acid}_2 \tag{3.7}$$

The corresponding equilibrium under this condition is

$$K^\circ_{1.2} = K^\circ_{A.1}/K^\circ_{A.2} = \frac{[\mathrm{base}_1][\mathrm{acid}_2]}{[\mathrm{acid}_1][\mathrm{base}_2]} \ . \tag{3.8}$$

The soil solution being an aqueous system, H_2O is the ever-present acceptor. For example, during the dissociation of an acid in the soil solution, H_3O^+ will be one of the dissociation products and the acid strength of the H acid relative to H_3O^+ will be a measurable parameter. In a dilute solution, the activity of the hydrated protons equals that of H_3O^+ and the pH value characterizes the H-ion activity.

Under these conditions, substituting for pH into Eq. (3.6) gives:

$$\mathrm{pH} - \mathrm{pK_A} = \log\frac{[\mathrm{base}]}{[\mathrm{acid}]} \ . \tag{3.9}$$

3.2 Precipitation-Dissolution

Under various physicochemical conditions, a process of dissolution and precipitation of minerals in water occurs. This reaction is more common for the natural systems and plays a lesser part in the fate of pollutants. A short discussion will be presented, however, since dissolution of a particular chemical in the soil water phase could affect the speciation of pollutants in the water phase.

The extent of the dissolution or precipitation reaction for systems that attain equilibrium can be estimated by considering the equilibrium constant. The precipitation of chemicals can only occur at supersaturation, since dissolution ends only at water saturation. In the soil medium the dissolution reaction is a heterogeneous chemical reaction and it is necessary to characterize the solubility in terms of a solubility product with an equilibrium constant which can characterize the solubility as well as predict how solution variables can change the solubility. The dissolution in water of an electrolyte comprising a cation A^{+n} and an anion B^{-m} behaves according to the following reaction:

$$A_mB_n(s) \rightleftharpoons mA^{+n}(aq) + nB^{-m}(aq) \ , \tag{3.10}$$

where (s) and (aq) stand for the solid and the aqueous phase, respectively. The equilibrium condition is

$$[A_mB_n(s)] \rightleftharpoons [A^{+n}(aq)]^m[B^{-m}(aq)]^n \ ,$$

which gives the coventional solubility expression

$$K_{so} \rightleftharpoons [A^{+n}(aq)]^m[B^{-m}(aq)]^n \ ,$$

where K_{so} is the solubility product, and [ ] activities. This formula assumes that the activity of the pure solid phase is set equal to unity and that the common standard-state convention for aqueous solutions is adopted. Although there are various features affecting the activity of the solid phase (the lattice energy, degree of hydration, solid solution formation, and the free energy of the surface), this activity is implicitly contained in the solubility equilibrium constant. Under these conditions, the solubility product is constant for varying compositions of the liquid phase under constant temperature and pressure and for a chemically pure and compositionally invariant solid phase. The solubility equilibrium is influenced by the ionic strength of the solution and, therefore, Stumm and Morgan (1981) recommended the use of an operational equilibrium constant, that is, a constant expressed as concentration quotients and valid for a medium of given ionic strength. The same authors consider that a possible test to determine whether a solution is over or undersaturated would be to check whether the dissolution free energy of the solid chemical is positive, negative or zero, and to compare the ion activity product (IAP) with K_{so}. The state of saturation of the solution with respect to a chemical in solid state is defined:

$$\begin{aligned} &\mathrm{IAP} > K_{so} \text{ (oversaturated)} \\ &\mathrm{IAP} = K_{so} \text{ (equilibrium, saturated)} \\ &\mathrm{IAP} < K_{so} \text{ (undersaturated)} \ . \end{aligned} \tag{3.11}$$

Oversaturation leads to the precipitation of the chemical; Fig. 3.2 (after Stumm and Morgan 1981) illustrates in a schematic way some of the domains of precipitation and solubility.

A chemical substance can exist in more than one crystalline form but, except at a transition temperature, only one is thermodynamically stable for a given chemical composition of the system. The metastable forms are more soluble

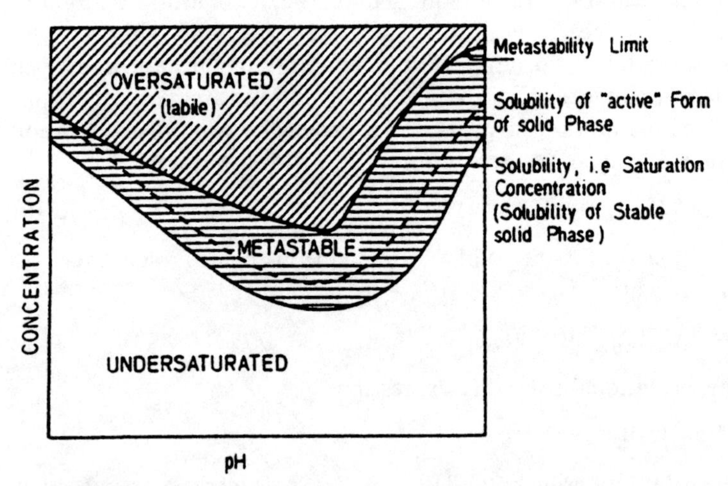

Fig. 3.2. Solubility and saturation. A schematic solubility diagram showing concentration range versus pH for supersaturated, metastable, saturated, and undersaturated solutions. (Stumm and Morgan 1981). An "active" form is very fine crystalline precipitate formed incipiently from strongly oversaturated solutions

than the stable ones. Similarly, the allotropic forms have different activities, the lowest being the stable one. Coprecipitation of inorganic trace elements, for example, as impurities in their oxide forms, may confound the solubility product approach by producing solid solutions with ill-defined stoichiometry.

Logarithmic diagrams are used to define the relationships among the activities of the various components involved in a precipitation/dissolution reaction at equilibrium. The solubility diagram is a graph of the logarithm of the ratio versus a well-defined soil solution parameter, such as pH. For a given value of the soil solution concentration, any parameter is treated as an independent variable, assuming that all solid phases are in their standard states. The solid that produces the largest value of the logarithm of the activity ratio is the one that is most stable and consequently the only one that will form at equilibrium.

It is not the aim of our book to treat solubility diagrams in depth, but the reader is directed to the books of Lindsay (1979), Stumm and Morgan (1981), and Sposito (1981) for further details.

3.3 Ligand Effect

A molecule or an ion is present in a solution in various forms called chemical species (or more concisely, species). Any combination of cations with molecules or anions which contains free pairs of electrons is called complex formation; it can be electrostatic, covalent, or a mixture of both. The cation will be the *central atom* and the anions or the molecules which form a coordination compound are the *ligands*. If a molecule contains more than one ligand, the atom forms a multidentate complex – such complexes are called chelates. In a broad classification of chemical reactions, the complex formation is a coordinative reaction, as are the acid-base and precipitation-dissolution reactions, and phenomenologically these are conceptually similar. Once it reaches the soil solution, the base ion enters a continuing search for a partner. The coordination reactions of a cation in the soil solution are exchange reactions with the coordinated water molecules, which are exchanged for some preferred ligands. The principle of hard and soft Lewis acids and bases (HSAB) proposed by Pearson (see Stumm and Morgan 1981) is useful to describe these reactions. A Lewis acid is any chemical species that employs an empty electronic orbital in initiating a reaction, while a Lewis base is any chemical species that employs a doubly occupied electronic orbital in initiating a reaction. Lewis acids and bases can be neutral molecules, simple or complex ions, or neutral or charged macromolecules. The proton and all of the metal cations of interest in soil solutions are Lewis acids. The Lewis bases include H_2O, oxyanions such as OH^-, COO^-, CO_3^{2-}, SO_4^{2-}, and $PO_4{}^{3-}$, and organic N, S, and P electron donors. A list of selected hard and soft Lewis acids and bases found in soil solution is presented in Table 3.1.

Table 3.1. Hard and soft Lewis acids and bases in soil solution. (Sposito 1981)

Lewis Acids
Hard acids
H^+, Li^+, Na^+, K^+, (Rb^+, Cs^+), Mg^{2+}, Ca^{2+}, Sr^{2+}, (Ba^{2+}), Ti^{4+}, Zr^{4+}, Cr^{3+}, Cr^{6+}, MoO^{3+}, Mn^{2+}, Mn^{3+}, Fe^{3+}, Co^{3+}, Al^{3+}, Si^{4+}, CO_2
Borderline acids
Fe^{2+}, Co^{2+}, Ni^{2+}, Cu^{2+}, Zn^{2+}, (Pb^{2+})
Soft Acids
Cu^+, Ag^+, Au^+, Cd^{2+}, Hg^+, Hg^{2+}, Ch_3Hg^+; pi-acceptors such as quinones; bulk metals
Lewis bases
Hard bases
NH_3, RNH_2, H_2O, OH^-, O^{2-}, ROH, CH_3COO^-, CO_3^{2-}, NO_3^-, PO_4^{3-}, SO_4^{2-}, F^-
Borderline bases
$C_6H_5NH_2$, C_2H_5N, N_2, NO_2^-, SO_3^{2-}, Br^-, (Cl^-)
Soft bases
C_2H_4, C_6H_6, R_3P $(RO)_3P$, R_3As, R_2S, RSH, $S_2O_3^-$, S^{2-}, I^-

R = organic molecular unit. ( ) indicates a tendency to softness.

The complex formed between a Lewis acid and base is called an *inner-sphere complex*. When the water molecule is interposed between the acid and the base we are dealing with an *outer-sphere* complex. A hard Lewis acid is a molecular unit of small size, high oxidation state, high electronegativity, and low polarizability, whereas a molecular unit of a relatively large size, characterized by low oxidation state, low electronegativity and high polarizability is a soft Lewis acid. A hard Lewis base is characterized by a high electronegativity and a low polarizability, while a soft Lewis base exhibits low electronegativity and high polarizability. Based on this theory, hard bases prefer to complex hard acids, and soft bases prefer to complex soft acids, under comparable conditions of acid-base strength. This rule is known as the HSAB Principle.

It is well known that the presence of foreign ligands in the soil solution leads to an increase of the solubility of coordinating ions. In the case of a complex formation with any ligand L_n or its protonated form HL, the total solubility is given by the expression

$$A_T \rightleftharpoons [A]_{free} + \Sigma\, [A_m H_K L_n (OH)_i] \tag{3.12}$$

where A represents the ion to be coordinated with different ligand types L_n and where all values of m, n, i, or $K \geq 0$ have to be considered in the summation.

If a compound $A_m L_n$ has a very small solubility product, K_{so}, the resultant molecular association exists in the solution as stable complexes. Stumm and Morgan (1981) show that this behavior is understandable if it is considered that the same forces that result in slightly soluble lattices are operative in the formation of soluble complexes.

3.4 Oxidation-Reduction Equilibrium

There is a conceptual analogy between acid-base and reduction-oxidation reactions. Similarly to the consideration of acids and bases as proton donors and proton acceptors, respectively, reductants and oxidants are regarded as electron donors and electron acceptors. Under this type of reaction are included those involving the transfer of electrons from donors to acceptors. The redox reaction between n_B mol of an oxidant A_{ox} and n_A mol of a reductant B_{red} can be written

$$n_B A_{ox} + n_A B_{red} \rightleftharpoons n_B A_{red} + n_A B_{ox} \ , \tag{3.13}$$

which is equivalent to the proton exchange equation.

Every redox reaction includes a reduction half-reaction and an oxidation half-reaction. A reduction half-reaction, for example, in which a chemical species accepts electrons, may be written in the form

$$aA_{ox} + bH^+ + e \rightleftharpoons zA_{red} + gH_2O \ , \tag{3.14}$$

where A represents a chemical species in any phase and the subscripts ox and red denote its oxidized and reduced states, respectively. The parameters a, b, z, and g are stoichiometric coefficients, H^+ and e denote the proton and the electron in the aqueous solution.

The designations oxidized and reduced for a given chemical species are provided with quantitative significance through the concept of *oxidation number*, the hypothetical valence that may be assigned to each atom. The oxidation number is denoted by a positive or negative Roman numeral and is computed according to a set of rules: (1) For a monoatomic species, the oxidation number is the same as the valence. (2) For a molecule, the sum of oxidation numbers of the constituent atoms must be equal to the net charge on the molecule, expressed in terms of the protonic charge. (3) For a chemical bond in a molecule, the shareable bonding electrons are assigned entirely to the more electronegative atom participating in the bond. If there is no difference in electronegativity between the two bonding atoms, the electrons are assigned equally. Considering C, Fe, Mn, N, O, and S as natural soil elements and As, Cr, Cu, and Pb as selected pollutants, Sposito (1981) listed some reduction half-reactions and their equilibrium constants in soil solution (Table 3.2).

The potential for redox reactions to occur in soil solution always exists, but the accomplishment of the reaction does not always occur; the redox reaction favored thermodynamically is not necessarily kinetically favored. Under such conditions, due to a slowness of the reactions, the equilibrium could be achieved in the presence of a catalysis process. In the soil medium, the microbial organisms may act as catalysts in the redox reactions and, under favorable conditions, their role will be effective. The redox reactions are controlled only by the thermodynamic processes and the soil microbial-induced catalysis is controlled only by the reaction kinetics.

Table 3.2. Selected reduction half-reactions in soil solution. (Sposito 1981)

Reaction[a]	log K_R
$\frac{1}{4}O_2(g) + H^+ + e = \frac{1}{2}H_2O(l)$	20.78
$\frac{1}{5}NO_3^- + \frac{6}{5}H^+ + e = \frac{1}{10}N_2(g) + \frac{3}{5}H_2O(l)$	21.06
$\frac{1}{3}NO_2^- + \frac{4}{3}H^+ + e = \frac{1}{6}N_2(g) + \frac{2}{3}H_2O(l)$	25.70
$\frac{1}{6}N_2(g) + \frac{4}{3}H^+ + e = \frac{1}{3}NH_4^+$	4.65
$\frac{1}{2}MnO_2(s) + 2H^+ + e = \frac{1}{2}Mn^{2+} + H_2O(l)$	20.69
$MnOOH(s) + 3H^+ + e = Mn^{2+} + 2H_2O(l)$	22.86
$\frac{1}{2}Mn_2O_3(s) + 3H^+ + e = Mn^{2+} + \frac{3}{2}H_2O(l)$	24.39
$\frac{1}{2}Mn_3O_4(s) + 4H^+ + e = \frac{3}{2}Mn^{2+} + 2H_2O(l)$	30.83
$\frac{1}{2}Mn_3O_4(s) + 4H^+ + \frac{3}{2}CO_3^{2-} + e = \frac{3}{2}MnCO_3(s) + 2H_2O(l)$	46.59
$Fe(OH)_3(s) + 3H^+ + e = Fe^{2+} + 3H_2O(l)$	15.87
$\frac{1}{2}Fe_3O_4(s) + 4H^+ + e = \frac{3}{2}Fe^{2+} + H_2O(l)$	20.79
$\frac{1}{2}HCOO^- + \frac{3}{2}H^+ + e = \frac{1}{2}HCOH + \frac{1}{2}H_2O(l)$	2.82
$\frac{1}{8}CO_2(g) + H^+ + e = \frac{1}{8}CH_4(g) + \frac{1}{4}H_2O(l)$	2.86
$\frac{1}{2}HCOH + H^+ + e = \frac{1}{2}CH_3OH$	3.92
$\frac{1}{2}CH_3OH + H^+ + e = \frac{1}{2}CH_4(g) + \frac{1}{2}H_2O(l)$	9.94
$\frac{1}{8}SO_4^{2-} + H^+ + e = \frac{1}{8}S^{2-} + \frac{1}{2}H_2O(l)$	2.52
$\frac{1}{8}SO_4^{2-} + \frac{9}{8}H^+ + e = \frac{1}{8}HS^- + \frac{1}{2}H_2O(l)$	4.26
$\frac{1}{8}SO_4^{2-} + \frac{5}{4}H^+ + e = \frac{1}{8}H_2S(g) + \frac{1}{2}H_2O(l)$	5.26
$\frac{1}{8}HSO_4^- + \frac{9}{8}H^+ + e = \frac{1}{8}H_2S(aq) + \frac{1}{2}H_2O(l)$	4.89
$\frac{1}{2}S(s) + e = \frac{1}{2}S^{2-}$	-8.0512

[a]Charged species are in aqueous solution; log K_R values refer to 298.15 K and 1 atm.

3.5 Effects of Mixed Solvents and Surfactants

Many of the potentially toxic organic chemicals reach the soil in a mixture of solvents or in formulations with dispersing agents (surfactants). The objective of such a formulation is to increase the solubility of the active compound in a solvent such as water. In a solvent which forms nearly ideal solutions with another, and when the solubility of a third substance is limited to a low value by its crystal stability, the logarithm of its solubility is nearly a linear function of the mole fraction composition of the solvent (Hartley and Graham-Bryce 1980). In the case of a pair of solvents in which the solubilities of the third substance differ greatly, these solvents – due to their different solvent powers – are generally much less soluble in one another and form miscible solutions which are far from linear. The amount of solute dissolved in a mixture of two equal amounts of the solvents is very much less than proportional to the amount expected to be dissolved by the more powerful solvent. In the case of a powerful organic solvent miscible with water, a more nearly linear slope for the log solubility vs. solvent composition relationship is obtained if the composition is plotted as volume fraction rather than mole fraction.

Yalkowsky and Roseman (1981) and Rubino and Yalkowsky (1987) suggest the following equations for relating the solubility (S_m) of a nonpolar solute in a binary mixture of an organic solvent and water to that in pure water (S_w):

$$\log S_m = \log S_w + \sigma_s \text{-} f_s \text{-} \quad (3.15)$$

with

$$\sigma_s = (A \log K_{ow} + B) \ ,$$

where f_s is the volume fraction of the cosolvent in the binary mixed solvent; A and B are empirical constants dependent on cosolvent properties and K_{ow} is the organic solvent-water partition coefficient of the solute.

Extrapolating this formula to a multiple solvents-solute case (complex mixture) Rao et al. (1989) express the solubility of a solute in a complex mixture as

$$\log S_m = \log S_w + \Sigma \sigma_i f_i \ , \quad (3.16)$$

where the subscript i indicates the i-th cosolvent. The solute mixtures of interest could comprise, for example, two hydrophobic organic chemicals, or one hydrophobic and one ionizable organic chemical. Crystalline salts of many organic acids and bases often have a maximum solubility in mixtures of water and water-miscible solvents. The ionic part of such a dissociable molecule requires a strongly polar solvent – such as water – to initiate dissociation. A mixture of water-miscible solvents hydrates and dissociates the ionic fraction of the solutes, and the water-miscible organic solvent accepts the remaining pollutants at a higher concentration than would either solvent alone. The deliberate use of a water-insoluble solvent as cosolvent in the formulation of toxic organic chemicals can lead to the increased solubility of hydrophobic organic chemicals in the aqueous phase and, consequently, to a potential increase in their transport from land surface to groundwater.

An increase in solubilization in water of nonpolar organic chemicals is obtained when surfactants are present in the water solution. However, the solvent power of surfactants is much greater than that of a simple cosolvent. The structure of the surfactant solution, above a rather well-defined critical concentration for micelles (cmc), is that of an ultrafine emulsion. The surfactant molecules are aggregated, forming a cluster of 20-200 units or more. The hydrophobic "tails" are oriented to the interior of the cluster and the hydrophilic "heads" are oriented to the exterior, in contact with the water phase. The interior retains a normal nonpolar solvent power for nonpolar crystalline solutes because it is not, on a molecular scale, diluted by water. The total solvent power is determined by the number of micelles, their size and structure, and is not proportional to surfactant concentration.

This type of dissolving action has been called solubilization despite the fact that it is not an appropriate name with regard to the mechanism involved. Hartley and Graham-Bryce (1980) showed that the solubility of a crystalline solute of low water solubility cannot increase continuously with the expansion of the micelle. The limitation comes from the fact that the micelle cannot increase indefinitely in size without modifications of its structure and properties due to reorientation of surfactant molecules. There is, therefore, a disadvantage, energetic and/or entropic, in indefinite expansion, and this sets a limit, in a complex way never exactly formulated, to the solubility. An example of the consequence of such a phenomenon is given by the presence in municipal disposal sites of surfactants which can, in this way, enhance the solubilization of hydrophobic organic substances and their transport through soils.

3.6 Temperature Effect of Solubility

The solubility in water of most inorganic and organic pollutants increases with temperature. The changes in the "real" concentration of a solute during changes in ambient temperature should be considered when we are dealing with the partitioning of the pollutants between the soil phases.

The temperature not only has a direct effect on chemical solubility in the soil aqueous phase but also has an indirect effect on the reactions previously described. The seasonal variation of temperature might affect the solubility of toxic chemicals in the soil solution; therefore, the observed solubility equilibria, if any, reflect only the solubility of the compound at a given time and ambient temperature.

3.7 Examples

So far, the theoretical aspects of the reactions affecting the solubility of pollutants in the soil solution have been discussed. In this part, selected examples will be presented only to illustrate the above reactions. It must, however, be remembered that these examples will not cover the whole spectrum of possibilities occurring in soils.

3.7.1 Acid Rains and Al Dissolution

Hydrogen ion regulation in soil solution is provided by numerous homogeneous and heterogeneous buffer systems. A buffer system is characterized by the range of pH within which it is efficient, relative to its acid- or base-neutralizing capacity. From the point of view of interaction with solid and dissolved bases, the most important acidic constituent is CO_2, which forms H_2CO_3 with water. In a natural environment, the pH of the soil solution does not generally correspond to an equilibrium state; this results in instability of pH values, which is reflected in short-term fluctuations, due to seasonal variations and to anthropogenic processes temporarily occurring in soils.

A classical example of a man-induced effect on the acid-base equilibrium in the soil liquid phase is the acid rain effect. As a result of atmospheric pollutants (e.g., HCL, HNO_3, H_2SO_4) the rainwater becomes acid. The genesis of an acid rainwater – as summarized by Zobrist and Stumm (1979) is presented in Fig. 3.3. If an acid rain reaches alkaline soils, the rain acidity is neutralized by the existing bases and the pH of the soil solution remains unchanged. In the case of an acid environment, where the buffering effect is lacking, the acid rain induces an increase in the acidity of the soil.

Dissolution of Al is the main consequence of acid atmospheric deposition in forest soils in large areas of northern and central Europe and North America (David and Driscoll 1984; Mulder et al. 1987). Increased ionic concentration of Al in soil solution has adverse effects on soil biology and crop development.

In a study with acid sandy soils, van Grunsven et al. (1992) observed that the logarithm of Al dissolution rates in individual soil samples followed an inverse linear relationship with pH. Together with pH, organic matter has been shown to influence Al dissolution. Figure 3.4 gives the results of an experiment designed by Hargrove (1986) on Al dissolution from mica-organic matter complexes showing that Al concentration in solution reaches a minimum value in a pH range of about 4–5. Above pH 5 the concentration of Al in solution

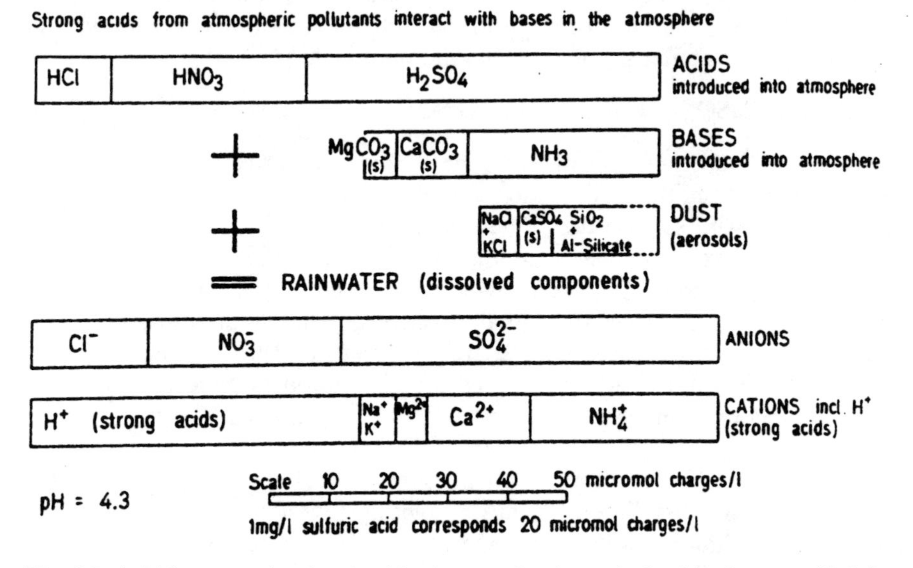

Fig. 3.3. Acid-base reaction involved in the genesis of a typical acid rainwater. (Zobrist and Stumm 1979)

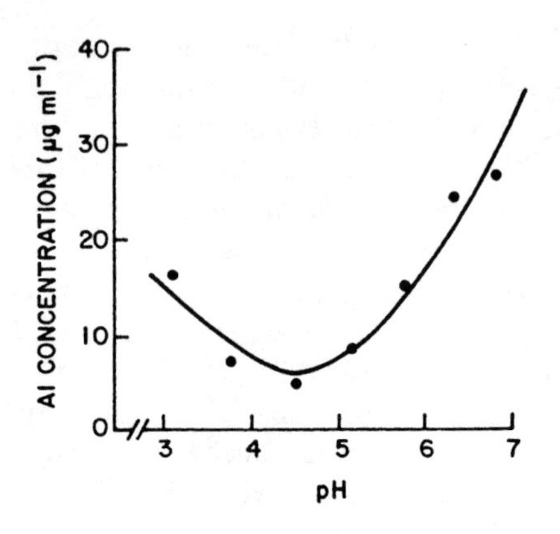

Fig. 3.4. Effect of pH on the dissolution of Al from Al-organic matter complexes. (Hargrove 1986)

increases dramatically. It was clearly observed that Al-organic matter is solubilized in the pH range 5–7. This experimental finding conforms with previous work on Al availability to plants.

3.7.2 Alkaline Soils and Gypsum Dissolution

Soil salinization, resulting from irrigation with saline and alkaline water or improper management of an irrigation-drainage system, could affect, besides the soil properties, the gypsum balance. In soils of which human action has enhanced the alkalinity, the distribution of gypsum – to be found in soils as a natural constituent or as added reclamation material – in the soil profile could be changed as a result of the effect of alkalinity on its dissolution.

The effective solubility of gypsum depends only on the soil solution and exchange phase composition. The dissolution exchange reaction

$$2NaX + CaSO_4 \rightleftharpoons Ca\,X_2 + 2Na^+ + SO_4^{2-} \quad (3.17)$$

(where X is one equivalent of exchanger) indicates that cation exchange will remove Ca^{2+} from the soil solution and allow additional gypsum dissolution.

Irrigation with sodic water brings an increase in the exchangeable Na^+ in the soil complex. Under such conditions, the dissolution of $CaSO_4$ will increase, to compensate for the Na^+ which is exchanged with Ca^{2+} in the soil. It has been reported (Oster 1982) that effective gypsum solubility is enhanced through ion-pair formation and electrolyte effects, and that under these conditions, the amount of water required for dissolution of gypsum in alkaline soils is likely to be much less than commonly inferred from gypsum solubility in water.

An example of the salt effect on gypsum dissolution is given in Fig. 3.5, which shows the dissolution kinetics of gypsum samples from Egypt in NaCl and $MgCl_2$ solution (Gobran and Miyamoto 1985). It appears that dissolved amounts at equilibrium are clearly related to the ionic composition of the solution (Fig. 3.5a). In fact, the dissolution rate is only apparently different; a

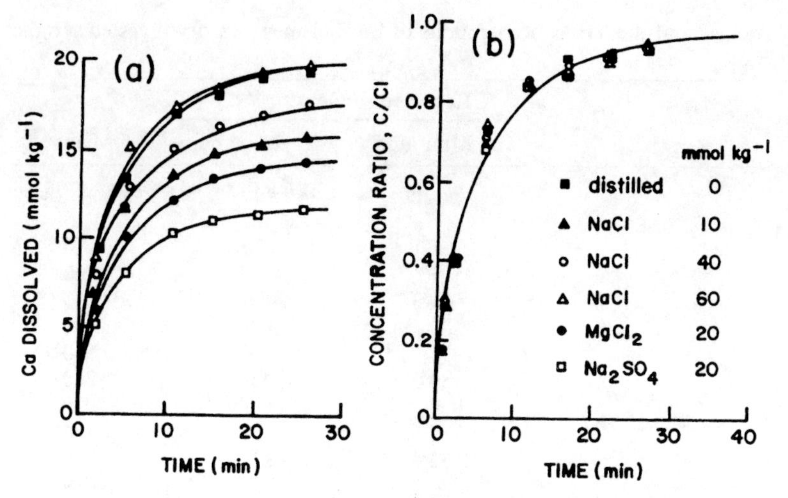

Fig. 3.5a,b. Kinetics of Ca dissolution from a gypsum sample. (After Gobran and Myamoto 1985). **a** In various salt solutions. **b** The ratio of dissolved Ca concentration (C) to its equilibrium value (C_{sat}) as a function of time

normalized representation allows all experimental points to fall on the same curve (Fig. 3.5b). This behavior was explained by Gupta et al. (1985) on the basis of the description proposed by Swartzendruber for dissolution kinetics:

$$\frac{dm}{dt} = KS(C'_{sat} - C) = \frac{KS}{C_{sat}}\left(1 - \frac{C}{C_{sat}}\right) , \tag{3.18}$$

where m is the amount of dissolved gypsum.

Clearly, it may be observed that the dissolution apparently increases with increasing salt concentration in the soil solution, as is expected from considering the ionic strength (Tanji 1969).

3.7.3 Sludge Effect on Potentially Toxic Inorganic Trace Element Solubility

Sludges from municipal origins are disposed of on land, in large amounts in small areas (around the water purification plants) or in relatively small amounts spread over large areas, where it is used as an additional organic manure in farming practices. The sludges of municipal origin are characterized by the presence of a large number of potentially toxic trace elements. The concentration ranges and median concentrations of the main elements found in dry, digested sewage sludges, as reported by Chaney (1983), are presented in Table 3.3. The above data encompass the results of sludge analysis in the US, Canada, and Europe. However, studies carried out in recent years have proved that the forms of potential toxic trace elements in the municipal sludges are affected by the wastewater treatment process.

Table 3.3. Ranges and median concentrations of trace elements in dry digested sewage sludges. (Chaney 1983)

Element	Reported range		Median
	Minimum	Maximum	
	...... mg/Kg dry sludge		
As	1.1	230	10
Cd	1	3410	10
Co	11.3	2490	30
Cu	84	17 000	800
Cr	10	99 000	500
F	80	33 500	260
Fe	1000	154 000	17 000
Hg	0.6	56	6
Mn	32	9870	260
Mo	0.1	214	4
Ni	2	5300	80
Pb	13	26 000	500
Sn	2.6	329	14
Se	1.7	17.2	5
Zn	101	49 000	1700

The soluble organics originating from sludges may retain the inorganic trace elements in a nonadsorbable form (mainly as anions or neutral species), thus raising the total dissolved trace element concentration in the solution above the dissolution capacity of the soil liquid phase. The inorganic forms associated with organic matter could include metallic particles, relatively pure precipitates (phosphates, carbonates, sulfides, or silicates), solid solutions resulting from coprecipitation with precipitates of Fe, Al as Ca, or as metal ions strongly adsorbed on surfaces of Fe, Al, or Ca minerals. Metals, for example, may be present in soils in many forms and the application of sewage sludge to soils may alter their speciation. The forms of metals in sludges, however, has been found to vary as a function of the sludge preparation process or technology and consequently their solubilities in soil solutions differ.

For the numerous complex-forming substances in soil, a general decreasing order of complex stability was stated by Scheffer and Schachstable (1982) as follows:

$$Fe^{3+} > Al^{3+} > Cu^{2+} > Pb^{2+} > Fe^{2+} > Ni^{2+} > Cd^{2+} > Zn^{2+} > Mn ,$$

humic acid complexes of Cu, Pb, Cd, and Zn have log K-values of 8.65, 8.35, 6.25, and 5.72, respectively, whereas for fulvic acid complexes of Cu, Pb, and Zn, log K-values are 4.0, 4.0, and 3.6, respectively.

Among the types of organic molecules involved in forming soluble complexes, the following may be included: oxalic, citric, malic, tartaric, and many other acids, and aliphatic and aromatic compounds (Stevenson and Ardakani

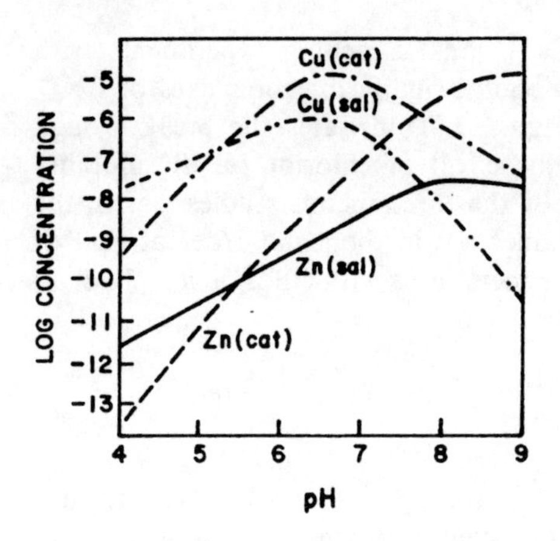

Fig. 3.6. Predicted concentrations of metal-ligand complexes as a function of pH in a solution containing 10^{-4} M catechol and salicylate. (McBride 1989)

1972). A hypothetical example given by McBride (1989) and reproduced in Fig. 3.6 illustrates the relative importance of these various organic acids as a function of the complexing metal and the pH. In this hypothetical example, McBride (1989) considers that 10^{-5} M total Cu and Zn are simultaneously present in solution with 10^{-4} M catechol and salicylic acid.

Speciation calculation predicts that an extremely small fraction of total soluble Zn actually forms complexes with these ligands at low pH, with salicylate dominating the speciation. For Cu^{2+}, important differences are noted; the fraction of the metal complexed by organics is much higher at low pH, and catechol is a weaker ligand than salicylate for pH values below 5.5. For both Cu^{2+} and Zn^{2+}, the speciation shifts in favor of catechol complexation at higher pH, and the fraction of Zn complexed with organics becomes significant above pH = 7.0. Figure 3.6 shows a decrease in the level of Cu complexed by catechol at high pH. This was attributed by McBride (1989) to the dominance of the Cu $(\text{catechol})_2^{2+}$ species over the Cu $(\text{catechol})_2^0$ species above pH 7.6, and the Cu is, in fact, predicted to be almost totally complexed with catechol at high pH.

McBride (1989) emphasized that the greater effect of catecholate than salicylate in this regard may indicate that the former ligand is a better electron donor, and the greater stability constant for the catecholate complex also suggests this.

All the reported results show that metal-organic complexes can modify the solubility of metals in the soil solution and consequently, their transport in the soil profile. This was assumed to be the explanation of the increased Cd transport in drainage water following sludge application (Lamy et al. 1993).

The sludge composition includes, besides the organic ligand, free salts which can form complexes with the potential pollutant trace elements. For example, in a gel filtration experiment reported by Baham et al. (1978), it was observed that Ca-chloro complexes were dominant among Cd-complexes, whereas Cu-fulvate complexes were the principal Cu-complex species in the system. Doner (1978) studied the mobilities of Cd, Cu, and Ni through soil columns as influenced by Cl^- and ClO_4^- ions. Leaching experiments with $NaClO_4$ were included for reference, because significant complexes of ClO_4^- with Cd, Cu, and Ni do not exist. Figure 3.7 illustrates the breakthrough curves obtained, which indicate that the effect of chloride on the mobility (e.g., solubility in the aqueous phase) of the three metals studies was in the order Cd $\gg$ Cu > Ni. The observed values are in the same order as that of the log K values for association of these metal ions with chloride ion: 1.98 $\gg$ 0.40 > − 0.40.

3.7.4 Redox Processes and Metal Solubility

As previously stated, the potentially toxic trace elements which can form a pollution hazard reach the soil as a consequence of sludge disposal practices and irrigation with reused water. The dissolution of trace elements into the soil liquid phase and their redistribution into the porous media between the land surface and the groundwater is also affected, besides the processes previously discussed, by the reduction and oxidation reactions occurring in the soil medium. The redox process could occur in the whole soil profile under saturation or drying conditions . In unsaturated soils the processes of reduction

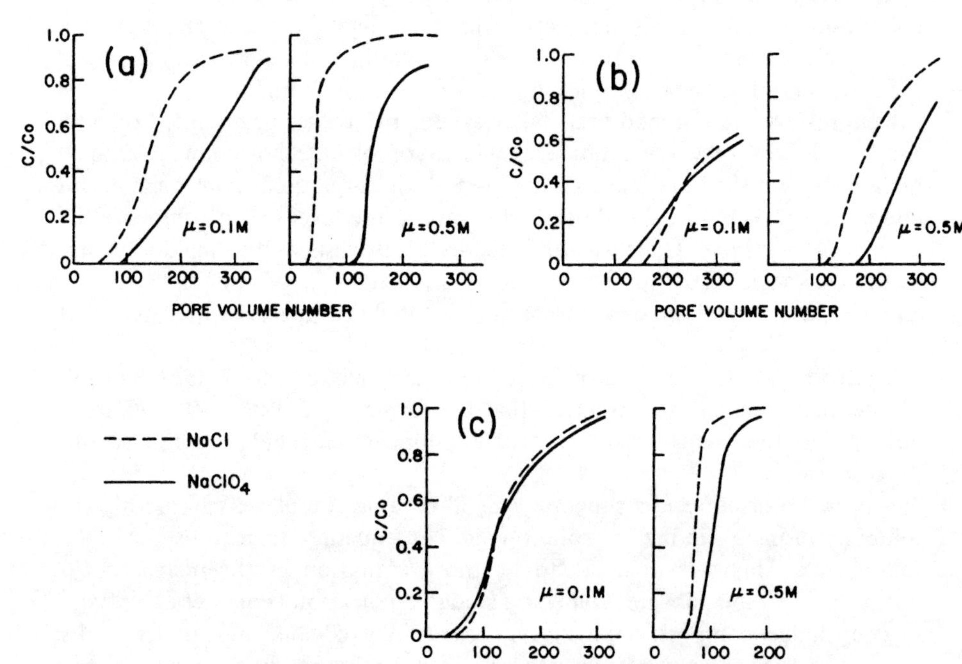

Fig. 3.7a–c. Transport of trace elements in a soil column. Breakthrough curves for Cd (**a**), Cu (**b**), and Ni (**c**) as affected by the presence of Cl^- and ClO_4^-. The greater the amount of complexed metal in solution, the more the breakthrough curve is shifted to the left. μ = concentration of NaCl and/or $NaClO_4$ solution. (After Doner 1978)

and oxidation proceed simultaneously at the aggregate soil level: reduction into the soil aggregate and oxidation at the aggregate surface.

It is known that the heavy metals are less soluble in an aqueous phase in their higher oxidation states. The reduction-oxidation in soils – enhanced by either biological processes or a chemical "catalysis" – affect the heavy-metal solubility. A biologically enhanced "redox effect" on Fe(II) solubility as well as the chemically enhanced oxidation of metals in the presence of Mn oxides is presented below.

The first case to be discussed is the effect of redox potential on the solubility of iron minerals. The approach of Lindsay (1979) serves as an example.

The electron activity in soils controls the ratio of Fe^{3+} to Fe^{2+} in solution according to the reaction.

$$Fe^{3+} + e^{-} \rightleftharpoons Fe^{2+} \quad \log K^{\circ} = 13.04 \ , \tag{3.19}$$

for which

$$\log \frac{[Fe^{2+}]}{[Fe^{3+}]} = 13.04 - pe \ . \tag{3.20}$$

When pe = 13.04, the ratio of Fe^{2+}/Fe^{3+} is unity. Changing pe by one unit changes the ratio of Fe^{2+}/Fe^{3+} by tenfold.

Figure 3.8 shows the activity of Fe^{2+} in equilibrium with soil-Fe as a function of redox and pH. The redox parameter (pe + pH) is used by Lindsay (1979) because it partitions the protons associated with the redox component of the reactions from those associated with pH. Lowering redox (Fig. 3.8) increases the activity of Fe^{2+} in equilibrium with soil-Fe. The lowest dashed line in this figure represents Fe^{2+} at (pe + pH) of 20.61, the equilibrium redox corresponding to 0.2 atm O_2(g). At this redox potential, the activities of Fe^{2+} and Fe^{3+} become equal at pH 7.57. Since the (pe + pH) of soils is generally lower than 20.61, the activity of Fe^{2+} in soils is most often greater than that of Fe^{3+} (see the dashed lines in Fig. 3.8).

The ability of Mn oxides to oxidize metals directly or to catalyze metal oxidation was emphasized by McBride (1989). This direct oxidation by O_2, catalyzed by metal oxides, provides a mechanism for lowering trace element solubility in the soil solution. Stumm and Morgan (1981) proved that the decrease of the Mn(II) concentration with time is an autocatalytic reaction (Fig.3.9). This autocatalytic type of reaction was explained by the authors as follows:

(1) The extent of Mn(II) removal during the oxygenation reaction is not accounted for by the stoichiometry of the oxidation reaction alone; that is, not all the Mn(II) removed from the solution [as determined by specific analysis for Mn(II)] is oxidized (as determined by measurement of the total oxidizing equivalents of the suspension). (2) As pointed out previously, the products of Mn(II) oxygenation are nonstoichiometric; they show various average degrees of oxidation ranging from about $MnO_{1.3}$ to $MnO_{1.9}$ (30–90% oxidation to MnO_2) under varying alkaline conditions. (3) The higher-valent manganese

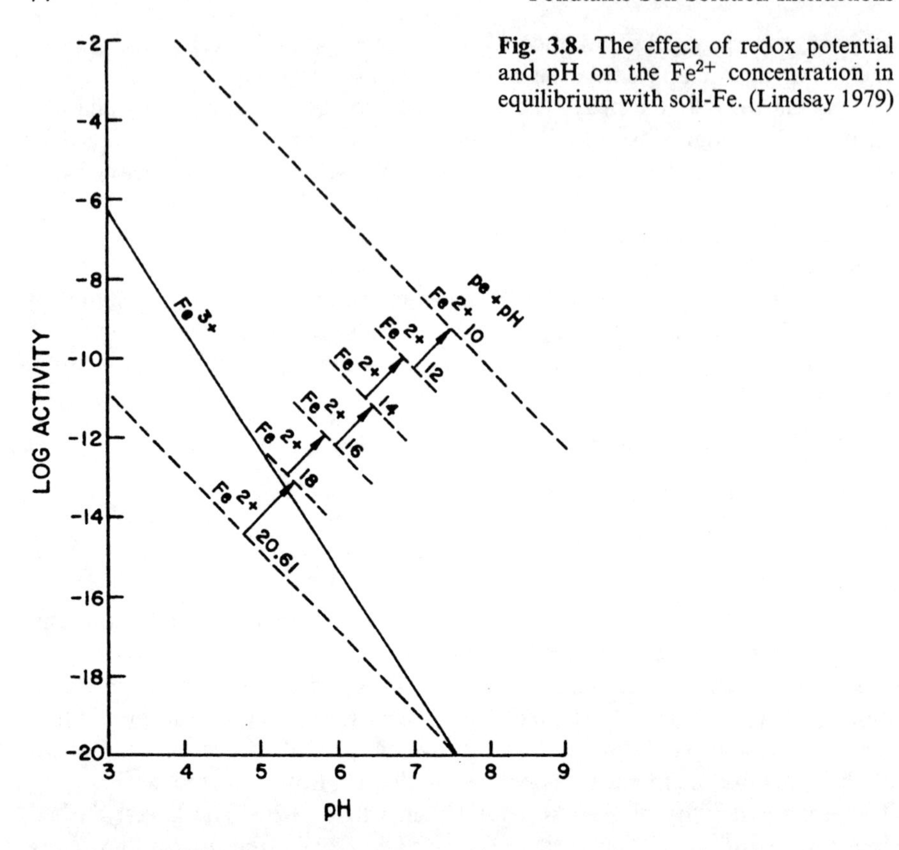

Fig. 3.8. The effect of redox potential and pH on the Fe^{2+} concentration in equilibrium with soil-Fe. (Lindsay 1979)

oxide suspensions show large sorption capacities for Mn^{2+} in slightly alkaline solutions. The relative proportions of Mn(IV) and Mn(II) in the solid phase depend strongly on pH and other variables. Mn oxides can reduce the solubility of metals by an oxidation process. The oxidation of Co(I) to Co(III) or Fe(I) to Fe(III) (Stumm and Morgan 1981) prove that in the absence of Mn-oxidizing organisms, Mn oxides could have an important role in reducing the concentration of Mn^{2+} in the soil solution.

3.7.5 Solubility of Organic Solvents Under a Waste-Disposal Site

The characteristics of a waste-disposal site include the presence of mixtures of inorganic and organic pollutants and of a mixture of water and polar and nonpolar organic solvents. The present example deals with the solubilities of several polyaromatic hydrocarbon chemicals in two water-polar organic mixtures. The data are taken from a paper by Yalkowsky (1986). In Fig. 3.10, the solubilities of naphtalene, anthracene, pyrene, and chrysene are plotted as fractions of solvent or cosolvent in binary and tertiary methanol-acetone-water systems. It may be observed that the solubilities of pollutants increase with

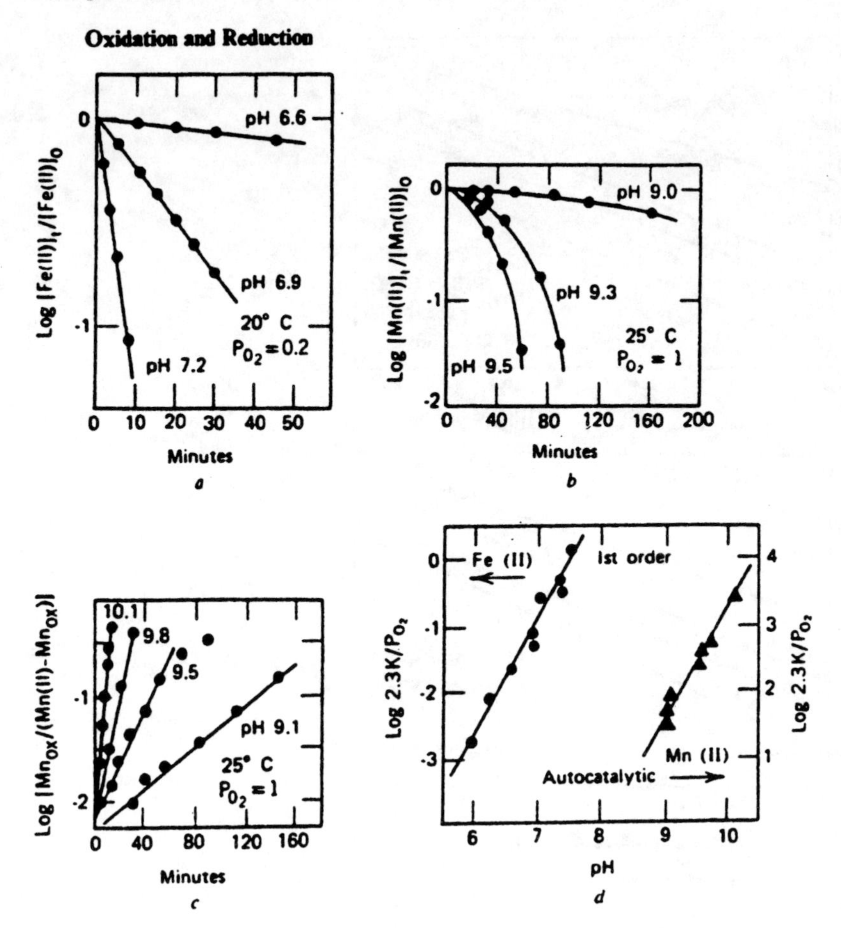

Fig. 3.9a–d. Oxydation of Fe(II) and Mn(II). (After Stumm and Morgan 1981.) **a** Oxydation of Fe(II) in bicarbonate solutions. **b** Removal of Mn(II) by oxydation in bicarbonate solutions. **c** Oxydation of Mn(II) in HCO_3^- solutions: autocatalytic plot. **d** Effect of pH on oxydation rates

increasing proportion of cosolvent in the water-cosolvent mixture. To prove this behavior in a ternary mixed solvent (methanol-acetone-water system), one of the cosolvent volumes was kept fixed at 0.5 while that of the second cosolvent and water varied in inverse relationship. Rao et al. (1989), reviewing some experiments from pharmaceutical industries, showed that a solvent (glyceryltriacetate) which is relatively insoluble in water was solubilized in water by propyl glycol or ethanol. The solubilized triacetate became an excellent cosolvent for other organic compounds from the drug industry. The use of an insoluble solvent as cosolvent could enhance the aqueous solubility of hydrophobic organic pollutants in water, leading to possible increases in soil and groundwater contamination.

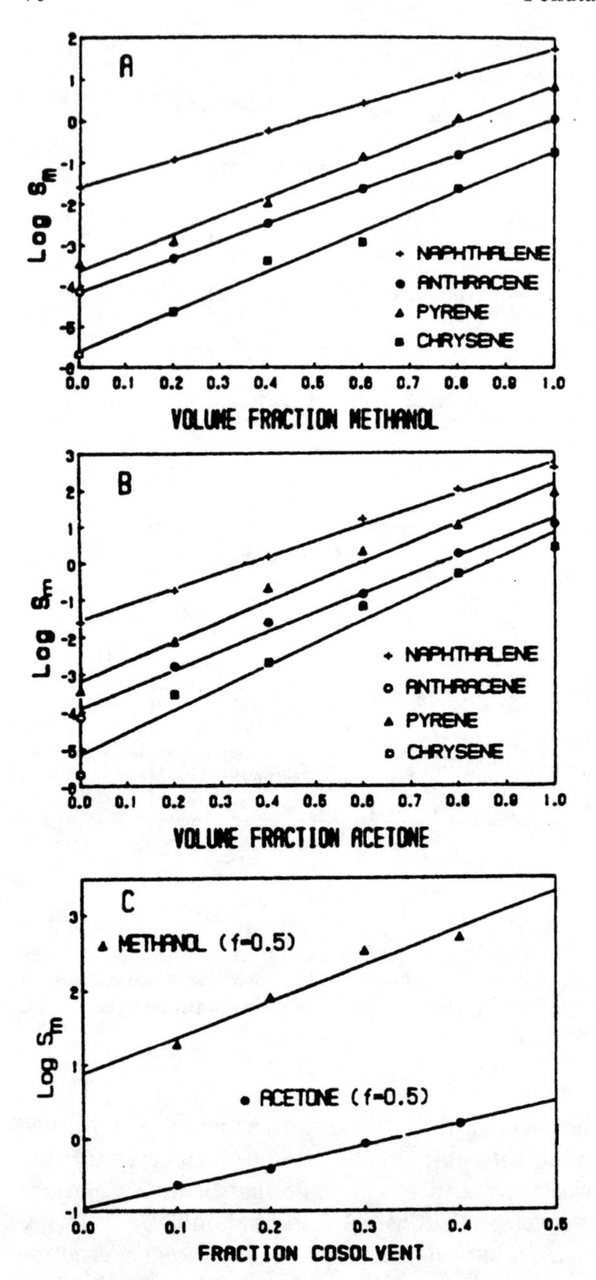

Fig. 3.10A-C. Solubilities of several polyaromatic hydrocarbon chemicals in (**A**) methanol-water and (**B**) acetone-water mixed solvent systems. Solubility data are presented in (**C**) for ternary mixed solvents (methanol-acetone-water) where one cosolvent volume fraction remains fixed at 0.5 while that of the second cosolvent varies with water. (Rao et al. 1989)

References

Baham J, Bell NB, Sposito G (1978) Gel filtration studies of trace metal-fulvic acid solutions extracted from sewage sludges. J Environ Qual 7: 181–188

Chaney RL (1983) Potential effect of waste constituents on the food chain. In: Parr JF, Marsh SB, Klay M (eds) Land treatment of hazardous wastes. Noyes Data Corp, Park Ridge, pp 152–210

David MR, Driscoll CT (1984) Aluminum speciation and equilibria in soil solutions of a Haplorthod in the Adironack Mountains. Geoderma 33: 297–318

Doner HE (1978) Chloride as a factor in mobilities of Ni(II), Cu(II), and Cd(II) in soil. Soil Sci Soc Am J 42: 882–885

Driessens FCM (1986) Ionic solid solutions in contact with aqueous solutions. In: Davis JA, Hayes KF (eds) Geochemical processes of mineral surface. ACS Symposium Series No 323. American Chemical Society, Washington, DC, pp 524–560

Gobran GR, Myamoto S (1985) Dissolution rate of gypsum in aqueous salt solutions. Soil Sci 140: 89–93

Gupta RK, Singh CP, Abrol IP (1985) Dissolution of gypsum in alkali soils. Soil Sci 140: 382–386

Hargrove WL (1986) The solubility of aluminum-organic matter and its implication in plant uptake aluminum. Soil Sci 142: 179–181

Hartley GS, Graham-Bryce IJ (1980) Physical principles of pesticide behavior, vol 1. Academic Press, London, 519 pp

Hendrickson LL, Corey RB (1981) Effect of equilibrium metal concentrations on apparent selectivity coefficients of soil complexes. Soil Sci 131: 163–171

Lamy I, Bourgeois S, Bermond A (1993) Soil cadmium mobility as a consequence of sewage sludge disposal. J Environ Qual 22: 731–727

Lewis GN, Randal M (1923) Thermodynamics and the free energy of chemical substances. McGraw Hill, New York

Lindsay WL (1979) Chemical equilibria in soils. John Wiley, New York, 449 pp

Mattigod SW, Sposito G, Page AL (1981) Factors affecting the solubilities of trace metals in soils. In: Chemistry of soil environment. ASA Spec Publ 40: 223–241

McBride MB (1989) Reactions controlling heavy metal solubility in soils. Adv Soil Sci 10: 1–47

Mulder J, van Grinsven JJM, van Breemen N (1987) Impacts of acid atmospheric deposition on woodlands soils. III. Aluminum chemistry. Soil Sci Soc Am J 51: 1640–1646

Oster JD (1982) Gypsum dissolution usage in irrigated agriculture – a review. Fertil Res 3: 79–89

Rao PSC, Lee LS, Nkedi-Kizza P, Yalkowsky SH (1989) Sorption and transport of organic pollutants at waste disposal sites. In: Gerstl Z, Chen Y, Mingelgrin U, Yaron B (eds) Toxic organic chemicals in porous media. Springer, Berlin Heidelberg New York, pp 176–193

Rubino JT, Yalkowsky SH (1987) Co-solvency and co-solvent polarity. J Pharmacol Res 4: 220–230

Scheffer F, Schachstable P (1982) Lehrbuch der Bodenkunde, 8th edn. Enke, Stuttgart, 444 pp

Sposito G (1981) The thermodynamics of soil solutions. Clarendon Press, Oxford

Stevenson FJ, Ardakani MS (1972) Organic matter reactions involving micronutrients in soils. In: Mortvedt JJ, Goldrand PM, Lindsay WL (eds) Micronutrients in agriculture. Soil Science Society of America, Madison, pp 78–144

Stumm W, Morgan JJ (1981) Aquatic chemistry. John Wiley, New York 624 pp

Swartzendruber D, Barber SA (1965) Dissolution of limestone particles in soils. Soil Sci 100: 287–291

Tanji KK (1969) Solubility of gypsum in aqueous electrolytes as affected by ionic strength. Environ Sci Technol 3: 656–661

van Grunsven HJM, Van Riensdijk WH, Otjes R, van Beemer N (1992) Rates of aluminum dilution in acid sandy soils observed in column experiments. J Environ Qual 21: 439–447

Yalkowsky SH (1986) Solubility of organic solutes mixed aqueous solvents. EPA Project Completion Report Cr 812581–01

Yalkowsky SH, Roseman TJ (1981) Techniques of solubilization of drugs. Marcel Dekker, New York

Zobrist J, Stumm W (1979) Forschung und Technik. Neue Zürcher Zeitung

CHAPTER 4

Volatilization into the Soil Atmosphere

Volatilization is defined as the transition of chemicals from the solid or liquid phase into the vapor phase. Each chemical has a characteristic saturation vapor pressure or vapor density, that varies with temperature. The potential volatility of a chemical is related to its inherent vapor pressure, but actual vaporization rates depend on the environmental conditions and all the other factors that control the behavior of the chemical at the solid-air-water interface. For surface deposits, the actual rates of loss, or the proportionality constants relating vapor pressure to volatilization rates, are dependent upon external conditions that affect movement away from the evaporating surface, such as turbulence, surface roughness, wind speed, etc. The rate at which a pollutant moves away from the surface is diffusion-controlled. Close to the evaporating surface there is relatively little movement of air, and the vaporized substance is transported from the surface through this stagnant air layer only by molecular diffusion. Since the thickness of the stagnant boundary layer depends on air flow rate and turbulence, vapor loss is influenced strongly by the type of soil cover and the atmospheric conditions, e.g., wind in the vicinity of the soil surface. In general, under a given set of conditions, as air turbulence increases, the volatilization rate increases.

As a result of volatilization, the actual concentration of a potentially toxic chemical in a specific soil site decreases, but the pollution hazard is not entirely eliminated. It is true that a substantial portion of the chemical involved could be transferred into the atmosphere, but part of the chemical in its gaseous phase could be transported downward to the groundwater, becoming a pollution hazard for the unsaturated zone. The volatilization of the pollutants in the soil from both liquid and solid phases has been discussed in recent years in reviews by Glotfelty and Schomburg (1989), Taylor and Spencer (1990), and Jayaweera and Mickkelsen (1991), and the reader is directed to these papers for a more detailed presentation of the process. Here, we will present only the basics of the volatilization process, with examples to illustrate the volatilization of both inorganic and organic pollutants from the soil medium.

4.1 The Volatilization Process

Two distinct processes govern the dispersion of pollutants into the atmosphere. The first is the vaporization of toxic molecules into the air from the residues present in the soil medium; the second is the dispersion of the resulting vapors into the soil gaseous phase or the overlying atmosphere, by diffusion and turbulent mixing. Despite the fact that the two processes are fundamentally different and controlled by different chemical and environmental factors, they are not wholly independent under natural conditions, and only by integrating their effects can the characterization of the volatilization process be done.

The vaporization process represents a phase change into vapor from the liquid or solid state. In their solid and liquid states, the potentially toxic molecules are held together by intermolecular forces.

The molar heats of fusion (ΔH_f), of vaporization (ΔH_v) and of sublimation (ΔH_s) are related according to the Born Haber cycle:

$$\Delta H_s = \Delta H_f + \Delta H_v \ . \tag{4.1}$$

Even at low temperature there may be enough energy to overcome the cohesive forces of the chemical and to escape from the solid or liquid state into the gaseous state.

The *vapor dispersion process* is described mathematically as a flux of vapor (IF) through any plane at a height z and is expressed by the equation

$$IF = kz \left(\frac{dp}{dz}\right) z \ , \tag{4.2}$$

where kz is the diffusion coefficient and dp/dz the gradient of vapor pressure at the height z.

The flow over the soil is turbulent and is described by the term, *eddy diffusion*, except for a shallow layer about 1 mm thick. In this region as well as into the soil matrix, the vapor dispersion may be essentially described by a molecular diffusion process.

The flux through the laminar layer (I_F) is expressed as:

$$I_F = D(p_s - p_d)/\delta \ , \tag{4.3}$$

where p_s and p_d are, respectively, the vapor pressure at the evaporating surface of the laminar layer of effective depth δ; D is the molecular diffusion coefficient of a particular compound.

Taylor and Spencer (1990) pointed out that the laminar layer can be regarded as the limiting distance above the surface to which the smallest eddies of the overland turbulent flow can penetrate. Thus, above this layer, transport takes place through eddy diffusion, and the corresponding vapor flux can be described, according to these authors, as:

$$I_F = k_z(p_1 - p_2)/(z_1 - z_2) \ , \quad (4.4)$$

where k_z is the main eddy diffusivity coefficient between the heights z_1 and z_2, and p_1 and p_2 are the corresponding pressures. The depth of the turbulent zone, often described as the *atmospheric boundary layer*, is several orders of magnitude greater than that of the laminar flow layer, being measured in meters rather than the millimeters of the latter. The difference between the two diffusion coefficients is also large, so that – except when the atmosphere is exceptionally stable – dispersion of the pollutant vapors in the turbulent zone outside the laminar layer is relatively rapid.

Despite these difficulties, the concept of separate regions of dilution by molecular diffusion and by turbulent mixing is of major importance in the theory of the exchange of gases at soil and plant surfaces. The concept is of particular value in the identification of the physical factors that impede the dispersion process under various soil and microclimate conditions.

4.2 Vapor Pressure and Vaporization Relationship

If a pollutant (in solid or liquid state) is placed in an initially empty vessel which is greater in volume than the pollutant itself, a portion of the chemical will evaporate so as to fill the remaining free space of the vessel with vapors. The pressure in the vessel at equilibrium is affected only by the temperature and is independent of the vessel volume. If any other gas is initially present in the vessel, the result will be the same (Dalton's Law). The pressure developed is called *vapor pressure* and characterizes any chemical in the liquid or solid state. Since the vapor pressure is a fundamental parameter in defining the vaporization of a chemical, many techniques have been developed for its measurement. Comprehensive reviews of these methods are found in the papers of Plimmer (1976) and Glotfelty and Schomburg (1989).

When the temperature increases, the proportion of molecules with energy in excess of the cohesive energy also increases and, consequently, an excess vapor pressure is observed. The Clausius-Clapeyron equation describes the variation of vapor pressure with temperature as follows:

$$d(\ln p)/dT = \Delta H_v/RT^2 \ , \quad (4.5)$$

where p is the vapor pressure, ΔH_v is the heat of vaporization (Pa m^{-3} mol^{-1}), R is the universal gas constant (8.314 J mol^{-1} K^{-1}), and T is the temperature (K).

Since the vapor pressure of potentially toxic chemicals is a key factor in controlling their dissipation in soil and from soil to the atmosphere, an accurate estimation of this value is required. In the case of chemicals with low vapor pressure reaching the land surface as a result of nonpoint disposal (e.g., pesticides in agricultural soils), their vapor pressure is sufficiently low to be below the detection limits of methods by which they may be accurately measured. This explains the discrepancies found in the literature in the reported values of vapor pressure (Glotfelty and Schomburg 1989).

This behavior may also be explained by the differences in the concentrations of pollutants on the land surface due to the fact that the soil is not a single-phase system, and equilibrium is usually achieved at considerably different concentrations in the various phases. Mackay (1979) showed that the partitioning between phases (corresponding to the maximization of entropy) can be expressed by equating the chemical potential in the respective phases. Thus, in partitioning between phases, matter flows from high chemical potential to low chemical potential. At equilibrium, the chemical potential is the same in all phases. However, since chemical potential is a difficult concept to use, Mackay (1979) reintroduced the much simpler concept of fugacity, which can be used quite simply to express the distribution of a pesticide among the various phases of the environment. It is important to underline again that high fugacity denotes a greater tendency for a chemical to escape from a phase. Fugacity is to mass diffusion as temperature is to heat diffusion: mass always flows from high to low fugacity just as heat flows from high to low temperature. Fugacity is linearly related to concentration,

$$C_i = f_i Z_i \ , \tag{4.6}$$

where i denotes the i-th phase, C is the concentration (mol m^{-3}) f is fugacity, and Z is the fugacity capacity (mol m^{-3} Pa^{-1}) of the phase (Mackay 1979). Fugacity capacity is like heat capacity. A phase with high Z will accept a large amount of chemical, just as a substance with high heat capacity will store a lot of heat. Each environmental phase has a Z value for each chemical at each temperature, and if the Z values are known, the equilibrium distribution between the phases can be determined. The fugacity concept in relation to transport processes will be discussed in Chapter 8.

The value of p_s in Eq. (4.3) is the maximum vapor pressure value that can be established at the inner surface of the laminar flow layer in air moving over a surface; it can be the saturation value only when the surface is uniformly covered by the pollutant. This situation is almost impossible to find – except at the very moment where the pollutant reaches the land surface. In reality, the pollutant is partially adsorbed on the soil solid phase or is dissolved in the soil solution, which reduces the vapor pressure below the equilibrium value of the pure compound.

Among the environmental factors affecting the vapor pressure value, adsorption, temperature, water solubility, and soil moisture content variations could be included. Since the processes of pollutant solubilization in water, and adsorption on soil constituents are discussed in Chaps. 3 and 5, at this point we will briefly mention their effect on the vapor pressure value in connection with the potential vaporization of the chemicals. More in this regard can be found in Taylor and Spencer's (1990) review on pesticide volatilization.

The vapor density of a chemical in the air reflects the solution concentration according to Henry's Law:

$$d = hc \ , \tag{4.7}$$

where d is the vapor density and c the solution concentration (both in mg l^{-1}) and h is Henry's coefficient. The vapor densities increase with residual concentration up to a limiting value, for the pure compound. When a pollutant in the gaseous phase passes through a soil layer, it is absorbed on the soil solid phase surface and its relative vapor density is reduced accordingly. It is clear that the gas adsorpton in soil will be affected by the soil constituent characteristics and this process will reduce the volatilization rate. An increase in the soil water content leads to a decrease of amount of gas adsorbed and, consequently to an increase of the volatilization rate.

4.3 Volatilization of Multicomponent Pollutants

In general, the soil pollutants reach the land surface not as pure components, but as a mixture of various potentially toxic chemicals. Each component in the mixture will have its own physicochemical characteristics. Complex mixtures of volatile organic compounds, such as petroleum products, typically contain components whose vapor pressures range over several orders of magnitude. The compositions of volatile organic mixtures in soil change with time due to volatilization of some of the components from the liquid phase via the soil vapor layer to the atmosphere. These changes in composition affect the physical properties of the remaining liquid mixture. Assuming that a mixture of components (W_L) was ideal, and that they were all well mixed, Woodrow et al. (1986) considered that the evaporation of each component (W_i) proceeds by a first-order process and is described by

$$W_i = W_L^o \exp(-k_i t) \ , \tag{4.8}$$

where k_i is the evaporation rate constant for the component i.

In a mixture, the more volatile components will evaporate sooner, resulting in a decreased evaporation rate for the mixture as its vapor pressure decreases. Woodrow et al.'s (1986) depiction of the evaporation of volatile components from a mixture is only one approach to this topic, and it confirms the previous development of Mackay's group (1979) on volatilization of organic components. Recently, Nye et al. (1994), studying the evaporation of single and multicomponent liquids through soil columns, emphasized the differences between the two cases. In the case of a pure liquid, three stages were identified for a gas moving through a column of length L:

1. A soil accumulation stage in which the soil sorbs the gas diffusing up the column.
2. A steady-state stage in which the vapor escapes to the atmosphere at a constant rate, after diffusing up the column under a constant concentration

gradient, $(C^{o}_{z=L} - C_{z=0})/L$. The concentration of gas sorbed will vary with z, and will be determined by the sorption isotherm. If the rate of sorption is slow, the time taken to arrive at a steady state will increase and in the limit no steady state will be reached before the supply is exhausted.

3. A soil depletion stage during which the gas sorbed by the soil diffuses to the atmosphere. When the sorption is slow, the initial rate of evaporation is relatively high (little sorption), but the time for complete evaporation is longer.

The same stages can be recognized when a mixture of volatile components evaporates, but with important variations. In the accumulation stage, the components are sorbed according to their volatility, diffusivity, and the competitive isotherm. No true steady state will be reached because the composition of the source liquid will be continuously changing, and the composition of the soil profile will be adjusting correspondingly. If the supply of source liquid is large compared with the soil's total sorption capacity, the adjustment will be slow, and evaporation will approach a steady state. In the depletion stage, the soil, already enriched in the less volatile components, will lose the volatile components slowly. The soil will still contain a small proportion of the more volatile components and will not lose them completely until all components are gone.

4.4 Examples

Volatilization of toxic chemicals from the soil body could lead, on the one hand, to an increase of the pollution level of the surrounding atmosphere and, on the other, to a vertical redistribution of the pollutants into the porous media, extending from the land surface to the groundwater.

In the present part we will use a few examples to illustrate the volatilization of pure inorganic and organic compounds and of multicomponent liquids from soils.

4.4.1 Ammonia Volatilization

Increasing use of fertilizers, and the increasing disposal of sewage sludges on agricultural lands raise the question of possible volatilization losses of NH_3 from soils. Ammonia volatilization is an example illustrating the behavior of volatile inorganic chemicals in soils under aerobic and anaerobic environmental conditions. It is recognized that ammonia volatilization is affected by wind, time and depth of incorporation, pH, temperature, soil moisture, soil cation exchange capacity (CEC), and base saturation. We will summarize two experiments dealing with the ammonia volatilization process: the first one under aerobic conditions from a sieved and steady enriched soil, and the second from a flooded soil system.

Summarizing the process of ammonia volatilization from soil, Feigin et al. (1991) emphasized that the relative concentrations of NH_3 and NH_4^+ in a solution depend on its pH, as indicated in the following equations:

$$NH_4^+ \rightleftharpoons NH_3(aq) + H^+ \tag{4.9}$$

$$\frac{[NH_3(aq)][H^+]}{[NH_4^+]} = K = 10^{-9.5} \tag{4.10}$$

$$\log\frac{[NH_3(aq)]}{[NH_4^+]} = -9.5 + pH \tag{4.11}$$

where $NH_3(aq)$ is NH_3 in solution and [ ] denotes concentration. According to these equations, the concentrations of $NH_3(aq)$ at pH = 5, 7 and 9 are 0.0036, 0.36, and 36%, respectively, of the total quantity of ammoniacal N in the soil solution. Loss of NH_3 from calcareous soils is considerably greater than from noncalcareous ones (Woldendorp 1968) and involves production of $(NH_4)_2CO_3$ (Fenn and Kissel 1973) or NH_4HCO_3 (Feagley and Hossner 1978). The fate of $(NH_4)_2SO_4$ added to a calcareous soil is represented by the overall reaction between $(NH_4)_2SO_4$ and $CaCO_3$.

$$CaCO_3 + (NH_4)_2SO_4 = 2NH_3 + CO_2 + H_2O + CaSO_4 \ . \tag{4.12}$$

In an experiment on factors affecting ammonia volatilization from sewage sludge-enriched soils, Donovan and Logan (1983) showed that it occurs in a manner similar to the reported volatilization from NH_3 fertilizers. These authors found that ammonia loss increased with increasing pH and temperature, as would be predicted from thermodynamic considerations. Loss from air-dry soil was much lower than from soil at moisture tension <1.5 MPa. The moisture added to the soil with the liquid sludge overshadows the effects that antecedent soil moisture has been shown to have on NH_3 volatilization from fertilizers. The type of sludge incorporated into the soil has an important role in NH_3 volatilization. Table 4.1 shows the percentage of ammonia volatilization from sludges of various preparations and origins. It may be seen that volatilization losses were greater from a lime-stabilized sludge, which had a pH of 12, than from the anaerobic and aerobic sludges. The N value of this sludge would decrease rapidly if the material were not incorporated immediately after application.

Donovan and Logan (1983) report that an initial soil moisture content corresponding to $\leq$ 1.5 MPa tension increased NH_3 volatilization compared with air-dried soil. Volatilization increased linearly with time elapsed after sludge incorporation. Ammonia loss was greater from soil at pH 7.5 than at pH 6.7 or 5.1. Volatilization decreased with decreasing temperature. When the sludge contained large sludge particles, NH_3 loss decreased. Figure 4.1 illustrates the effect of the soil moisture content on ammonia volatilization.

Table 4.1. Ammonia-N volatilized as percentage of NH_3–N applied and tests of significance[a] for several sludges. (After Donovan and Logan 1983)

Sludge origin	Type of sludge	NH_3 volatilized over 24 h as percentage of total NH_3-N applied	Test of line equality	Duncan's multiple range test
Columbus	Anaerobically digested	8.3	a	a
Dewatered Columbus	Anaerobically digested	7.8	a	a
Medina	Anaerobically digested	0.4	a	a
Columbus compost	Primary	None detected	a	a
Ashland	Aerobically digested lime-stabilized	15.8	b	b

[a]Means followed by the same letter are not significantly different at the 0.05 level.

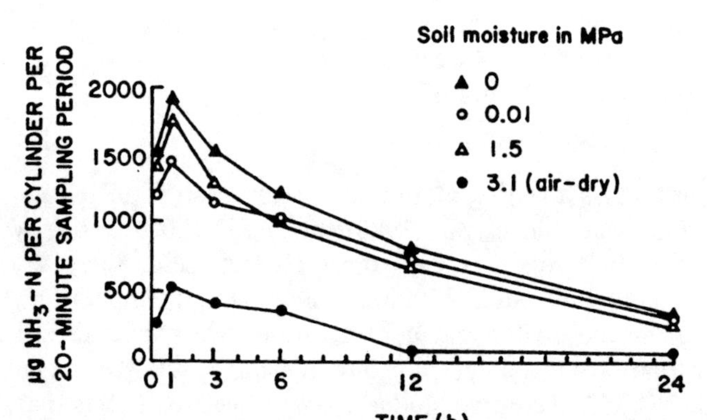

Fig. 4.1. Ammonia-N volatilization as a function of time for sewage sludge applied to soils at 0, 0.01, and 1.5 MPa, and air-dry initial moisture levels. *Horizontal bars* indicate the amount collected during the 20-min sampling period. (Donovan and Logan 1983)

Ammonia volatilization from flooded soil systems involves a more complex pathway, in the terrestrial-atmospheric nitrogen (N) cycle. In a comprehensive review on this topic, Jayaweera and Mikkelsen (1991) showed that a variety of water, soil, biological, and environmental factors and management practices influence the kinetics and extent of NH_3 volatilization from flooded soil systems. Ammoniacal N concentration, pH, P_{CO_2}, alkalinity, buffering capacity, temperature, depth, turbulence, and biotic activity are several floodwater characteristics that influence NH_3 volatilization. The NH_4^+ concentration in floodwater is influenced by N management practices such as source, timing and method of application, and water depth, as well as biotic activity. The dominant soil factors affecting NH_3 volatilization are soil pH, redox status, cation exchange characteristics, $CaCO_3$ content, soil texture, biotic acitivity, and fluxes affecting adsorption and desorption of NH_4^+ at the soil-water interface. Atmospheric conditions such as wind speed, P_{NH_3}, air temperature, and solar radiation also influence NH_3 volatilization. According to Jayaweera and Mikkelsen (1991), the rate of NH_3 volatilization is principally a function of two parameters, the NH_3 (aq) concentration in floodwater and the volatilization rate constant for NH_3.

4.4.2 Pesticide Volatilization

Since it is a major pathway of pesticide losses, the volatilization of pesticides has been the subject of a large number of studies through the years. Both the

Table 4.2. Saturation vapor pressures and densities of selected pesticides (As summarized by Taylor and Spencer 1990)

Pesticide	Molecular wt. (g)	Temperature (°C)	Vapor pressure (mPa)	Vapor density (μg/L)
Alachlor	270	25	2.9	3.2 (−1)[a]
Atrazine	216	25	9.0 (−2)	8.0 (−3)
Bromacil	261	25	2.9 (−2)	3.0 (−3)
Carbofuran	221	25	1.1	1.0 (−1)
Chlorpropham	214	25	1.3	1.2 (−1)
DDT	354	25	4.5 (−2)	6.0 (−3)
Diazinon	304	25	1.6 (+1)	2.0
Dicamba	221	20	5.0 (+2)	45
Dieldrin	381	25	6.8 (−1)	1.0 (−1)
Diuron	233	25	2.1 (−2)	2.0 (−3)
EPTC	189	20	1.1 (+3)	8.3 (+1)
EPTC	189	25	2.8 (+3)	2.2 (+2)
Fenitrothion	277	20	8.0 (−1)	9.1 (−2)
Heptachlor	373	20	2.2 (+1)	3.3
Lindane	291	25	8.6	1.0
Linuron	249	20	1.1	1.2 (−1)
Malathion	330	20	1.3	1.8 (−1)
Methyl parathion	263	25	2.4	2.5 (−1)
Metolachlor	284	20	1.7	2.0 (−1)
Monuron	199	25	2.3 (−2)	2.0 (−3)
Parathion	291	25	1.3	1.5 (−1)
Picloram	241	20	7.4 (−4)	7.3 (−5)
Prometon	225	25	8.3 (−1)	1.0
Simazine	202	25	2.0 (−3)	1.7 (−4)
Triallate	305	20	2.6 (+1)	3.2
Trifluralin	335	20	1.5 (+1)	2.0

[a]Numbers in parentheses represent power of ten factors.

theoretical and practical aspects of pesticide volatilization were reviewed a few years ago by Taylor and Spencer (1990).

Pesticides are characterized by a range of saturation vapor pressures and densities (Table 4.2) and, therefore, evaporate from the soil medium at differing rates and in differing proportions of the amount incorporated in the soil. The volatilization is controlled, not only by the properties of the molecules, but also by the properties of the soil and environmental factors, such as humidity and temperature. This statement is supported by the results of Spencer and Cliath (1970, 1973) on the vapor density of lindane and dieldrin, as presented in Fig. 4.2a,b. When pesticide concentrations in soils are below the saturation level, the temperature relation becomes more complex. A more detailed analysis of the results presented in Fig. 4.2b shows that, below the saturation level, the vapor density increased with temperature more slowly than that of the pure compound. Spencer and Cliath (1970) explain this behavior by the combination of solubility relationships, Henry's law, and free energy changes during adsorption.

Pesticide volatilization is also affected by the soil constituents and the soil moisture status. From the results reported by Spencer (1970) on the effect of organic matter and clay content on vapor density of dieldrin, it was observed that its vapor densities over both wet and dry soils were inversely related to the soil organic matter content (Table 4.3). In the case of dieldrin, the clay content of soils, which was unrelated to the organic matter content, had only a minor effect. This finding was confirmed by other experiments (Chiou 1989) and, based on this, it may be emphasized that soil organic matter content and soil

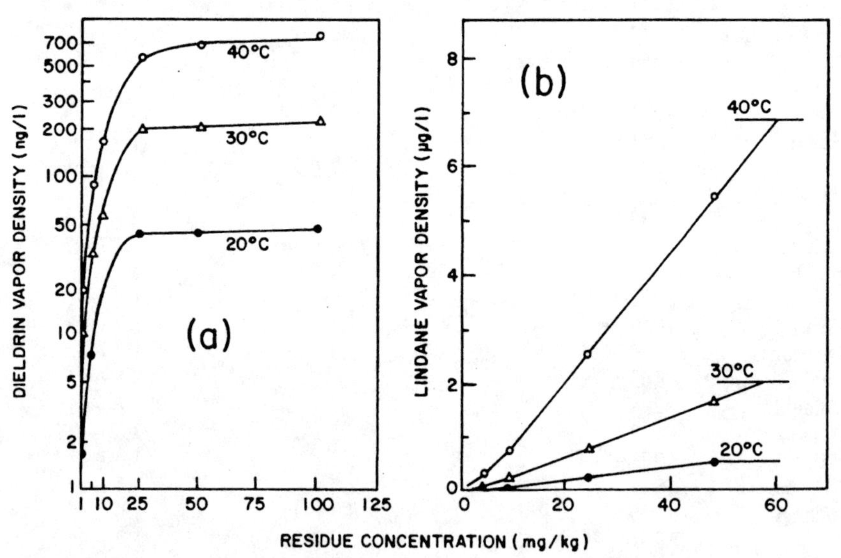

Fig. 4.2a,b. Vapor density of dieldrin (**a**) and lindane (**b**) in air equilibrated at 20, 30 and 40 °C with residues adsorbed on Gila silt loam soil at a moisture content of **a** 0.10 $kg\,kg^{-1}$ and **b** 0.0394 $kg\,kg^{-1}$. (Spencer and Cliath 1969, 1970)

Table 4.3. Effect of organic matter and clay content on vapor density of dieldrin (at 10 mg/kg dieldrin and 30 C; Spencer 1970)

Soil type	Organic matter (g/kg)	Clay (%)	Vapor density (mg/L)	
			Wet[a]	Dry[b]
Rosita very fine sandy loam	1.9	16.3	175	1.7
Imperial clay	2.0	67.6	200	0.9
Gilat silt loam	5.8	18.4	52	0.7
Kentwood sandy loam	16.2	10.0	32	0.4
Linne clay loam	24.1	33.4	32	0.6

[a]Wet – approximately 0.2 Mpa matrix suction.
[b]Dry – equilibrated at 50% relative humidity. All soils contain 10 ppm dieldrin: all temperatures 30 °C.

organic matter partition coefficients are of primary importance in describing the rate of volatilization in soils for compounds having a high affinity for the organic matter.

Over the years, retardation of pesticide volatilization in dry soils was noted. This is illustrated by the results of Spencer (1970) with trifluralin and dieldrin (Fig. 4.3a,b). These show that even where the concentrations in a moist soil of dieldrin and trifluralin were so high as to give vapor densities approaching those of pure compounds, reduction of the soil water content caused large

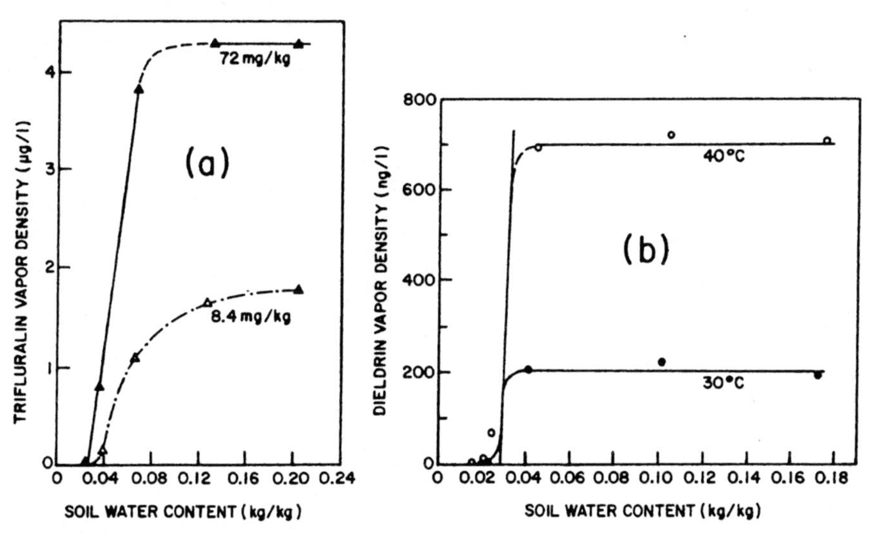

Fig. 4.3a,b. Vapor density of trifluralin (**a**) and dieldrin (**b**) as affected by temperature, initial concentration in the soil and soil-water content. **a** Vapor density of trifluralin in air equilibrated at 30 °C with Gila silt loam containing initially 8.4 and 72 mg kg^{-1} trifluralin. (Spencer and Cliath 1974). **b** Vapor density of dieldrin in air equilibrated at 30 and 40 °C with Gila silt loam containing initially 100 mg/kg dieldrin. (After Spencer and Cliath 1969, 1973)

Table 4.4. Field measurements of volatilization. (As summarized by Taylor and Spencer 1990)

Pesticide	Exp. no.	Usage	Surface	Season, place	Fraction volatilized in period
Alachlor	1	Surface	Fallow silt loam	May 1981, Maryland	26% in 24 d
Atrazine	2	Surface	Fallow silt loam	May 1981, Maryland	2.4% in 24 d
Chlorpropham	3	Surface	Soil under soybean	May 1976, Maryland	49% in 50 d
Chlorpropham	4	Microencap	Soil under soybean	May 1976, Maryland	20% in 50 d
Chlordane	5	Surface	Moist fallow silt loam	Aug. 1975, Maryland	50% in 50 h
Chlordane	6	Surface	Dry fallow sandy loam	June 1978, Maryland	2% in 50 h
Dacthal	7	Surface	Moist fallow silt loam	Aug. 1975, Maryland	2% in 34 h
DDT	8	Incorp. to 9 cm, then surface	Moist fallow silt loam	Oct. 1968, Louisiana	44%in 120 d
DDT	9	Cotton field	Foliage, 45% ground cover	Sept. 1976, Mississippi	21% in 5 d 58% in 11 d
Dieldrin	10	Surface	Moist soil fallow	Sept. 1969, Louisiana	20% in 50 d
Dieldrin	11	Incorp. to 7.5 cm	Soil under corn	May 1969, Ohio	4% in 170 d
Dieldrin	12	Vegetation	Short orchardgrass	July 1973, Maryland	90% in 30 d
EPTC	13	Surface in irrigation water	Soil under 25 cm alfalfa	May 1977, California	74% in 52 h
Heptachlor	14	Incorp. to 7.5 cm	Soil under corn	May 1969, Ohio	7% in 170 d
Heptachlor	15	Surface	Moist fallow silt loam	Aug. 1975, Maryland	50% in 6 h 90% in 6 d
Heptachlor	16	Surface	Dry fallow sandy loam	June 1978, Maryland	40% in 50 h
Heptachlor	17	Vegetation	Short orchadgrass	July 1973, Maryland	90% in 7 d
Lindane	18	Surface	Moist fallow silt loam	June 1977, Maryland	50% in 6 h 90% in 6 d
Lindane	19	Surface	Dry fallow sandy loam	June 1978, Maryland	12% in 50 h
Photodieldrin	20	Vegetation	Short orchardgrass	July 1973, Maryland	5% d^{-1}
Simazine	21	Surface	Fallow silt loam	May 1981, Maryland	1.1% in 24 d
Toxaphene	22	Cotton crop	Foliage	Sept. 1974, Mississippi	25% in 5 d
Toxaphene	23	Cotton field	Foliage 45% groundcover	Aug. 1976, Mississippi	53% in 33 d

Toxaphene	24	Surface	Fallow silt loam	May 1981, Maryland	33% in 24 d
Trifluralin	25	Incorp. to 2.5 cm	Soybean on sandy loam 0.5% o.m.	June 1973, Georgia	22% in 120 d
Trifluralin	26	Incorp. to 7.5 cm	Soybean on loam 4% o.m.	May 1973, New York	3.4% in 90 d
Trifluralin	27	Surface	Moist fallow silt loam	Aug. 1975, Maryland	50% in 7.5 h 90% in 7 d
Trifluralin	28	Surface	Moist fallow	June 1977, Maryland	87% in 50 h
Trifluralin	29	Surface	Dry fallow sandy loam	June 1978, Maryland	25% IN 50 h
2,4-D	30	Surface	20-cm high spring wheat	June 1981, Saskatchewan	21% in 5 d

reductions in the equilibrium vapor densities. Similar results were reported by the same group for DDT, DDE isomers, and lindane (Spencer and Cliath 1970, 1973). In practical terms, in the Gila soil, a moisture content of 0.0394 kg/kg is in equilibrium with air of a relative humidity of 94%. The moisture contents at which the equilibrium vapor densities of pesticides are reduced to low values are thus easily reached in surface soils drying in moderately dry air.

The above data demonstrate the dominant role of soil moisture in controlling the volatilization of soil-incorporated pesticides. When soil moisture moves to the surface as a result of surface evaporation, the volatilization of pesticides is controlled by the "wick effect", and reflects the rate of replenishment by upward transport of the chemical in the flow of soil moisture (Taylor and Spencer 1990).

The above experiments on volatilization were carried out in the laboratory under controlled conditions. There have also been field experiments which confirmed the validity of the laboratory experiments, with small deviations caused by scale and condition differences. Table 4.4 reproduces information on field experiments on pesticide volatilization, as summarized by Taylor and Spencer (1990).

The data in Table 4.4 show a wide range of observed volatilization rates extending from 90% losses within a few days to small losses over weeks or months. Direct measurement of pesticide volatilization in the field is, however, complicated, and requires a large number of samples to be analyzed in order to obtain a statistically meaningful result. The data obtained by field measurements confirm, however, the trends in volatilization processes of pesticides from soil to the atmosphere as determined in the laboratory under controlled environmental conditions.

4.4.3 Petroleum Products Volatilization

In recent years, pollution of the unsaturated zone by petroleum products has aroused concern, particularly in the light of reports on groundwater quality. Crude oil and petroleum products are, typically, mixtures of compounds, mainly hydrocarbons, the hydrocarbon group itself having various carbon structures (e.g., *n*-paraffin: C_{11} to C_{32}, paraffins; 1-ring to 6-ring cycle paraffins; aromatics: benzene, toluene; C_8 to C_{11} aromatics, etc.) The relative ratios among the hydrocarbon groups and the composition of each group are determined by the origin of the crude oil and the distillation procedure.

Two nonsynthetic petroleum hydrocarbon mixtures, kerosene and weed oil, will be used as examples to illustrate the volatilization of multicomponent pollutants from soil. Weed oil is characterized by the presence of light hydrocarbons (C_9 to C_{12}) since the kerosene component encompasses a larger range of light and heavy components (C_9 to C_{15}). Consequently, Beacon weed oil will illustrate the case of evaporation immediately after application, while the experiments with kerosene will illustrate the changes with time in the pollutant composition, due to volatilization of the light components.

Field evaporation of Beacon weed oil, as reported by Woodrow et al. (1986) could be described by an exponential function of time:

$$\ln R = 1.8817 - 0.8124t \quad (r^2 = 0.96) ,$$

where R is the remaining amount (mg cm^{-2}) and t is time (h).

The authors explain that the exponential evaporation of the oil in the field was probably partly due to increasing temperature and wind speed during the test period. The half-life of evaporation of the Beacon weed oil was 0.85 h, which was equivalent to a transport to the atmosphere of 328 kg/ha when the initial oil deposit was 6.56 mg cm^{-2}. Figure 4.4 exhibits the qualitative examination of the gas chromatograms of air and surface residues and also indicates that loss from soil was dependent primarily on oil component volatility, with no apparent decomposition occurring on the soil or in the air.

In a series of experiments on kerosene behavior in soils carried out by Yaron's research group (Yaron et al. 1989; Galin et al. 1990a,b; Fine and Yaron 1993; Jarsjö et al. 1994) the effects of soil composition and aggregation on kerosene volatilization were reported. Figure 4.5 shows the relative total volatilization of kerosene from dune sand, loamy sand, and silty loam soils. A significant difference was found between the silty loam soils, on the one hand,

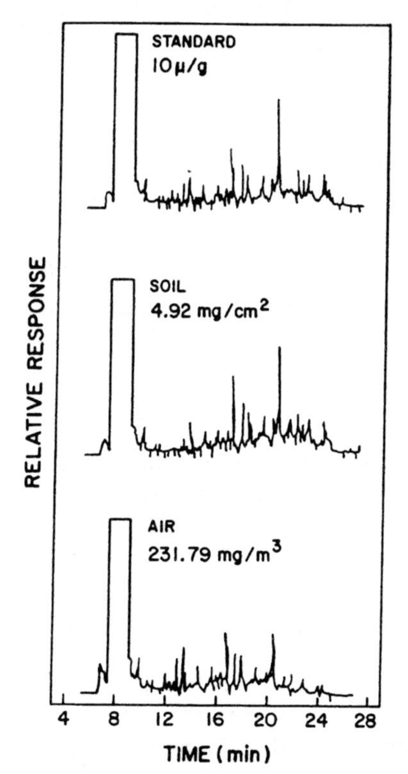

Fig. 4.4. Gas chromatograms of beacon oil standard and of residue on soil and in air for the same postspray interval. (Woodrow et al. 1986)

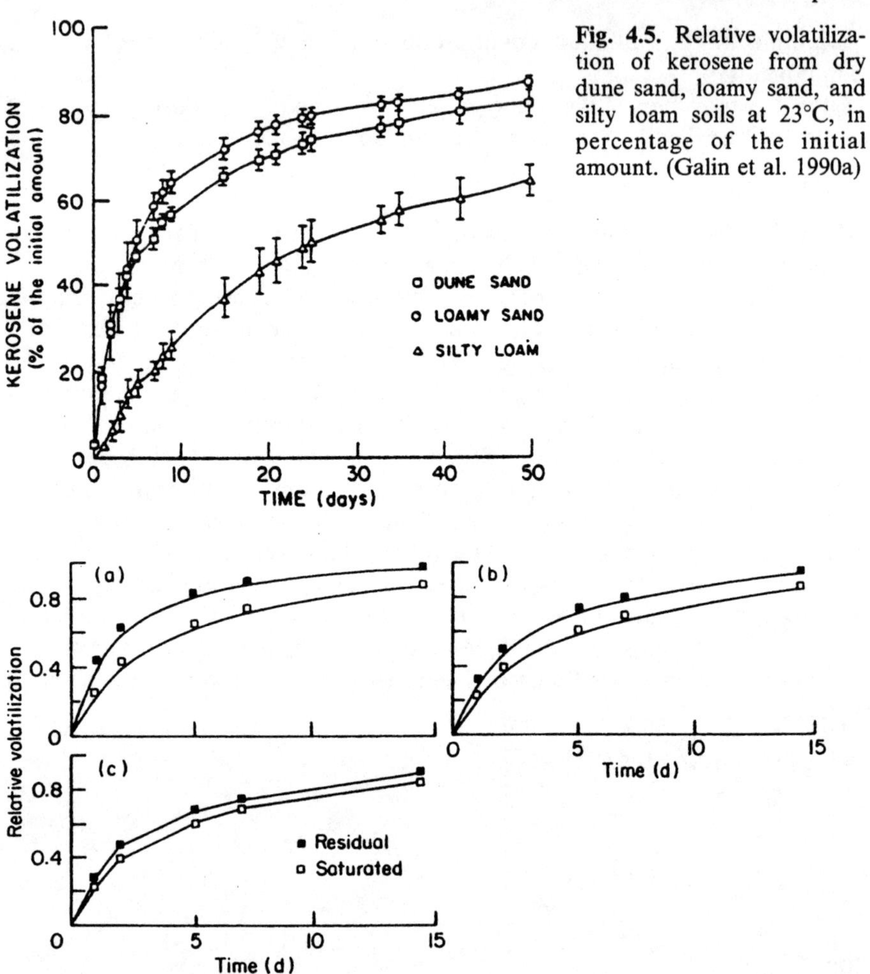

Fig. 4.5. Relative volatilization of kerosene from dry dune sand, loamy sand, and silty loam soils at 23°C, in percentage of the initial amount. (Galin et al. 1990a)

Fig. 4.6a–c. Relative volatilization of kerosene from (**a**) coarse, (**b**) medium, and (**c**) fine sand, as proportions of the initial amount. (Galin et al. 1990b)

and the dune sand and loamy sand soils, on the other. This behavior may be explained by the effect of pore-size distribution in each soil investigated.

The initial amount of kerosene applied to coarse, medium, and fine sands significantly affected the rate of volatilization for the three inert materials used in the experiment (Fig. 4.6). The greatest effect of initial concentration on the evaporation rate was observed in the coarse sand. When the initial amount of kerosene was equivalent to the residual saturation capacity, the evaporative surface decreased (due to rapid filling of the pores by the liquid kerosene) and a decrease in volatilization rate occurred. In fine sand, the free evaporative surface in the porous medium remained similar when the initial amount of kerosene applied ranged between residual and saturated, and this was reflected by a similar rate of volatilization. In the medium sand the effect of initial concentration on the rate of volatilization was intermediate.

The differential volatilization of the kerosene components led to a change over time of the chemical composition of the remaining petroleum product. Figure 4.7 shows gas chromatograms of fresh kerosene and the kerosene recovered from coarse, medium, and fine sands after 1, 5, and 14 days of incubation. The amount of kerosene applied to the three sands was equivalent to the saturated retention capacity. Note the loss of the more volatile hydrocarbons by evaporation in all the sands studied 14 days after application, and the lack of resemblance to the original kerosene. It should be emphasized that all the changes were caused by evaporation, and no biodegradation of the kerosene occurred, since the water content (<0.1%) was too low to support significant biological activity.

The pore size of the sands affected the chemical composition of the remaining kerosene. The C_9–C_{12} fractions disappeared completely 14 days after their incorporation, except from saturated fine sand, where 5% of the initial C_{12} applied was still found. About 40% of C_{13} was recovered from all three sands, when the initial amount applied was equivalent to the saturated retention capacity. When the amount of kerosene applied was equivalent to the

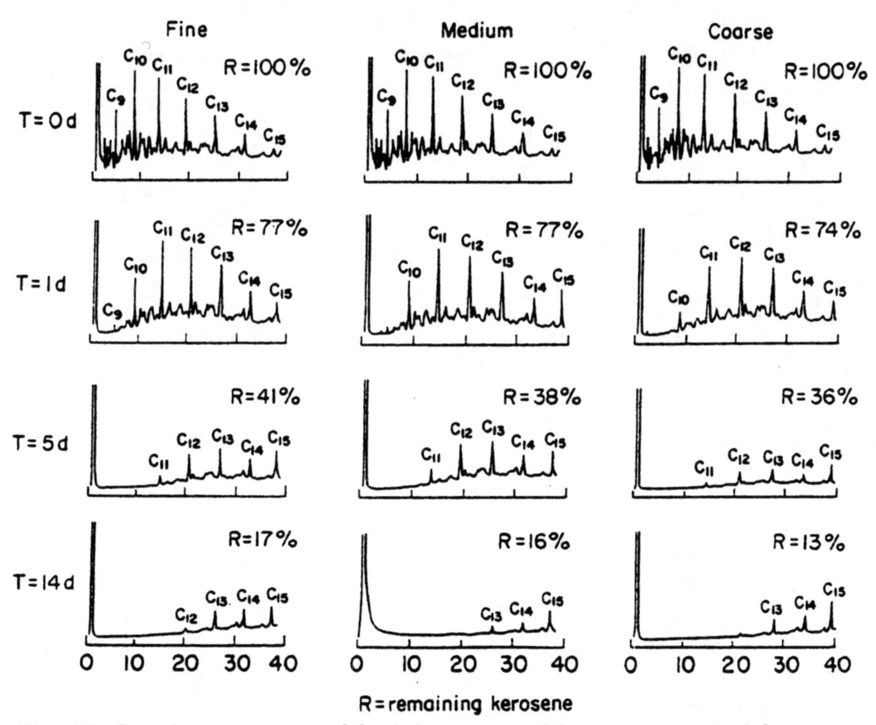

Fig. 4.7. Gas chromatograms of fresh kerosene and kerosene recovered from coarse, medium, and fine sands after 1, 5, and 14 days of incorporation. The initial amount of kerosene applied was equivalent to the saturated retention value. (Galin et al. 1990a)

residual capacity, only 20% of C_{13} was recovered from the fine sand, and the kerosene was completely volatilized from the medium and coarse sands. Values of 10–40% of the initial amount of C_{14}–C_{15} were recovered after 14 days of incubation in the saturated sands. When the initial concentration was equivalent to the residual capacity, only about 20% of the above components were recovered.

The relative importance of the individual components of the kerosene changed with time owing to evaporation of the more volatile hydrocarbons. Figure 4.8 shows, for example, the relative concentration of the selected fractions C_{10}, C_{12}, and C_{15} during 14 days of kerosene incubation in a coarse, medium, and fine sand, when the initial amount was equivalent to the residual and saturated retention capacities. As a general pattern it was observed that, after this period of time, liquid-phase hydrocarbon was still present in the porous medium. The importance of the volatile fraction (C_{10}, C_{12}) diminished and that of the heavy fractions increased for all the sands. The relative rate of decrease of C_{10} and C_{12} and the relative increase of C_{15} were greater in the coarse sand than in the fine sand. The medium sand exhibited intermediate kinetics. In all the experiments, the relative decrease or increase was more pronounced in the sands where kerosene was applied at the residual retention capacity than at the saturated retention capacity.

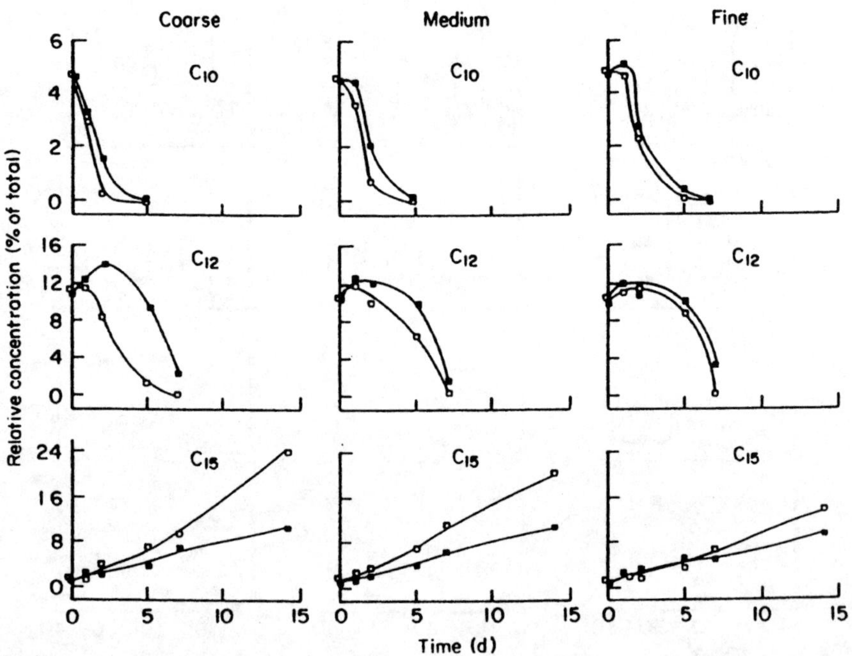

Fig. 4.8. The relative concentration of the C_{10}, C_{12}, and C_{15} components of a kerosene during 14 days of incubation in coarse, medium and fine sand (in percentage of total amount of kerosene), at (□) residual and saturation capacity (■). (Galin et al. 1990b)

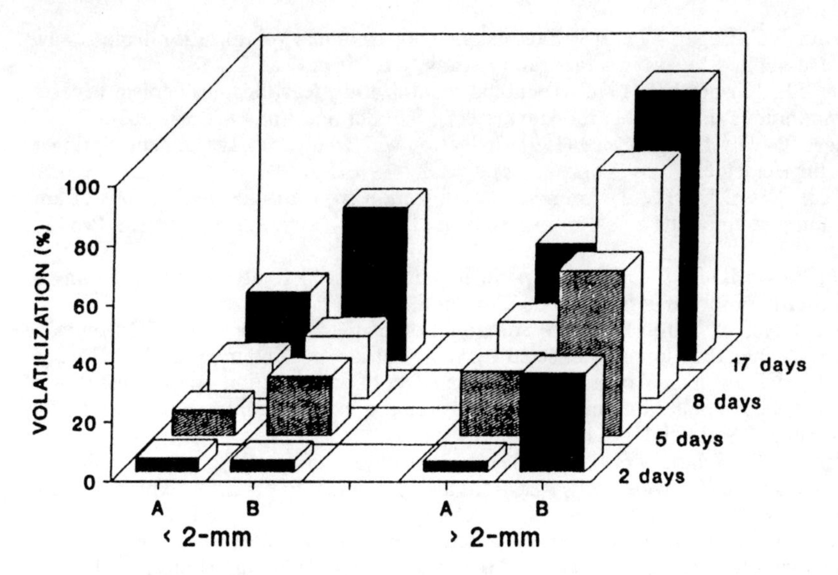

Fig. 4.9. Volatilization of kerosene from the golan vertisol as affected by aggregate size. (Fine and Yaron 1993). Free (*A*) and enhanced venting (*B*)

Fine and Yaron (1993) studied the effect of soil aggregation on kerosene volatilization by comparing the rate of the process in two size fractions of Golan vertisol: the <1-mm fraction and the 2–5-mm aggregates. The total porosity of these two aggregate fractions was very similar: 53 and 55% of the total soil volume, respectively. The respective kerosene retention capacity (KRC) was, however, different in these two size fractions: 8.7 ± 0.25 and 4.1 ± 1.1 mg/100 g soil, respectively. The difference in retention of the heavy fraction seems to emanate from the effect of pore size distribution on the rate of kerosene flow through the soil and on the kinetics of kerosene retention in soil micropores. The two soil separates differed also in air permeability, their respective Ka values being 0.0812 ± 0.009 cm^2 and 0.145 ± 0.011 cm^2. Figure 4.9 shows that the rate of kerosene removal from the two soil separates is related to their permeabilities to air (effect of venting). The more permeable to air the soil is, the more it releases kerosene. From the above examples, it might be observed that the soil aggregation status and the pore-size distribution are soil parameters which affect the volatilization of petroleum products from soils.

References

Chiou TC (1989) Theoretical considerations on the partition uptake of nonionic organic compounds by soil organic matter. In: Shawhney BL, Brown K (eds) Reaction and movement of organic chemicals in soils. SSSA Spec Publ 22: 1–31

Donovan WC, Logan TJ (1983) Factors affecting ammonia volatilization from sewage sludge applied to soil in a laboratory study. J Environ Qual 12: 584–590
Feagley SE, Hosner LR (1978) Ammonia volatilization reactions mechanism between ammonium sulfate and carbonate systems. Soil Sci Soc Am J 42: 364–367
Feigin A, Ravina I, Shalhevet J (1991) Irrigation with treated sewage effluent. Springer, Berlin Heidelberg New York, 224 pp
Fenn LB, Kissel DE (1973) Ammonia volatilization from surface applications of ammonium compounds on calcareous soils. I. General theory. Soil Sci Soc Am Proc 37: 855–859
Fine P, Yaron B (1993) Outdoor experiments on enhanced volatilization by venting of kerosene components from soil. J Contam Hydrol 12: 335–374
Galin T, Gerstl Z, Yaron B (1990a) Soil pollution by petroleum products, III. Kerosene stability in soil columns as affected by volatilization. J Contam Hydrol 5: 375–385
Galin T, McDowell C, Yaron B (1990b) The effect of volatilization on the mass flow of a nonaqueous pollutant liquid mixture in an inert porous medium. Experiments with kerosene. J Soil Sci 41: 631–641
Glotfelty DE, Schomburg CJ (1989) Volatilization of pesticides from soils. In: Shawhney BL, Brown K (eds) Reactions and movement of organic chemicals in soils. SSSA Spec Publ 22: 181–209
Jarsjo J, Destouni G, Yaron B (1994) Retention and volatilization of kerosene: Laboratory experiments on glacial and post-glacial soils. J Contam Hydrology 17: 167–185
Jayaweera GR, Mikkelsen DS (1991) Assessment of ammonia volatilization from flooded soil systems. Adv Agron 45: 303–357
Mackay D (1979) Finding fugacity feasible. Environ Sci Technol 13: 1218–1223
Nye PH, Yaron B, Galin T, Gerstl Z (1994) Volatilization of multicomponent liquid through dry soils. Testing a model. Soil Sci Soc Am J 58: 269–278
Plimmer JR (1976) Volatility. In: Kearney PC, Kaufman DD (eds) Herbicides: chemistry, degradation and mode of action, 2nd edn, vol 2. Marcel Dekker, New York, pp 891–934
Spencer WF (1970) Distribution of pesticides between soil, water and air. In: Pesticides in the soil: ecology, degradation and movement. Michigan State University, East Lansing, pp 120–128
Spencer WF, Cliath MM (1969) Vapor density of dieldrin. Environ Sci Technol 3: 670–674
Spencer WF, Cliath MM (1970) Desorption of lindane from soil as related to vapor density. Soil Sci Soc Am Proc 34: 574–578
Spencer WF, Cliath MM (1973) Pesticide volatilization as related to water loss from soil. J Environ Qual 2: 284–289
Taylor AW, Spencer WF (1990) Volatilization and vapor transport processes. In: Cheng HH (ed) Pesticides in the soil environment. SSSA Book Ser 2. Soil Science Society of America, Madison, pp 213–369
Woldendorp JW (1968) Losses of soil nitrogen. Stickstoff 12: 32–46
Woodrow JE, Selber JN, Yong-Hwe K (1986) Measured and calculated evaporation losses of two petroleum hydrocarbon herbicide mixtures under laboratory and field conditions. Environ Sci Technol 20: 783–786
Yaron B, Sutherland P, Galin T, Acher AJ (1989) Soil and groundwater pollution of petroleum products. II. Adsorption-desorption of kerosene vapors as affected by type of soil and its moisture content. J Contam Hydrol 4: 347–358

CHAPTER 5

Retention of Pollutants on and Within the Soil Solid Phase

Pollutants retained on and within the soil solid phase have reached the soil directly as solute, water-immiscible liquid, suspended particles, or in the gaseous phase. Pollutant retention is controlled by the physicochemical and physical properties of the soil solid phase by the properties of the pollutants themselves, and by environmental factors such as temperature and soil moisture content. Since, under natural conditions, we are dealing, in general, not with a single pollutant but with a mixture of pollutants and natural organic and inorganic compounds, the competition between the compounds for the soil adsorption sites will control their retention on or in the soil solid phase. For quantifying the retention of pollutants in the soil solid phase, an equilibrium should be determined; consequently, major attention should be given to the kinetics of the process.

The capacity of the soil solid phase to retain and release both toxic chemical species and pathogenic microorganisms is an essential part of its function as a filter and reservoir of pollutants. An understanding of the nature of this phenomenon is a prerequisite for any attempt to quantity and predict pollutant behavior in soil. The retention of pollutants on and in the soil solid phase is the result of a physicochemical process, adsorption on the surface, chemical reactions with the solid phase, and mechanical trapping of the pollutants in the solid-phase pores. In the present section we will deal separately with the surface adsorption-release and with the mechanical retention of the pollutants. Examples illustrating both processes will follow.

5.1 Surface Adsorption

Adsorption is defined as the excess concentration of pollutants at the soil-solid interface compared with that in the bulk solution or the gaseous phase, regardless of the nature of the interface region or of the interaction between the adsorbate and the solid surface which causes the excess. Adsorption removes a compound from the bulk phase and, thereby, greatly affects its behavior in the soil environment. Due to some hysteresis effects, explained among other possibilities by a possible bound residue formation, the release of the compound from the solid phase to the liquid or gaseous phase does not always affect the amount of pollutants in the soil medium. Surface adsorption is due to electrical charges and nonionized functional groups on mineral and organic constituents.

5.1.1 Adsorption of Ionic Pollutants

Adsorption of charged ionic pollutants on the surface of the soil solid phase is subject to a combination of chemical binding forces and the electric field at the interface that is implicitly controlled by the adsorption itself. It was shown above (Chap. 1.1.5) that the soil solid phase has a net charge which, in contact with the liquid or gaseous phase, is faced by one or more layers of counter or co-ions which have a net charge equal to and separated from the surface charge. Electrical neutrality on the colloidal surface requires that an equal amount of charge of the opposite sign must accumulate in the liquid phase near the charged surface. For a negatively charged surface, this means that positively charged cations are thus electrostatically attracted to the charged surface.

At the same time, due to diffusion forces, the cations are also drawn back towards the equilibrating solution. The distribution of cations in a "diffuse layer" is established so that the concentration of cation increases towards the surface, the concentration increasing from a value equal to that of the equilibrating solution to a higher value, principally determined by the magnitude of the surface charge. On the other hand, ions of the same sign (anions) are repelled by such a surface with diffusion forces acting in an opposite direction, such that there is a deficit of anions near the surface.

Electric double layer theory. The overall pattern as discussed above is known as a *diffuse double layer (DDL)*. The theory of the diffuse double layer was developed about 100 years ago by Gouy and Chapman. The Gouy-Chapman theory assumes that the exchangeable cations exist as point charges, that the colloid surfaces are planar and infinite in extent, and that the surface charge is uniformly, distributed over the entire colloid surface. Despite the fact that the above assumption does not correspond to the soil conditions, it works "surprisingly well" for the soil colloids, and this fact could be explained only by mutual cancellation of the errors. Stern (1924) and Grahame (1947) refined the Gouy-Chapman theory by recognizing that counterions are unlikely to approach the surface more closely than the ionic radii of anions and the hydrated radii of cations. They also introduced the concept of binding energy for specific adsorption. Fig. 5.1 shows a schematic representation of the Gouy-Chapman and Stern-Grahame models, illustrating the fundamental differences between the two models. Detailed critical presentation of the diffuse double-layer theories can be found in Greenland and Hayes (1981), Sposito (1984), Sparks (1986) and Bolt et al. (1991). A short presentation of the evolution of the diffuse double-layer theory, without going into the mathematical analysis, follows.

The Gouy-Chapman model considers that the charge is uniformly spread over the surface. The space charge in the solution is considered to be built up of a nonuniform distribution of point charges. The solvent is treated as a continuous medium, influencing the double layer only through its dielectric constant which is assumed invariant with position in the double layer. Further, it is assumed that ions and surface are involved only in electrostatic interactions. The present derivation is for a flat surface, infinite in size. The double-layer theory applies equally well to rounded or spherical surfaces (Overbeek 1952). Stern's model (1924) considers that the region near the surface consists of a layer of ions adsorbed at the surface and forming a compact double layer, the Stern layer, and a diffuse double layer, the Gouy-Chapman layer (Fig. 5.1). In the Stern model the surface charge is balanced by the charge in solution which is distributed between the Stern layer at a distance, d, from the surface and a diffuse layer which has a Boltzmann distribution. The total surface charge, σ, is therefore due to the charge in the two layers

$$\sigma = -(\sigma_1 + \sigma_2) \ , \tag{5.1}$$

σ_1 is the Stern layer charge and σ_2 is the diffuse layer charge.

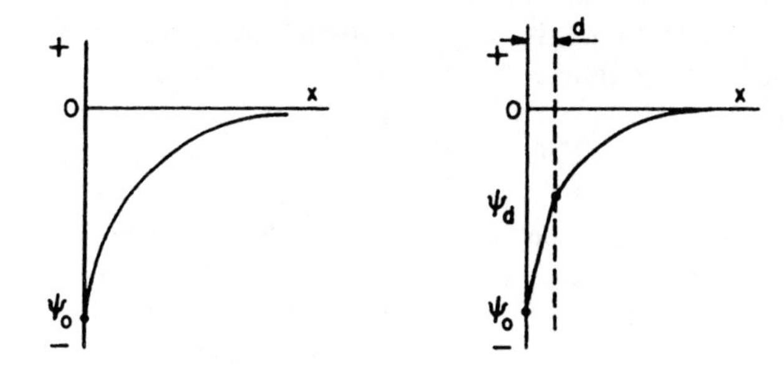

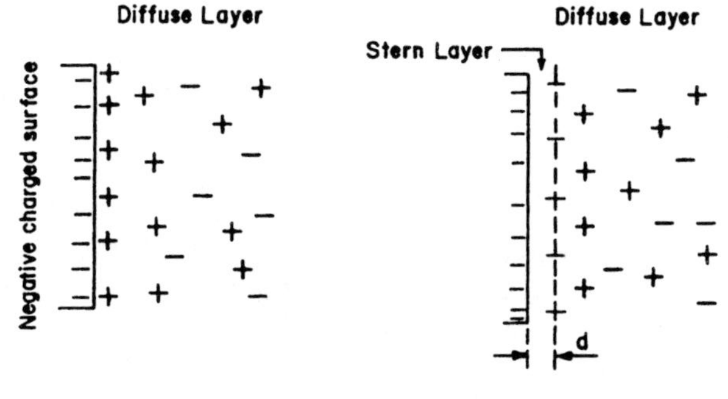

Fig. 5.1a,b. Distribution of electrical charges and potential in double layer according to **a** Gouy-Chapman theory and **b** Stern theory. ψ_o and ψ_d are surface and Stern potential, respectively, *d* is the thickness of the Stern layer

A development of the diffuse double-layer theory considers also the interactions between the two flat layers of the Gouy-Chapman model. The double-layer charge is, however, only slightly affected when the distance between the two plates is large. Further development of the theory includes that of Grahame (1947), which suggests that specifically adsorbed anions are adsorbed into the Stern layer when they lose their water of hydration, whereas the hydrated cations are only electrostatically attracted to the surface. Bolt (1955) later introduced the effects of ion size, dielectric saturation, polarization energies of the ion, Coulombic interaction of the ions, and the short-range repulsion between the ions into the Gouy-Chapman theory. However, the simple Gouy-Chapman theory gives fairly reliable results for colloids with a charge density not exceeding 0.2–0.3 C m^{-2}, especially if the colloid under consideration has a constant charge density. The Gouy-Chapman equation expresses the fundamental relation between the point charge, the electrolyte concentration, and the surface electrical potential. The Gouy-Chapman model gives an invaluable answer – at least qualitatively – to a series of surface processes occurring in the soil system. For example, pending some corrections, this theory explains the exchange capacity concept for the range of surface charge density normally encountered in soil clays.

The double-layer theory has found its application in the soil-pollutant system, in connection with ion exchange process, when the pollutants have charges opposite to that of the surface. The excess ions of opposite charge – over those of like charges – are called exchangeable ions because any ion can be replaced by an ion of the same charge by altering the chemical composition of the equilibrium solution. The cation exchange capacity under a given set of solution conditions is defined by the maximum number of counterions present in the electric double layer per unit mass of exchangers. Expressions of the cation exchange capacity can be derived from the theory describing the distribution of ions near charged surfaces (Bolt et al. 1991). In a variable-charge soil system, the surface potential is controlled by the adsorbed cations, and depends on the activity of these ions in the equilibrium solution. This must be taken into account for expressing the cation exchange capacity.

Generally, the soils contain surfaces with a mixture of both constant and variable charges and the total net surface charge density becomes

$$\sigma = \sigma_p + \sigma_v \ , \tag{5.2}$$

where σ_p and σ_v represent the constant and the variable surface charge density, respectively.

Cation exchange and selectivity. Cation exchange and selectivity form one of the most important processes which control the fate of ionically charged pollutants in soils. It involves both the cationic concentration in solution and the cation dimensions, on the one hand, and the configuration of the exchange sites, on the other. For cationic molecules the retention properties follow the relation

$$\frac{[A_{soln}]}{[B_{soln}]} = K_K \frac{[A_{ads}]}{[B_{ads}]} , \tag{5.3}$$

where K is a selectivity coefficient which expresses the inequality of the activity ratios of the cationic molecules, A and B, in solution (soln) and adsorbed (ads) phases. Since Kerr (1928) was the first to propose this equation, the exchange coefficient is called the Kerr exchange coefficient and is denoted K_K. A simple way to show the selectivity phenomenon is to plot the molar fraction adsorbed versus the molar fraction in solutions in a square diagram (Fripiat et al. 1971).

When cations of differing charge are exchanged, Eq. (5.3) is modified. In a case of an exchange between bivalent and monovalent cations, for example, the exchange coefficient is expressed by the equation

$$\frac{[M^+{}_{ads}]^2}{[M^{2+}{}_{ads}]} \times \frac{[M^{2+}{}_{soln}]}{[M^+{}_{soln}]^2} = K_K , \tag{5.3a}$$

where M^+ and M^{2+} are mono- and bivalent cations and all terms represent molar quantities.

This equation was modified later and the modification most used at present is that suggested by Gapon (1933), who approached the process as an exchange of mole equivalents of electric charge rather than of moles, the Gapon relation is

$$KG = \frac{[M^+{}_{ads}]^2}{[\frac{1}{2}M^{2+}{}_{ads}]} \times \frac{[M^{2+}{}_{sol}]}{[M^+{}_{sol}]^2} , \tag{5.4}$$

where the solute concentration is measured as activities and the adsorption is measured on an equivalent basis.

The exchange properties of negatively charged surfaces in soils do not affect different cations which have the same valence, in the same manner. This is due to differences in size and polarizability among the cations themselves, to the structural characteristics of the surfaces where the cations are held, and to differences in the surface charge distribution. In general, the preference of soil minerals, for instance, for monovalent cations decreases in the order

$$Cs^+ > Rb^+ > K^+ > NH_4^+ > Na^+ > Li^+ ,$$

and this is known as the lyotropic series, indicating the greater attraction of the surfaces for less hydrated cations. The ammonium ion is more strongly preferred than the anhydrous H^+ or Na^+ in 1:2 clay minerals since it may form NH oxygen links in the hexagonal holes of the Si–O sheets, and may also link them to adjacent oxygen planes in the interlayer space by an O.H–N–H.O bond. This mechanism occurs only in the case of minerals which have isomorphic substitution in the tetrahedral sheet of the layer. Selectivity of the divalent alkaline earth cations is less pronounced. Trivalent cations such as Al^{3+} octahedrally coordinated to water molecules would link more strongly than hydrated Ca^{2+} ions.

Cation selectivity on soil organic matter is related mainly to the disposition of the acidic groups in soil organic compounds. Multivalent cations are preferred to monovalent cations and transitional group metals to the strongly basic metals. In the case of soil organomineral complexes, it is mentioned that the CEC of the complex is invariably less than the sum of the CECs of the components, and that the pattern of cation selectivity is also considerably modified (Greenland and Hayes 1981).

The selectivity equations characterizing the exchange between two cations are specific for saline soils, and describe the sodium adsorption on the soil solid phase. The US Salinity Lab. (1954) derived two equations from the Gapon equation for the molecular adsorption of sodic solution of the soils. The first describes the sodium adsorption ratio (SAR), based on the Na content in the soil solution, as follows:

$$SAR = \frac{[Na^{+}]}{([Mg^{2+}] + [Ca^{2+}])^{1/2}} , \quad (5.5)$$

where the solution concentration is in $\mu mol\, l^{-1}$. The exchangeable sodium fraction in soils, known as ESR, is expressed by the formula:

$$ESR = \frac{ES}{CEC - ES} , \quad (5.6)$$

where ES is the exchangeable Na in $cmol\, kg^{-1}$ and CEC is the cation exchange capacity in $cmol\, kg^{-1}$.

The processes involved in the adsorption of metal cations in soils include exchange or Coulombic adsorption and specific adsorption by the solid phase components of the soil. The heavy metal cations take part in exchange reactions with negatively charged surfaces of clay minerals as follows:

$$Clay - Ca + M^{2+} \rightleftharpoons Clay - M + Ca^{2+} . \quad (5.7)$$

Cationic adsorption is affected by the pH, and in an acid environment ($pH < 5.5$) some of the heavy metals do not compete with Ca^{2+} for the mineral adsorption sites. When the pH becomes greater than 5.5, the heavy metal adsorption increases abruptly and the reaction becomes irreversible. Table 5.1 shows, for example, the effects of pH on the adsorption of selected heavy metals on the goethite surface.

Cationic organic pollutants (e.g., pesticides) compete with the mineral ions for the same adsorption sites. At a low pH level, an organic cationic molecule is much more strongly adsorbed on the soil surface than a mineral ion with the same valency. At a moderate pH, the mineral ion adsorption is favored over that of cationic organic molecules. In general, the charge density of the soil solid phase is a determinant factor in adsorption of cationic organic molecules. Their adsorption is affected, however, by their molecular configuration (Mortland 1970). Similarly to the organic cation, the basic organic molecules could be adsorbed by clays, by cationic adsorption. However, this adsorption mechanism is conditioned by the acidity of the medium.

Table 5.1. Adsorption of heavy metals on goethite as a function of pH. Data expressed as percent of initial amount of metallic cation in solution. (Quirk and Posner 1975)

Metal	pK_1 (approx.)	pH							
		4.7	5.2	5.5	5.9	6.4	7.2	7.5	8.0
Cu	8.0	17	55	75	90	–	–	–	–
Pb	8.0	–	43	56	75	–	–	–	–
Zn	9.0	–	–	–	13	22	68	–	–
Cd	9.5	–	–	–	–	23	44	53	–
Co	9.5	–	–	–	–	–	39	54	78

Living pathogenic organisms could also be adsorbed on soil solid surfaces by cation exchange mechanisms. Multivalent cations are necessary for certain bacteriophages to be adsorbed on colloidal charged surfaces and dependence on cation-mediated sorption has been reported for many virus-cell pairs. Using a positively charged resin as a model for the exchange with a negatively charged bacterial cell, Daniels (1980) proposed a formal mechanism of adsorption from liquid suspensions. He considered that the large complex structure of the bacterial cell can behave as either a cation or an anion and react, respectively, with the charged groups of an anionic or a cationic exchange resin. The exchange resin can be represented as a large polymeric network that assumes either a positive or negative charge in association with small, dissociable counterions of opposite charge.

The exchange between a negatively charged bacterial cell and a positively charged ion-exchange resin is shown in Eq. (5.8)

$$\underbrace{\mathrm{H{-}\overset{\displaystyle R}{\overset{|}{C}}{-}COO^-}}_{\substack{\text{Surface of Bacterial Cells}\\ \mathrm{NH_2}}}\mathrm{H^+} + \overbrace{\mathrm{Cl^-(H_3C)_3N^+{-}R'}}^{\text{Resin}} \rightarrow \mathrm{H{-}\underset{\displaystyle NH_2}{\overset{\displaystyle R}{C}}{-}COO^-{-}N^+(H_3C)_3{-}R' + H^+Cl^-} \quad . \tag{5.8}$$

$$\underbrace{\mathrm{H{-}\overset{\displaystyle R}{C}{-}COO^-}}_{\substack{\text{Surface of Bacterial Cells}\\ \mathrm{NH_2}}}\mathrm{H^+} + \mathrm{M^{2+}} + \overbrace{\mathrm{R''{-}\underset{\displaystyle O}{\overset{\displaystyle O}{S}}{-}O^-\,H^+}}^{\text{Resin}} \rightarrow \mathrm{H{-}\underset{\displaystyle NH_2}{\overset{\displaystyle R}{C}}{-}COO^-\,M^{2+}\,O{-}\underset{\displaystyle O}{\overset{\displaystyle O}{S}}{-}R'' + 2(H^+, Cl^-)} \quad . \tag{5.9}$$

Cation exchange between a negatively charged bacterial cell and a negatively charged (cation) ion-exchange resin is possible if a multivalent cation, M^{2+}, can act as a bridge between the cell charge and the cation, as shown in Eq. (5.9).

Since in dealing with soil pollution problems we must consider also the fate of pathogens in the soil media, the understanding of the adsorption of microorganisms on soil aggregates is absolutely necessary.

Negative adsorption. When a charged solid surface faces an ion – of like charge – in an aqueous suspension, the ion is repelled from the surface by Coulomb forces. This phenomenon is called negative adsorption. The Coulomb repulsion produces a region in the aqueous solution that is relatively depleted of the anion and an equivalent region far from the surface that is relatively enriched. Sposito (1984) characterized this strictly macroscopic phenomenon quantitatively through the definition of the *relative surface excess* of an ion, i, in a suspension:

$$\Gamma_i^{(w)} = \frac{n_i - M_w m_i}{S}\,, \tag{5.10}$$

where n_i is total moles of ion i in the suspension per unit mass of solid, M_w is the total mass of water in the suspension per unit mass of solid, m_i is the molality of ion i in the supernatant solution, and S is the specific surface area of the suspended solid. Thus $\Gamma_i^{(w)}$ is the excess moles of the ion (per unit area of suspended solid) relative to an aqueous solution containing M_w kilograms of water and the ion at molality m_i.

If an anion approaches a charged surface, it is subject to attraction by positively charged sites on the surface, or repulsion by negatively charged ones. Since clay minerals in soils are normally negatively charged, anions tend to be repelled from the mineral surfaces. If a dilute neutral solution of KCl, for example, is added to a dry clay, the equilibrium Cl^- concentration in the bulk solution will be greater than the Cl^- concentration in the solution originally added to the clay. This process, occurring when an anion is added to a dry colloid with no adsorption capacity for the anion at the prevailing pH, is called negative adsorption, and is related to the unequal ion distribution in the diffuse double layer charged colloids. Anion negative adsorption is affected by the anion charge, concentration, pH, the presence of other anions, and the nature and charge of the surface. Negative adsorption decreases as the soil pH decreases, and when anions can be adsorbed by positively charged soils or soil colloids. The greater the negative charge of the surface, the greater is the anion negative adsorption. Acidic toxic organic compounds in their anionic form are expected to be repelled by negatively charged clay surfaces. Some of the soil clays, amorphous materials, also have some pH-dependent positive charge at low pH values. Still, under such conditions, most acidic organics are in the molecular form in the soil, so that significant adsorption by formation of anionic bonds is improbable.

5.1.2 Adsorption of Nonionic Pollutants

The adsorption of nonionic pollutants on the soil solid phase surfaces is subjected to a series of mechanisms such as protonation, water bridging, cation bridging, ligand exchange, hydrogen bonding, and van der Waals interactions. In this class of pollutants are hundreds of compounds belonging to chemically different groups, such as chlorinated hydrocarbons, organophosphates, carbamates, ureas, anilines and anilides, amides, etc. The great differences between the properties of these groups and among compounds within a group, which are reflected in the variability of the adsorption mechanism. That has been demonstrated for all the nonionic organic compounds, is the predominant role of the soil organic colloid in the adsorption process. Details of nonionic pollutant adsorption on soils and clays can be found in the reviews of Mortland (1970), Calvet (1989), Hassett and Banwart (1989), Hayes and Mingelgrin (1991), in the books of Theng (1974) or edited by Greenland and Hayes (1981) or Saltzman and Yaron (1986). Here, we will give only a general view of the mechanism involved in the process.

Hassett and Banwart (1989) consider that the sorption of nonpolar organics by soils is due to enthalpy-related and entropy-related adsorption forces. They consider that sorption occurs when the free energy of the sorption reaction is negative: $\Delta G = \Delta H - T\Delta S$ with $\Delta G < 0$. The free energy of a sorption reaction can be negative because of either enthalpy term, the entropy term, or contributions from both. The enthalpy term is primarily a function of the difference in the bonding between the adsorbing surface and the sorbate (solute) and that between the solvent (water) and the solute. The entropy term is related to the increase or decrease in the ordering of the system upon sorption.

For a chemical reaction at equilibrium, the free energy (ΔG) can be calculated from the equilibrium constant (K). The enthalpy change (ΔH) can be calculated from the variation of K with temperature, and ΔS calculated as the difference:

$$\Delta S = (\Delta H - \Delta G)/T \ .$$

1. *Enthalpy-related adsorption forces* include the following processes:

Hydrogen bonding which refers to the electrostatic interaction between a hydrogen atom covalently bound to one electronegative atom (e.g., oxygen) and to another electronegative atom or group of atoms in a molecule. The hydrogen atom may thus be regarded as a bridge between electronegative atoms. Generally, this bonding is conceived of as an induced dipole phenomenon. The H bond is considered, generally, as the asymmetrical distribution of the first electron of the H atom induced by various electronegative atoms.

Ligand exchange processes involve replacement of one or more ligands by the adsorbing species. In some instances the ligand exchange processes can be regarded as condensation reactions (e.g., between a carboxyl group and a hydroxy aluminum surface). Under some conditions, ligand exchange reactions are very likely to be involved when humic substances interact with the clay material.

Protonation mechanisms include a Coulomb-electrostatic force resulting from charged surfaces. The development of surface acidity by the solid surface of the soils offers the possibility that solutes having proton-selective organic functional groups can be adsorbed through a protonation reaction.

πBonds occur as a result of the overlapping of π orbitals when they are perpendicular to aromatic rings. This mechanism can be used to explain the bonding of alkenes, alkylenes, and aromatic compounds to soil organic matter.

London-van der Waals forces are generally multipole (dipole-dipole or dipole-induced dipole) interactions produced by correlation between fluctuating induced multipole (principal dipole) moments in two nearly uncharged polar molecules. Despite the fact that the time-average induced multipole in each molecule is zero, the correlation between the two induced moments does not average zero. As a result, an attractive interaction between the two is produced at very small molecular distances.

The van der Waals forces also include the dispersion forces that arise from correlations between the movement of electrons in one molecule and those of neighboring molecules. Under such conditions, even a molecule with no permanent dipole moment will form an instantaneous dipole as a result of fluctuations in the arrangements of its electron cloud. This instantaneous dipole will polarize the charge of another molecule to give a second induced dipole, and this will result in a mutual dipole-dipole attraction. All molecules are subject to attraction by dispersion forces whether or not more specific interactions between ions (or dipoles) occur. Although the momentary dipoles and induced dipoles are constantly changing positions, the net result is a weak attraction. Sposito (1984) pointed out that the van der Waals dispersion interaction between two molecules is necessarily very weak, but when many groups of atoms in a polymeric structure interact simultaneously, the van der Waals components are additive.

Chemisorption denotes the situation in which an actual chemical bond is formed between the molecule and the surface atoms. Hassett and Banwart (1989), covering this topic with regard to the adsorption of nonpolar compounds showed that a molecule undergoing chemisorption may lose its identity as the atoms are rearranged, forming new compounds at the demand of the unsatisfied valences of the surface atoms. The enthalpy of chemisorption is much greater than that of physical adsorption. The basis of much catalytic activity at surfaces is that chemisorption may organize molecules into forms that can readily undergo reactions. It is often difficult to distinguish between chemisorption and physical sorption because a chemisorbed layer may have a physically sorbed layer upon it.

2. *Entropy-related adsorption force*. This force, denoted as hydrophobic sorption (more generally solvophobic interaction), is the partitioning of nonpolar organics out of the polar aqueous phase onto hydrophobic surfaces in the soil where they are retained through dispersion forces. Figure 5.2 shows a

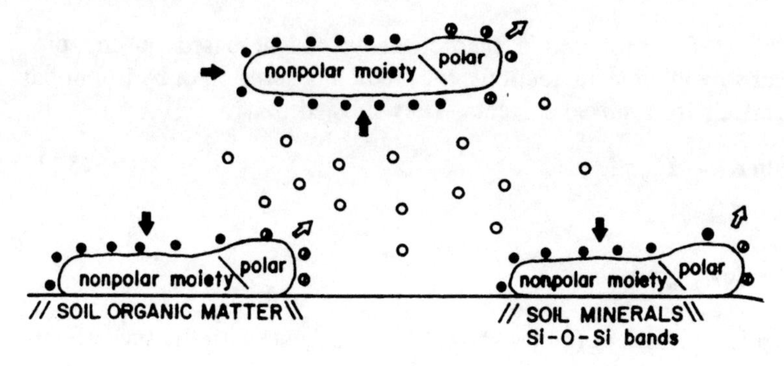

Fig. 5.2. Diagrammatic model of forces contributing to the sorption of nonpolar (hydrophobic) organics. (After Horvath and Melander 1978 in Hassett and Banwart 1989)

diagrammatic model of the forces contributing to the sorption of hydrophobic organics. The major feature of hydrophobic sorption, is the weak interaction between the solute and the solvent. The primary force in hydrophobic sorption appears to be the large entropy change resulting from the removal of the solute from solution. The entropy change is largely due to the destruction of the cavity occupied by the solute in the solvent and the destruction of the structured water shell surrounding the solvated organic. The hydrophobic surfaces in soils are essentially present in the soil organic matter, but may also be $-S_i-O-S_i$ bonds at mineral surfaces, which have their isomorphic substitutions located in the octahedral sheet of the layer.

Hydrophobic sorption, being an entropy-driven process, provides the major contribution to sorption of hydrophobic pollutants on soil surfaces.

5.1.3 Adsorption of Complex Mixtures

A system characterized by the presence of multiple solutes and multiple solvents has been termed a complex mixture (Rao et al. 1989). Since soil pollution under a waste disposal site generally occurs as complex mixtures, an understanding of the adsorption process in these conditions is required. In general, this situation mainly involves organic pollutants, and the sorption might occur from complex solvents containing partially miscible or completely miscible organic solvents.

Rao et al. (1985) proposed a theoretical approach, based on the predominance of solvophobic interactions, for predicting sorption of hydrophobic organic chemicals from mixed solvents. They showed that

$$\ln K^{m} = \ln K^{w} - \Sigma\, \alpha_i \sigma_i f_i \ , \qquad (5.11)$$

where

$$\sigma_i = [(\Delta\gamma_i \, HSA/(kT)] \ .$$

K is the sorption coefficient (moles solvent kg^{-1} sorbent) with the superscripts w and m indicating values for sorption from water and mixed solvents, respectively; σ_i is a dimensionless term unique to each solvent-sorbate combination; $\Delta\gamma_i$ is the differential interfacial free energy (J nm^{-2}) at the solvent-sorbate interface; HSA is the sorbate hydrocarbonaceous surface area (nm^2); k is the Boltzmann constant (J/K); T is the thermodynamic temperature (K); f_i is the volume fraction of the i-th consolvent; and α_i is an empirical constant.

Equation (5.11) suggests that with increasing volume fraction (f_i) of a completely miscible organic solvent in a binary mixed solvent, the hydrophobic organic solvent sorption coefficient (K^m) decreases exponentially; this is based on the assumption that solubility and sorption coefficient are inversely related.

When the hydrophobic organic compound is adsorbed onto soils, from a multiphasic (water and organic) solvent, the liquid-ligand partitioning of the compound and its adsorption by the soil directly from the aqueous phase should be considered. This behavior is explained by the fact that in unsaturated and saturated soils, water is likely to be the wetting phase. In this case, the wetting phase will completely or partially coat the soil surface, increasing its sorbing capacity due to the fact that the wetting phase will serve as additional sink.

A special case is that of multisorbate mixture adsorption from aqueous solutions. Rao et al. (1989) showed that for hydrophobic organics at least four mechanisms of adsorption should be considered. The first mechanism involves the sorption of the neutral molecular species from the aqueous phase; this is similar to hydrophobic sorption. A second mechanism of interest comprises the specific interactions of the dissociated (ionic) species with various functional groups on the sorbent surface. Several models that have been developed for predicting the ion exchange of inorganic ions may be useful in predicting this type of sorption. A third sorption mechanism, known as molecular ion pairing, involves the transfer of the organic ions, together with inorganic counterions, from the aqueous phase to the organic surface phase. A fourth mechanism involves the transfer of the organic ions from the aqueous phase to the organic surface, with the counterions remaining in the electric double layer of the aqueous phase. While all these mechanisms are plausible, further research is needed to explicitly define and describe the sorption processes involved. The relative contribution of each of these mechanisms will depend upon: (1) the extent of compound dissociation as a function of pK_a and solution pH; (2) the ionic charge status of the soil as a function of the pH and of the point of zero net charge; and (3) the ionic strength and composition of the aqueous phase.

5.1.4 Adsorption Isotherms

It was shown that adsorption is the process leading to a net accumulation of a compound at the interface of two contiguous phases. In our case, the adsorbent is the soil solid phase and the sorbate can be in a gaseous or a liquid phase (as a solute). The adsorption data should be considered as valid only when an equilibrium state has been achieved under controlled environmental conditions.

When the measured adsorption data are plotted against the concentration value of the adsorbate at equilibrium, a graph is obtained called the *adsorption isotherm*. A useful relationship between the shape of the adsorption isotherm and the adsorption mechanism for a solute-solvent adsorbent system is based on the classification of Giles et al. (1960) as follows:

The *S-curve* isotherm is characterized by an initial slope that increases with the concentration of a substance in the soil solution. This suggests that the relative affinity of the soil solid phase for the solute at low concentration is less than the affinity of the soil for the solvent.

The *L-curve* isotherm is characterized by an initial slope that does not increase with the concentration of the substance in the soil solution. This behavior points out the high relative affinity of the soil solid phase at low concentration and a decrease of the free adsorbing surface.

The *H-curve* isotherm is characterized by a large initial slope which indicates the high affinity of the soil solid phase for the adsorbate.

The *C-curve* isotherm exhibits an initial slope that remains independent of the substance concentration in the solution under the possible experimental condition. This type of isotherm looks like a constant partitioning of a solute between the solvent and the adsorbing surface. It may be due to a proportional increase of the adsorbing surface as the surface excess of an adsorbate increase.

Depending on their molecular properties and the properties of the solvent, both inorganic and organic pollutants exhibit adsorption isotherms fitting one of Giles' isotherm classifications. Giles' adsorption isotherms for organic toxic compounds are illustrated in Fig. 5.3. These isotherms can describe the adsorption mechanism of potentially toxic organic compounds as controlled by the properties of the adsorbent, properties, and concentration of the solute, and characteristics of the solvent.

Weber and Miller (1989), summarizing the published data on 230 adsorption isotherms for adsorption from aqueous solution, found the following distribution in the Giles classification: 16% S-group, 64% L-group, 12% H-group, and 8% C-group. The fact that L-type isotherms are the most common for the sorption of organic chemicals on soils and soil materials might be explained by the fact that soils are heterogeneous mixtures of adsorbents, each

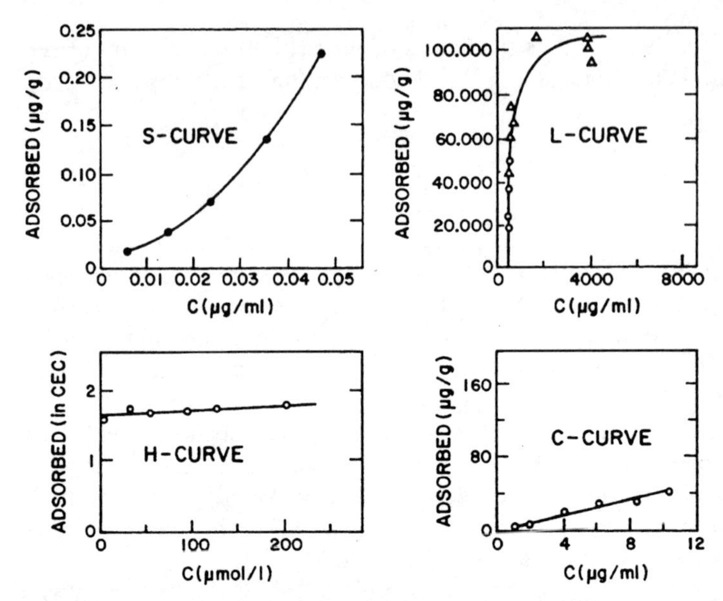

Fig. 5.3. Examples of adsorption isotherms. *S-type* aldrin on oven-dry kaolinite from aqueous solution. *L-type* parathion on oven-dry attapulgite from hexane solution. *H-type* methylene blue at pH 6 on montmorillonite from aqueous solution. *C-type* parathion on clay soil (RH = 50%) from hexane solution

of which sorbs the organic chemicals to a different extent. The adsorbing soil may affect the type and the slope of the adsorption isotherm. Weber et al. (1986) pointed out the case of fluoridone, which exhibits an S-type sorption isotherm on soils with low organic matter content and high montmorillonite content, and an L-type sorption isotherm on soils with moderate organic matter content and mixed mineralogy.

The mathematical description of these isotherms almost invariably involved adsorption model described by Langmuir, Freundlich, or Brunauer, Emmett and Teller (known as the BET-model). Typical adsorption isotherms as described by the above models are shown in Fig. 5.4.

The *Langmuir equation*, derived to describe the adsorption of gases on solids, assumes that the adsorbed entity is attached to the surface at definite homogeneous localized sites, forming a monolayer. It is also assumed that the heat of adsorption is constant over the entire monolayer, that there is no lateral interaction between the adsorbed species, that equilibrium is reached, and that the energy of adsorption is independent of temperature. The Langmuir equation is:

$$\frac{\chi}{m} = \frac{KCb}{1 + KC} = \frac{Kb}{\frac{1}{C} + K} , \tag{5.12}$$

where C is the equilibrium concentration of the adsorbate, χ/m is the mass of adsorbate per unit mass of adsorbent, K is a constant related to the bonding

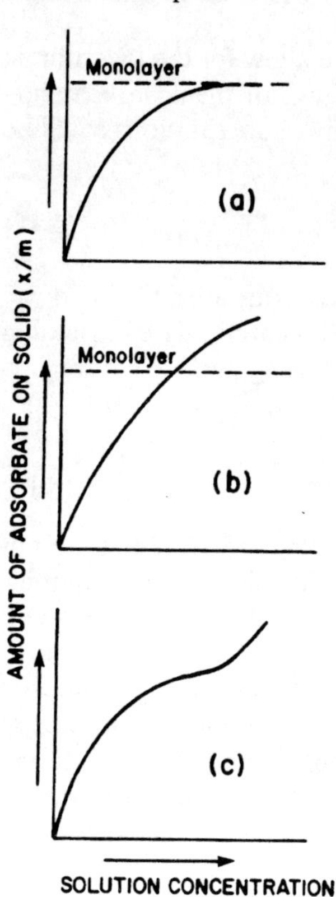

Fig. 5.4a–c. Typical adsorption isotherms described by the **a** Langmuir, **b** Freundlich, and **c** BET equations

strength, and b is the maximum amount of adsorbate that can be adsorbed. The best way to determine these parameters is to plot the distribution coefficient (K_d), which is the ratio between the amount adsorbed per unit mass of adsorbent (χ/m), and the concentration in solution (C)

$$K_d = \frac{\chi/m}{C} \ . \tag{5.13}$$

If we multiply Eq. (5.12) by $1/C + K$ and solve for K_d, we can observe that the Langmuir equation becomes linear and is expressed as

$$K_d = bK - K\chi/m \ . \tag{5.14}$$

If a straight line is obtained when K_d is plotted against χ/m, the Langmuir equation is numerically applicable. The observation of a straight line is not a proof of Langmuir adsorption, since complex precipitation reactions, which occur in particular cases, are depicted by the same type of curve (Veith and Sposito 1977). Corrections have been introduced to the Langmuir equation, in the course of time, to overcome the problems of heterogeneous sites, coupled adsorption-desorption reactions and adsorption of trace elements (organic and inorganic) on soils.

The Freundlich equation was empirically derived to allow for the logarithmic decrease in adsorption energy with increasing coverage of the adsorbent surface. Freundlich found that adsorption data for many dilute solutions could be fitted by the expression

$$\frac{\chi}{m} = KC^{1/n} , \tag{5.15}$$

where K and n are empirical constants and the other terms are defined above.

The Freundlich equation can, however, be derived theoretically by assuming that the decrease in energy of adsorption with increasing surface coverage is due to the surface heterogeneity (Fripiat et al. 1971).

The linear form of the Freundlich equation is:

$$\log \chi/m = 1/n \log C + \log K . \tag{5.16}$$

The main limitation of the Freundlich equation is the fact that it does not predict a maximum adsorption capacity. However, in spite of its limitations, the Freundlich equation is widely used for describing pollutant adsorption on the soil solid phase.

The BET equation describes the phenomenon of multilayer adsorption which is characteristic of physical or van der Waals interactions. In the case of gas adsorption, for example, multilayer adsorption merges directly into capillary condensation when the vapor pressure approaches its saturation value and often proceeds with no apparent limit. The BET equation has the form

$$\frac{P}{V(P_0 - P)} = \frac{1}{V_m C_h} + \frac{(C_h - 1)P}{V_m C_h P_0} , \tag{5.17}$$

where P is the equilibrium pressure at which a volume V of gas is adsorbed; P_0 is the saturation pressure of the gas; V_m the volume of gas corresponding to an adsorbed monomolecular layer, and C_h is a constant related to the heat of adsorption of the gas on the solid in question. If a plot of $P/[P_0 - P]$ against P/P_0 results in a straight line, the effective surface area of the solid can be calculated after V_m has been determined, either from the slope of the line $(C - 1)/V_m C$, or from the intercept, $1/V_m C$.

5.1.5 Kinetics of Adsorption

For a long time it has been observed that the adsorption of pollutants on the soil solid phase is not necessarily instantaneous. The study of sorption from a kinetic perspective can lead to a better understanding of the mechanism of the process. Details of the kinetics of soil chemical processes can be seen in a book by Sparks (1989) and in Pignatello's (1989) review. Boyd et al. (1947) showed that the ion exchange process is diffusion-controlled and that the reaction rate is limited by mass transfer phenomena that are either film (FD)- or particle diffusion (PD)-controlled.

In the case of soil cation exchange, charge-compensation cations are held in the soil solid phase as follows: within crystals in interlayer positions (mica and smectites), in structural holes (feldspars), or on surfaces in cleavages and faults of the crystals and on external surfaces of clays, clay minerals, and organic matter.

Cations held on external surfaces are immediately accessible to the soil solution. Once removed to this phase, they move to a region of smaller concentration. This movement is controlled by diffusion and the diffusion coefficient (D) through soil can be calculated using the Nye and Tinker (1977) development which shows that

$$D = D_e \Theta f \frac{dC_{solution}}{dC_{sol}} , \qquad (5.18)$$

where D_e is the diffusion coefficient in water, Θ the water content of the soil, f is the "impedance" factor related to soil tortuosity, and C the cation concentration in the soil and in the solution (mass/volume). The last term represents the buffer capacity of the system and it is a specific constant for the soil. Two cation exchange properties are involved: the number of exchange sites occupied by the cation investigated and the selectivity of the cation relative to the concentration of the exchanging cation.

Cations held on the external surfaces of clays exhibit a quite rapid diffusion process, but are subject to an additional restriction compared with the first category, due to the fact that the arrival rate of ingoing cations at the exchange site is much slower than the release rate of the outgoing cations. The rate-controlling step is the influx of exchanging cation to negatively charged sites. Sparks (1986) defined the following concurrent processes which take place for Na^+ and K^+ exchange in vermiculite: (1) diffusion of Na^+ ions with Cl^- through the solution film that surrounds the vermiculitic particles (FD); (2) diffusion of Na^+ ions through a hydrated interlayer space of the vermiculite particle (PD); (3) chemical reaction exchange of Na^+ ions for K^+ ions from the particle surface (CR); (4) diffusion of displaced K^+ ions through the hydrated interlayer space of the vermiculite particle (PD); and (5) diffusion of displaced K^+ ions with Cl^- through the solution film away from the particle (FD).

The Weber diagram (Fig. 5.5) shows the rate-determining steps in a heterogeneous soil system. For actual chemical reaction exchange to occur, ions must be transported to the active fixed sites of the particles. The film of water adhering to and surrounding the particle, and the hydrated interlayer spaces in the particle, are both zones of low concentration which are constantly being depleted by ion adsorption to the sites. The decrease in concentration of ions in these interfacial zones is then compensated by ion diffusion from the bulk solution. Thus, in most soil and soil-constituent systems, either PD and/or FD may be rate-limiting.

The characteristic period of ion exchange in soils ranges from a few seconds to days; this is due to the properties of inorganic and organic constituents of the soil solid phase, on the one hand, and to the properties of the pollutants –

ion charge and radius – on the other. The slowly exchangeable cations are situated on exchange sites in interlayer spaces of the soil minerals (e.g., mica), or in cages and channels of soil organic matter and exchange out into solution by diffusive flux. Ion exchange occurs when a driving force, such as a chemical potential gradient, is maintained between solid and solution or when access to sites is freely maintained by the use of a hydrated and less preferred cation for exchange.

As was previously shown, in the case of most organic nonionic compounds of interest, e.g., environmental contaminants, the driving force for adsorption consists of entropy changes and relatively weak enthalpic (bonding) forces. Hydrophobic bonding to the soil organic phase may be the most important interaction for neutral molecules. The sorption of these molecules by soils has generally been characterized by an initial rapid rate followed by a much slower approach to an apparent equilibrium. The initial reaction(s) have been associated with diffusion of the organic toxic compounds to and from the surface of the sorbent, while the slower reaction(s) have been related to particle diffusion control of the movement of molecules into and out of micropores of the sorbent. Diffusion will be seen to be an important factor determining the release of pollutants (see Chap. 6).

5.1.6 Factors Affecting Adsorption

The adsorption process (mechanism, extent, and rate) is controlled by environmental factors such as soil constituents, soil moisture content, and temperature.

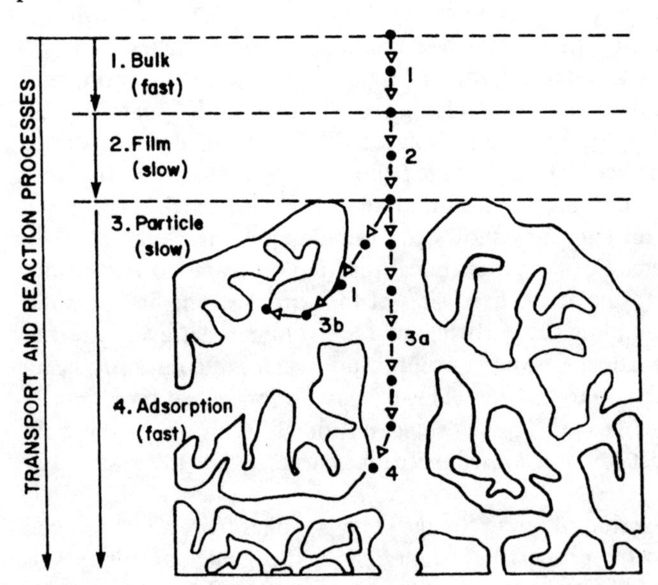

Fig. 5.5. Rate-determining steps in heterogeneous soil systems. (Weber 1984)

Independently of the molecular properties of the pollutants, the soil constituents form a major factor in controlling the adsorption process. The type of clay mineral, for example, has significant effect on both rate and extent of adsorption, mainly for cations.

All soil components contribute in some measure to cation exchange, but the extent and nature of the contribution could be altered by pH, soil-water complexes, and the electrolytes in the soil solution. At the same time, all soil constituents, each one in a different manner, contribute to the surface adsorption of nonpolar compounds, but this adsorption is much affected by the shape and size of the adsorbed molecule.

The mineralogical composition of the soil is one of the major factors in defining the rate and extent of the ion exchange and this is related to the structural properties of the clay materials. In the case of kaolinite, the tetrahedral layers of adjacent clay sheets are tightly held by hydrogen bonds and only planar external surface sites are available for exchange. In the case of smectite (montmorillonite), clay particles are able to swell, if there is adequate hydration, following a rapid passage of ions into the interlayer space. The vermiculite mineral is characterized by a more restructured interlayer space since the region between layers of silicate is selective for certain types of cations such as K^+ or NH_4^+ (Sparks and Huang 1985).

Cation exchange is also affected by the particle size fractions of soil. It has been observed, for example, that of the total Ca-Na content of a sand layer, 90% comprises particles of 0.12–0.20 mm and only 10% on the 0.20–0.50-mm sand fraction (Kennedy and Brown 1965). Similar behavior was observed on silt materials. The exchange rate (Ba-K) on medium and coarse silt, for example, has been found to diminish with increasing particle size. The low exchange rate in the coarse fraction was explained by a slow diffusion of ions into the particle (Malcolm and Kennedy 1970).

Carboxyl and phenolic hydroxyl functional groups contribute most to the cation exchange capacity of humus. Uranic acids in polysaccharides and the acidic amino acids (e.g., aspartic and glutamic acids) and carboxy-terminal structures in peptides can contribute to the negative charge and CEC of soil organic matter under appropriate pH conditions. The basic amino acids lysine, arginine, and histidine are positively charged at pH = 6.0, and these, as well as the amino terminal groups in peptides and polypeptides, can be expected to be the principal contributors to positive charges in soil organic materials, in the appropriate pH environment (Talibuden 1981).

The cation exchange capacity (CEC) of the organic-mineral complexes is less than the sum of those of the two components. The reduction of the cation exchange capacity of the organic-mineral complexes is due in great part to the changes in the properties of humic substances.

A considerable contribution to the understanding of the relative importance of the roles of mineral and organic colloids on organic pollutant adsorption was made through studies with separated soil fractions and well-defined model materials. One approach was the study of adsorption by isolated soil fractions (e.g., humic acid, clay); another was the comparison of adsorption before and after organic matter removal from soil samples. The second approach was used by Saltzman et al. (1972) to assess the relative importance of soil colloids in parathion uptake. Although the removal of organic matter from soils by oxidation with hydrogen peroxide could affect the properties of the adsorbent, the results obtained may be considered to provide qualitative information about the role of the properties of the adsorbent. These results emphasize that parathion has a greater affinity for organic adsorptive surfaces than for mineral ones. However, the important finding suggested in this type of work is that adsorption was dependent on the type of association between organic and mineral colloids, which determines the nature and the magnitude of the adsorptive surfaces (Gaillardon et al. 1977).

Using model materials, the behavior of montmorillonite, humic acids, and their mixtures as adsorbents was studied for terbutryn, a herbicide of the s-triazine group. In contrast to the parathion case, the picture obtained from this work is complicated by the pH-dependence of adsorption, because terbutryn is a weakly basic compound that could protonate and be adsorbed as a cation. In a slightly acid medium (pH 5.6–6.0), the humic acids were the main adsorbent, but a decrease in pH increased the adsorption by the clay fraction and by the clay-humic acid mixtures. It is suggested that in acid conditions the mixtures have a synergistic effect on adsorption. A hysteresis in desorption was observed for the humic acids and the mixtures, but not for the clays (see Chap. 6). This work also pointed out that the organic material-mineral colloids relationship, rather than isolated parameters, must be considered in the assessment of the adsorption of organic pollutants by soils.

Although the importance of organic matter in organic pollutants adsorption has been well established, the properties of the organic colloids which are relevant to adsorption have not yet been thoroughly characterized. The available information shows that these properties could be: the proportions of humic acid, fulvic acid, and humin; the presence of active groups such as carboxyl, hydroxyl, carbonyl, methoxy, and amino; and high cation exchange capacity and surface area (Weed and Weber 1974; Burchill et al. 1981).

In contrast to the organic matter, the role of mineral colloids (mainly the clays) as adsorbents for toxic organics, has been studied intensively. The main properties affecting the adsorptive capacity of clays are considered to be the available surface area and the cation exchange capacity, as well as the saturating cation, the hydration status, and the surface acidity, which is related to the two preceding properties and to the clay structure (Greenland 1965). The degree of pesticide adsorption from aqueous solutions by clays is variable, with a general trend of increasing adsorption as we progress from nonionic, nonpolar molecules to more polar pesticides, weak bases, and cationic compounds.

Although amorphous oxides and hydroxides of iron, aluminum, and silica can adsorb pesticides, very little information concerning this interaction is available. Al and Fe hydroxides, adsorbing mainly pesticides, increase the adsorptive capacity of montmorillonites (Terce and Calvet 1977). Huang et al.

(1984) studied the relative importance of organic matter, sesquioxides, and different particle size fractions of soils in the adsorption of atrazine. They found that aluminum, iron oxides, and probably other mineral compounds present in soil fractions ranging from clay to sand provided adsorption sites for atrazine. The removal of aluminum and iron oxides from soils significantly decreased the amount of atrazine adsorbed and changed the adsorption kinetics. However, the amount of extractable Al and Fe in the different particle-size fractions was not proportional to the extent of adsorption, suggesting that different forms of sesquioxides could have different reactivities for atrazine. The adsorption capacity of sesquioxide components was attributed to their high specific surface area and proton donor functional groups.

In addition to the properties of soil components, the soil environment influences the distribution of synthetic organic chemicals among the soil phases. The environment of a specific soil is determined not only by its intrinsic properties, but also by external factors, especially climatic conditions and agricultural practices. These factors affect the extent and timing of changes in the soil environment.

The soil moisture content affects adsorption in several ways, (Calvet 1984). Pesticides, for example, are usually transported to the adsorbing surfaces by water. The moisture content determines the accessibility of the adsorption sites, and water affects the surface properties of the adsorbent. Toxicity and chemical analyses of DDT, gamma-BHC, and dieldrin applied on mud blocks showed that bioactivity closely followed any changes in humidity, that the rate of diffusion of the insecticides into the mud blocks increased with humidity, and that the insecticides sorbed were reactivated by increased moisture content. The explanation for this behavior is that there is competition for adsorption sites between water and these nonionic pesticides. Preferential adsorption of the more polar water molecules by the mud blocks hindered insecticide adsorption at high humidity; the competition was less at low moisture contents, so that there was increased insecticide adsorption. The insecticides sorbed under low humidity conditions were desorbed when humidity increased, thereby increasing their diffusion both to the surface and into the mud blocks. Negative relationships between pesticide adsorption and soil moisture content have often been reported (e.g., Ashton and Sheets 1959; Yaron and Saltzman 1978).

Studies of pesticide-clay-water systems have been remarkably useful for achieving an understanding of the effect of water on pesticide adsorption (Green 1974; Burchill et al. 1981). Although the hydration of clays and the properties of adsorbed water are not yet fully understood, it is generally accepted that water molecules are attracted by the clay surfaces, mainly by the exchangeable cations, and form hydration shells. Adsorbed water provides adsorption sites for pesticide molecules. An important feature of water associated with clay surfaces is its increased dissociation, giving the surfaces a slightly acidic character. Generally, a negative relationship exists between the surface acidity of clays and their water content.

The effect of hydration of the soil organic matter in relation to pesticide adsorption has been studied less. It has been suggested that hydration influences the molecular shape of humic substances, and thus accessibility for pesticides. Strong, sometimes irreversible, retention of pesticides by hydrated humic substances could be explained by the penetration and trapping of pesticides into the internal structure of the swollen humic substances. The hydrated exchangeable cations and some dissociated functional groups, as well as water held by various polar groups of the humic substances, could also provide adsorption sites. At low moisture contents, the hydrophobic portions of the organic matter structures could bind hydrophobic, nonionic pesticides (Burchill et al. 1981).

As the adsorption processes are exothermic, changes in soil temperature could have a direct effect on the phase distribution of pesticides. Adsorption usually increases as the temperature decreases, and desorption is favored by increasing temperature. Temperature could indirectly influence adsorption by its effect on pesticide-water interactions. The complex relationship among adsorbent, adsorbate and solvent, as affected by temperature was described by Mills and Biggar (1969). The adsorption of lindane (1,2,3,4,5,6-hexachlorocyclohexane) and its beta-isomer by a peaty muck, a clay soil, Ca-bentonite, and silica gel decreased as the temperature of the systems increased. The authors suggested that this adsorption-temperature relationship reflects not only the influence of energy on the adsorption process, but also the change in solubility of the adsorbate. The activity, and hence the chemical potential, of a solute in solution is more or less dependent on its solubility, as affected by temperature and solvent. Mills and Biggar (1969), considered that changes in activity in solution are important, as the difference between the activity in solution and on the adsorbent is the driving force in the adsorption process. The authors assumed that the change in activity in solution with temperature is related to the change in the reduced concentration, which is the ratio between the actual concentration of the solute at a given temperature and its solubility at the same temperature. Adsorption isotherms obtained by using the reduced concentration showed, in contrast to normal adsorption isotherms, an increase in adsorption with increasing temperature, suggesting that the heat effect involved in the adsorption process was mainly that involved in the solubility of the solute. Similar results, emphasizing the significant influence of temperature on adsorption through its solubility effect, have been obtained by Yaron and Saltzman (1978) for parathion adsorption by soils, and by Yamane and Green (1972) for atrazine and ametryne adsorption by soils and montmorillonite.

5.2 Nonadsorptive Retention

These processes have been somewhat neglected until now, despite the fact that they provide important mechanisms for pollutant retention in the soil medium.

5.2.1 Pollutant Precipitation

If adsorption was defined as a net accumulation at an interface, precipitation can be defined as an accumulation of a substance to form a new bulk solid phase. Sposito (1984), defining this process, showed that both of these concepts imply a loss of material from an aqueous solution phase, but one of them is inherently two-dimensional and the other inherently three-dimensional. The chemical bonds formed in both cases can be very similar, and mixed precipitates can be inhomogeneous solids with one component restricted to a thin outer layer because of poor diffusion. In soils, the problem of differentiating adsorption from precipitation is made especially severe by the facts that new solid phases can precipitate homogeneously onto the surfaces of existing solid phases and that weathering solids may provide host surfaces for the more stable phases into which they transform chemically.

Precipitation takes place when the solubility limits are reached and it happens on a microscale between and within soil aggregates. In the presence of lamellar charged particles with impurities, precipitation of cationic pollutants, for example, might occur even at concentrations below saturation with respect to the theoretical solubility coefficient of the solvent. The precipitation is never complete and the precipitates have their own solubility products, and thus ions are always present in the supernatant liquid.

Heavy and transition metals, for example, are present as hydrated cations in water at neutral pH values and at this stage they behave like acids, due to water molecule dissociation in the hydration shell of the cation. The "acidity" of hydrated cations depends on the pK_a, values of which are presented in Table 5.2. From the data presented in Table 5.2, it can be deduced that the lower the pK_a value of the metal the lower will be the pH at which precipitation takes place.

There is a relationship between the solubility of a metal in water, the amount of precipitation, and the pH. Sposito (1984) demonstrates, however, that an observation, according to which an ion activity product in a soil solution is larger than the corresponding solubility product constant, is not prima

Table 5.2. pK_a values of the major heavy metal pollutants. (de Boodt 1991)

Heavy metal	pK_a
Cd^{2+}	8.7
Co^{2+}	8.9
Cr^{2+}	6.5
Cu^{2+}	6.7
Mn^{2+}	10.6
Ni^{2+}	8.9
Pb^{2+}	7.3
Zn^{2+}	7.6
Fe^{3+}	3.0

facie evidence of precipitation. Consider a solid which precipitates according to the reverse of the reaction

$$M_aL_b(s) \rightleftharpoons aM^{m+}(aq) + bL^-(aq) \ , \qquad (5.20)$$

where M is a metal and L is a ligand that precipitates to form the solid $M_aL_b(s)$, and a and b are stoichiometric coefficients subject to the constraint of electroneutrality. The ion activity product (IAP) can be larger than the corresponding solubility product constant for the solid if the activity of the solid is greater than unity, which might occur when the precipitate comprises particles of radius $< 1\ \mu m$. Sposito (1984) explains the behavior of heavy and transition metals by the fact that the surface energy of these very small particles is large enough to contribute importantly to the Gibbs energy of the precipitate and, therefore, to increase its activity relative to that in the standard state, where the interfacial energy component of the Gibbs energy is, by definition, negligible. An additional process of precipitate formation could occur in the presence of nucleating agents when the rate of heterogeneous precipitation from a supersaturated soil solution is large, and the mechanism is solid formation.

5.2.2 Trapping

Trapping is an additional form of nonadsorptive retention of pollutants in soils and occurs in the case of water-immiscible fluid compounds or of pollutants adsorbed on suspended particles.

The water-immiscible fluids hinder each other's transport in the soil pore space until a minimum degree of saturation is reached. However, there are some cases where the soil pore geometry permits the flow of the nonwetting fluid at a level greater than the degree of saturation, leaving behind an enclave of water-immiscible liquid. The trapped immiscible liquids will remain in the unsaturated zone for an indefinite time (Schwille 1984), serving as a source of contamination which will decrease in magnitude as a result of abiotic processes such as volatilization or dissolution in the water phase. The dominance of capillarity explains the capillary trapping of organic liquid in water-wet porous media previously saturated with water. The trapped organic liquid remains behind as small, immobilized, disconnected pockets of liquid, sometimes called blobs or ganglia, no longer connected to the main body of organic liquid. This condition is usually referred to as residual organic liquid saturation. The degree of soil saturation for an immiscible liquid can be expressed as the utilization of pore space by the liquid and air phases (Van Dam 1967; Schwille 1984) or as the organic liquid content of the medium in ml.

If the organic liquid saturation is measured as the volume of organic liquid per unit void volume, measured over a representative elementary volume of the porous medium, we can write the expression

$$S_o = \frac{V_{\text{organic liquid}}}{V_{\text{voids}}} , \tag{5.21}$$

which is the fraction of the pore space occupied by the organic liquid. The subscript o indicates the organic liquid.

The residual saturation at which the organic liquid becomes discontinuous and immobile is defined by

$$S_{or} = \frac{V_{\text{discontinuous organic liquid}}}{V_{\text{voids}}} , \tag{5.22}$$

where the subscript r indicates residual. In the saturated zone the water saturation is given by

$$S_w = 1.0 - S_o . \tag{5.23}$$

The extent of the trapping process is determined primarily by the physical properties of the vadose zone. If the organic liquids are characterized by a high boiling point and a low solubility in water, they will remain trapped in the unsaturated zone. In this particular case, the porous medium behaves like an inert material and the organic liquid's behavior depends only on its own properties, with no interactions occurring between the liquid and the solid phases.

Pollutants – adsorbed on suspended particles – can also be retained in the soil pores by trapping, and thus create a source of future contamination of the soil medium. The origin of these contaminated suspended particles is the land disposal of water and wastewater, for irrigation or waste disposal. Vinten et al. (1983) proved that the vertical retention of contaminated suspended particles in soils is controlled by the soil porosity and pore size distribution. A schematic representation of the fate of a contaminated substance reaching the soil in suspension is presented in Fig. 5.6.

Three distinct ways in which mass transfer of contaminant can occur are considered:

1. As the contaminated suspension passes through the soil, the contaminant in solution is adsorbed by the soil.
2. As a result of this, desorption of contaminant from the suspended solids (SS) phase occurs.
3. As the suspension passes through the soil, deposition of particles occurs, taking the contaminant with it.

The suspended solid particle size and the volume of effluent also have to be considered in defining the solid deposition along a soil profile under irrigation with sewage effluent (waste disposal site). The coarse fraction of suspended solids is retained in the upper layer of a soil profile; the finer colloidal fraction is mobile and its deposition is controlled by the soil porosity. When the diameter of the suspended solids is greater than the diameter of the soil pores, the contaminated suspended solid is retained.

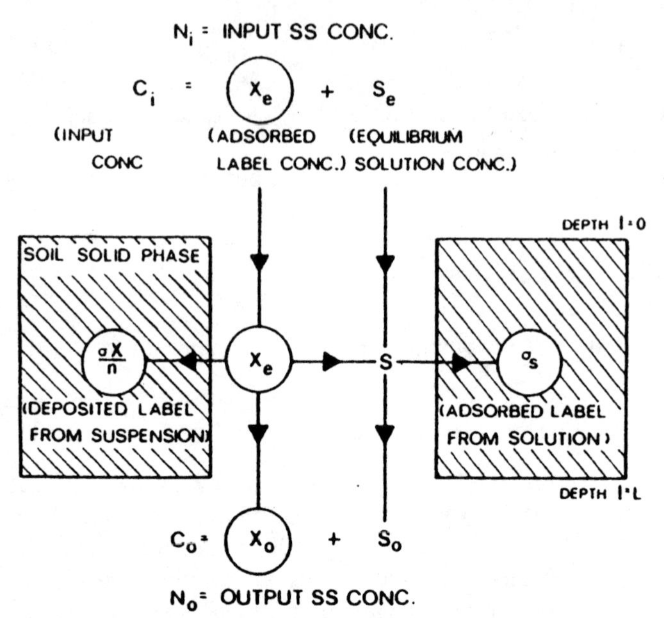

Fig. 5.6. Diagrammatic representation of transport of labeled suspension through soil. *SS* Suspended solids. (After Vinten et al. 1983)

An additional form of trapping is the retention of pollutants on living organisms (e.g., roots, microbial population, earthworms, etc.) and on solid wastes dispersed on land.

Living organisms have the capacity to adsorb chemicals on their surfaces and to release these wholly or partially when the living activity of their cells ceases. Vertically oriented earthworm burrows exist in many agricultural sites. Earthworm activity induces changes in the characteristics of the burrow walls, increasing the adsorption capacity of narrow surfaces around the walls forming enclaves of higher retention capacity for toxic organic molecules reaching the soil. A similar situation occurs when the toxic organic chemicals are adsorbed directly by the plant roots, leading to enclaves of higher concentration of pollutants along the root material. In both cases, the retention of pollutants on living organisms will be different from – usually greater than – that on the surrounding soil solid phase and should be considered as part of the soil spatial variability in defining the extent of pollution in the soil medium.

Sludge and manure spreading have been generally accepted as a disposal practice on a particular soil that recycles beneficial plant nutrients. Other wastes with high adsorption capacities such as activated carbon from the food-processing industry, and additional organic wastes of various origins, could be incorporated in soil during the disposal procedure. The adsorption of organic toxic molecules, in general, is much greater on these incorporated organic materials than on the soil solids, so that they form small enclaves of high pollutant concentration.

All the processes described above illustrate the importance of nonadsorptive retention of pollutants in spatially variable field conditions.

5.3 Examples

The following examples will illustrate some aspects of the adsorption and nonadsorptive retention of pollutants on soil solid phase, as affected by the properties of the chemicals and of the soil constituents, as well as by environmental factors such as soil moisture content and temperature. The examples are grouped according, not to their retention mechanism, but to their pollution hazard capacity. Considerations in the grouping include data on their adsorption on soil solid phase as the result of irrigation with sewage water or of the disposal of heavy metals and of crop protection by pesticide application.

5.3.1 Pollutants Adsorption

Cation exchange under irrigation with sewage saline water is the main process leading to the deterioration of soils as a result of an exchange process between Na^+ from the irrigation water and the soil-saturating cations. When sodium-containing water first flows through a soil, the relationship between the SAR of water and ESP of soil shows no significant correlation. Thomas and Yaron (1968), in a column experiment carried out on a series of Texan soils of differing mineralogy, proved that the total electrolyte concentration of the sodic water influenced the rate of sodium adsorption, but at equilibrium the exchangeable sodium percentage (ESP) in the soil was influenced more by the soil mineralogy and the cationic composition of the water than by the total electrolyte concentration (Table 5.3).

Table 5.3. Effect of irrigation water, SAR, and Na concentration in water on the ESP of soils of different mineralogy. (Thomas and Yaron 1968)

Soil	Clay distribution[a]		Clay mineralogy		SAR	Conc	ESP
	<0.2 μ	2–0.2 μ	<0.2 μ	0.2–2.0 μ		mEq L^{-1}	
Burleson	71	29	$M_1Mi_3K_3$	$M_2Mi_2K_2Q_2$	28	11.0	18.9
						33.0	20.4
Houston Black	78	22	M_1	$M_2K_2Mi_2Q_2$	28	11.0	16.0
						33.0	16.0
Miller	49	51	$M_1Mi_2K_3$	$M_2Mi_2K_2Q_{2F}3$	28	11.0	25.8
						33.0	37.9
Pullman	41	59	$Mi_2K_2M_2$	$Mi_2K_2Q_2F_3$	28	11.0	31.3
						33.0	32.5

[a] % of clay content

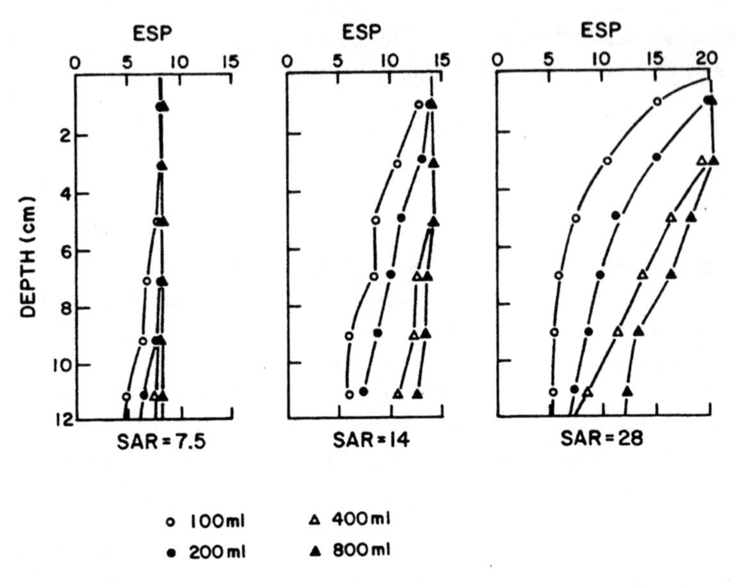

Fig. 5.7. Exchangeable sodium percentage along a burleson soil column as a function of the sodium adsorption ratio of the irrigation water. The values were obtained by percolating the soil columns with sodic water (total electrolyte concentration 11 $\mathrm{mEq\,L^{-1}}$). Each *curve* corresponds to a given applied volume of solution. (After Thomas and Yaron 1968)

Typical data for the adsorption of sodium are shown in Fig. 5.7, which shows the results obtained on Burleson soil. Three synthetic solutions with a total electrolyte concentration of 11 $\mathrm{mEq\,L^{-1}}$ and SAR values of 7.5, 14.0, and 28.0 were passed through the soil column. At equilibrium the solution with $\mathrm{SAR} = 7.5$ gave an ESP of 8.1, the solution with $\mathrm{SAR} = 14.0$ an ESP of 13.2 and the solution with $\mathrm{SAR} = 28.0$ an ESP of 18.0. It can be seen from Fig. 5.7 that the quantity of solution that had to be passed through columns of Burleson soil in order to achieve a constant Na^+ content in the entire profile increased with the SAR of the solution. In other words, for a given applied volume of solution, the greater the SAR the smaller is the depth where a constant ESP is achieved. For example: for a volume of 400 ml, the depth is between 9 and 10 cm, 5 and 6 cm, and 3 and 4 cm for SAR of 7.5, 14.0, and 28.0, respectively. One hundred ml of solution with an SAR of 7.5 gave a constant Na^+ content for half the length of the column, whereas the solutions with SAR of 14.0 and 28.0 gave constant Na contents for less than 20% of the column length.

Pesticide retention in soils is controlled by the molecular properties of the compound, by the soil constituents, and by environmental factors.

Diquat and paraquat organocation pesticides are used to illustrate the effects of type of clay and of the resident inorganic cation on the clay surfaces on the adsorption of the chemicals. Table 5.4 shows the adsorption of these cationic pesticides by homoionic exchange in montmorillonite, vermiculite, illite,

Table 5.4. Influence of the resident cation on the adsorption of diquat (II) and paraquat (III) by homoionic-exchanged clays. (Hayes and Mingelgrin 1991)

Clay	Resident inorganic cation	CEC (μEq g^{-1})	Organic adsorption (μEq g^{-1})		Enthalpy change ΔH (kJ mol^{-1})	
			Diquat (II)	Paraquat (III)	Diquat (II)	Paraquat (III)
Montmorillonite	Na^+	999	950	970	–32	–48
	K^+	850	920	940	–31	–51
	Ca^{2+}	950	940	950	–36	–55
Vermiculite (Libby)	Na^+	1440	1210	1130	+26	+42
	K^+	ND	270	230	+1.6	–0.9
	Ca^{2+}	1130	(870)[a]	(630)[a]	(–9.2)[a]	(–5.2)[a]
Illite	Na^+	230	240	230	–13	–22
	K^+	ND	220	220	–6.8	–17
	Ca^{2+}	240	230	220	–16	–24
Kaolinite	Na^+	30	36	36	–5.0	–11
	K^+	ND	35	34	+1.3	–4.9
	Ca^{2+}	60	36	36	–18	–25

ND Not determined.
( )[a]Data for incomplete cation exchange.

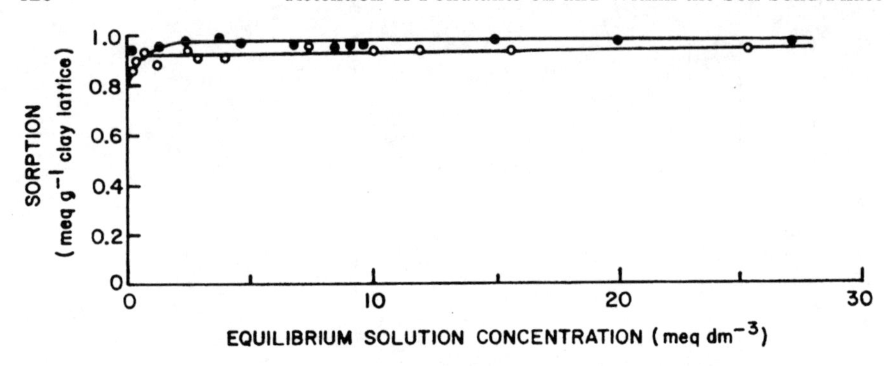

Fig. 5.8. Isotherms for the adsorption of diquat (II) and paraquat (III) by Na^+-montmorillonite. (After Hayes and Mingelgrin 1991)

and kaolinite clays. It may be seen from these data that the cation can have a significant influence on the extent and the energetics of the adsorption processes. The cation exchange capacities (CEC) of the clays can be seen to vary according to the nature of the resident cation. However, exchange of the resident cations by both of the organocations was essentially complete in the cases of adsorption by the kaolinite, montmorillonite, and illite clays, although there was a slight preference for paraquat over diquat in the cases of the homoionic montmorillonite clays. These affinities were reversed in the cases of adsorption by the Na^+-vermiculite and illite clays.

The isotherms were of the high-affinity type for the adsorption of the two organocations by all of the clays. Figure 5.8 shows the isotherms obtained for the adsorption of paraquat and diquat by Na^+-montmorillonite, and the shapes shown are characteristic also of those obtained for adsorption by kaolinite and illite. Adsorption by Na^+-vermiculite was also of the high-affinity type, but the isotherm was more rounded prior to the attainment of the plateau level.

Considerable attention has been given to the adsorption of diquat and paraquat by humic acid. Hayes and Mingelgrin (1991), on the basis of data presented by Burns et al. (1973a,b), explain how the exchangeable cations on humic macromolecules can influence the adsorption of paraquat (Fig. 5.9). Adsorption decreased in the order Ca^{2+}-humate > H^+-humic acid > Na^+-humate. Adsorption by the Na^+-humate approached the cation exchange capacity of the polyelectrolyte. This can, in part, be explained by the preference of ion exchangers for mineral ions of higher valence. However, the abilities of the organocations to penetrate towards adsorption sites in the interior of the macromolecules was more important in this instance than considerations of valence. Adsorbed organocations bridged negative charges within strands and between strands, and the interstrand binding caused the macromolecules to shrink, water was excluded from the macromolecular matrix, and eventually precipitation of the paraquat-humate complex occurred. As was observed for

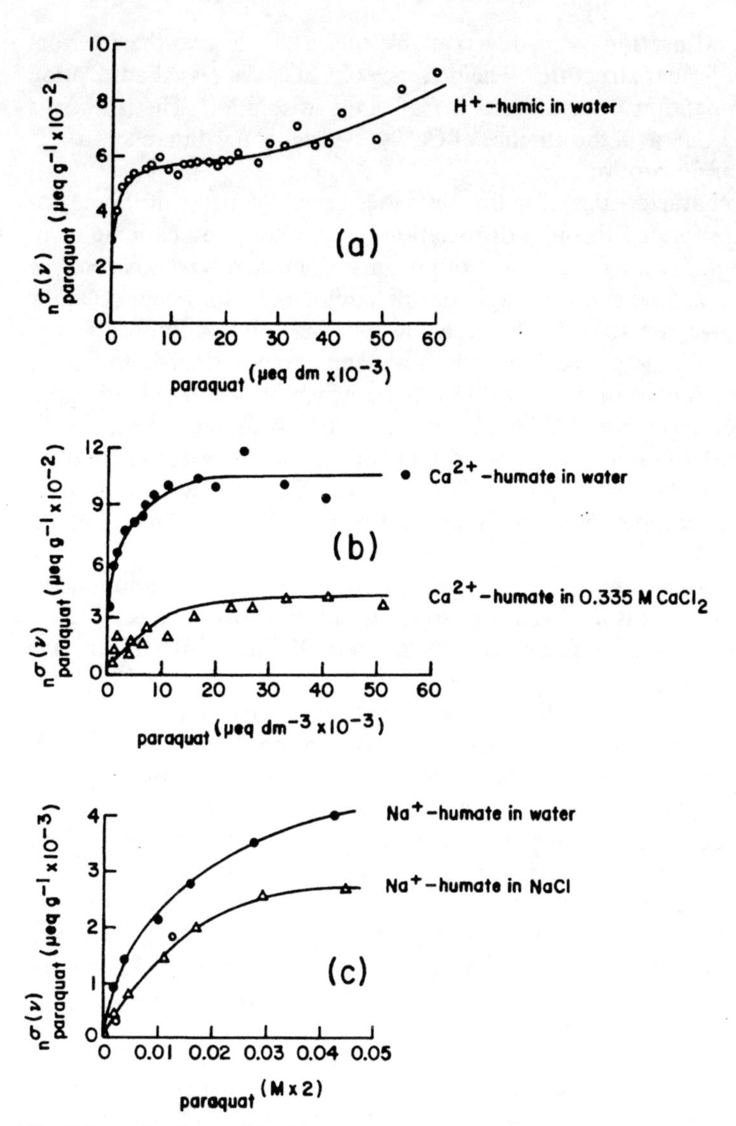

Fig. 5.9a–c. Isotherms for the adsorption of paraquat (II) **a** by H^+-humic acid; **b** by Ca^2-humate in water and in 0.335M $CaCl_2$ solution, and **c** by Na^+-humate solutions in water and dilute NaCl solutions. (Burns et al. 1973a)

Na^+- montmorillonite, uptake of paraquat by the Na^+-humate approached the CEC value of the polyelectrolyte.

Adsorption by the Ca^{2+}- and H^+-exchanged humic acids was significantly less than the CEC of the preparations. Comparisons of the isotherms in Fig. 5.9a and b indicate that high-affinity adsorption of paraquat by the H^+-humic acid in water and by Ca^{2+}-humate in water were similar, and representative of adsorption by readily available sites at or close to the exteriors of the macromolecular structures. The presence of $CaCl_2$ depressed adsorption because of competition from the excess Ca^{2+}-ions in solution. The isotherms for the H^+- and Ca^{2+}- (in the absence of $CaCl_2$) systems show differences at the plateau levels of adsorption.

The charge characteristics of many pesticides are pH-dependent. Some anionic species are formed through dissociation of a proton, and cationic compounds may be formed by the uptake of protons. Compounds with carboxylic acid groups are characteristic of anionizable compounds, although phenolic groups could give rise to anionic species in alkaline soil conditions. The s-triazine family of soil-applied biocidal compounds are considered to be the classical representatives of cationizable species, which have widespread applications as pre-emergent weed killers. Several compounds containing carboxylic acid groups enter the soil when sprayed on plant cover, and some are applied directly to the soil. These act as neutral molecules at pH values well below their pK_a values, but become increasingly anionic as the pK_a is reached and exceeded.

The s-triazine herbicides can be used as an example of ionizable compounds. Table 5.5 shows the adsorption from aqueous solution of a series of compounds from the s-triazine family on Na-montmorillonite as affected by the solution pH.

Gaillardon et al. (1977) studied the relative effects of pH on the adsorption of an additional s-triazine (terbutryne) on a Ca-montmorillonite, humic acids and a mixture of the two soil constituents (Fig. 5.10). The authors conclude that the effect of pH on the adsorption of terbutryne by humic acids and by a Ca-montmorillonite or mixtures of the two shows a certain similarity with its effect on that of clay alone. Around the neighborhood of neutral pH, only humic acids adsorb terbutryne. In an acid environment, the isotherms of terbutryne adsorption by humic acids and montmorillonite are types L and S, respectively, and reflect a difference between the adsorbent and the adsorbed molecules in their interactions; in the case of mixtures, isotherms of adsorption

Table 5.5. Adsorption from aqueous solutions of s-triazine compound by Na^+-montmorillonite. [Hayes and Mingelgrin (1991) after Jordan 1970; Weber 1970]

Compound	pK_a	Water solubility (ppm)	Adsorption (μmol g^{-1})			
			pH = 2	pH = 3	pH = 4	pH = 5
Atrazine (XI)	1.68	33	275	200	115	70
Atratone (XIII)	4.20	1654	410	450	475	400
Ametryne (XIV)	3.12	193	520	610	650	560
Propazine (XV)	1.85	9	150	110	20	20
Hydroxypropazine (XVI)	5.20	310	150	220	240	245
Prometone (XVII)	4.30	750	290	380	400	350
Prometryne (XVIII)	3.08	48	460	490	490	415

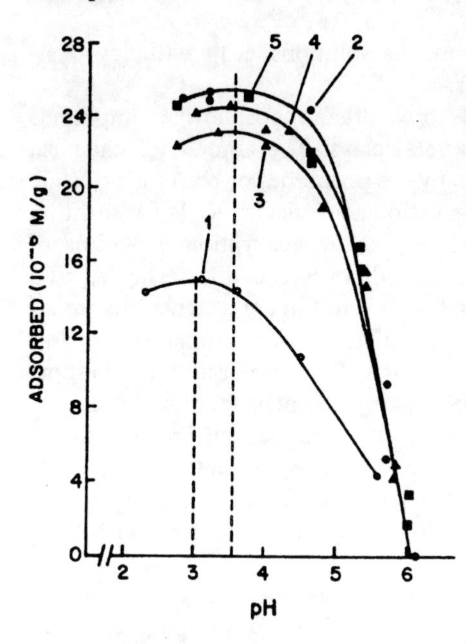

Fig. 5.10. pH Effect on the adsorption of terbutyryne by the humic acids (*1*), Ca-montmorillonite (*2*), and their mixture containing 25% (*3*), 50% (*4*), and 75% (*5*) clay. (After Gaillardon et al. 1977)

are different and reveal a synergistic effect which suggests interaction between the colloids.

Hayes and Mingelgrin (1991) emphasize that under low pH conditions, sorptive molecules can take place through association of the triazine with carboxyl groups in the humic substances followed by electrostatic binding to the conjugate carboxylate base of the proton donor.

The behavior of organophosphorus pesticides will be used as an example of the adsorption of nonpolar toxic chemicals on soil surfaces. The results obtained on this aspect by one of the authors (Yaron 1978a,b; Yaron and Saltzman 1978; Gerstl and Yaron 1981) are summarized below.

The organophosphorus pesticides are part of the phosphoric acid ester group, with a general formula of the type

$$(RO)_2\overset{O(S)}{\overset{\|}{P}}-\underset{(S)}{O}-X \qquad (5.24)$$

where R is an alkyl group. In contact with clay surfaces, such organic nonionic polar molecules are retained at the surface. These esters are stable at neutral or acidic pH but are susceptible to hydrolysis in the presence of alkalies, where the P–O–X ester bond breaks down. The rate of the process is related to the nature of the constituent X, to the presence of catalytic agents, and to pH and temperature. The organophosphorus compounds are characterized by a low solubility in water as compared with their solubility in organic solvents.

Consequently, the adsorption from aqueous solutions dealt with low concentrations in the equilibrium solutions.

Table 5.6 shows the percentage of some organophosphorus compounds adsorbed from aqueous solution by several clays. The amount of chemical adsorbed on the clay surfaces is affected by the properties of both the pesticide and the clay surface area. The saturating cation also affects the clay adsorption capacity for the pesticide. Figure 5.11 shows the adsorption isotherms of parathion for a montmorillonite saturated with various cations. The sorption sequence for parathion ($Al^{3+} > Na^{+} > Ca^{2+}$) is not in agreement with any of the ionic series obtained when considering different ionic properties, such as size, volume, hydration, or energy. This shows that the parathion-montmorillonite interaction in aqueous suspensions is very complex; factors such as clay dispersion, steric effects, and hydration shell are determinant in the sorption processes. Organophosphorus adsorption on clays is described by the Freundlich empirical equation and the K constants for parathion sorption by several clays are: 3 for Ca-kaolinite, 125 for Ca-montmorillonite, and 145 for Ca-attapulgite.

A temperature effect is observed when the organophosphorus pesticides are adsorbed from aqueous solutions by clays. In the case of attapulgite, Gerstl and Yaron (1978) showed that the amount of parathion adsorbed on the mineral in a dehydrated system decreased with increasing temperature. The effect of temperature can be explained as being due to an increase in pesticide solubility with increasing temperature, or simply to the fact that adsorption, being an exothermic process, decreases with increasing temperature. By introducing a correction factor based on "escaping tendency of the solute" theory, described in Chapter 4 as fugacity, it may be possible to distinguish between the influence of temperature on the adsorption energy and the solubility effect.

Table 5.6. Percent of adsorbed organophosphorus pesticides by clays from aqueous solutions. (Yaron 1978)

Insecticides	Solution equilibrium concentration ppm	Clay		
		Ca-montmorillonite	Ca-kaolinite	Attapulgite
Parathion ($C_{10}H_{14}NO_5PS$)	6.5	73	14	87
Pirimiphos-methyl ($C_{11}H_{20}N_3O_3PS$)	16.0	94	30	[a]
Pirimiphos-ethyl ($C_{13}H_{24}N_3O_3PS$)	16.0	92	75	[a]
Menazon ($C_6H_{12}N_5O_2PS_2$)	16.0	16	5	[a]

[a]Not determined

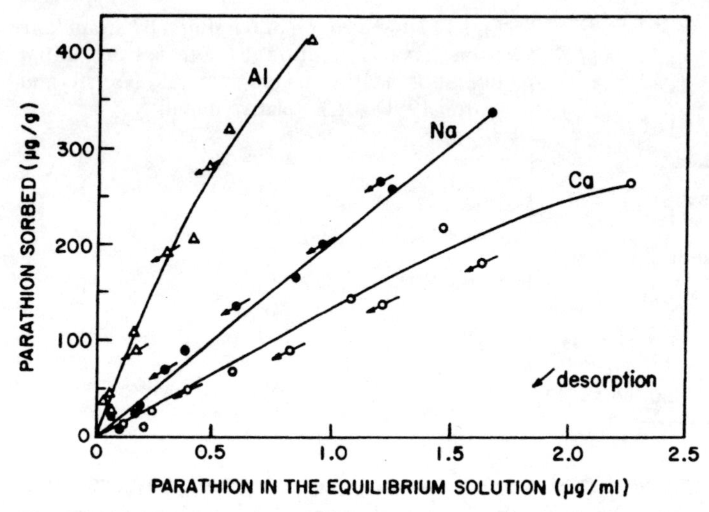

Fig. 5.11. Adsorption of parathion by montmorillonite from aqueous solutions. (After Yaron and Saltzman 1978)

The maximum adsorption capacity of clays for organophosphorus pesticides is attained only when the reaction with the chemical occurs in an appropriate organic solvent. It is known that in adsorption at a solid-liquid interface, when the liquid is a mixture of two components (e.g., pesticide and solvent), preferential or selective adsorption occurs. Consequently, in order to obtain a maximum adsorption capacity, careful consideration has to be given to the selection of the proper solvent. The maximum adsorption capacity on a clay surface is affected by the type of clay and the saturating cation. Taking parathion as a typical example, the amount of pesticide adsorbed from a hexane solution on Ca-saturated clay is about 10% on montmorillonite, 8% on attapulgite, and less than 0.5% on kaolinite.

The effect of saturating cation on paraffin adsorption from hexane solution is reflected in the following series:

Montmorillonite:

Mg – 13.5%, Fe – 11%, Ca – 10.5%, Cu – 5.5%, Al – 5.5%, Zn – 3.5%, Na – 3.5%.

Kaolinite:

Al – 0.5%, Ca – 0.45%, Na – 0.35%.

The hydration status of the clay affects its adsorption capacity for the organophosphorus compounds. Figure 5.12 shows, for example, the decrease in the amount of parathion adsorbed from hexane solution by an attapulgite, as affected by the hydration water of the mineral (equilibrated previously up to a relative humidity of 98%). The effect of water on adsorption may be explained as follows: in a dry attapulgite-parathion-hexane system, the slightly polar parathion molecules compete effectively with apolar hexane molecules for the adsorption sites. In partially hydrated systems, parathion molecules cannot

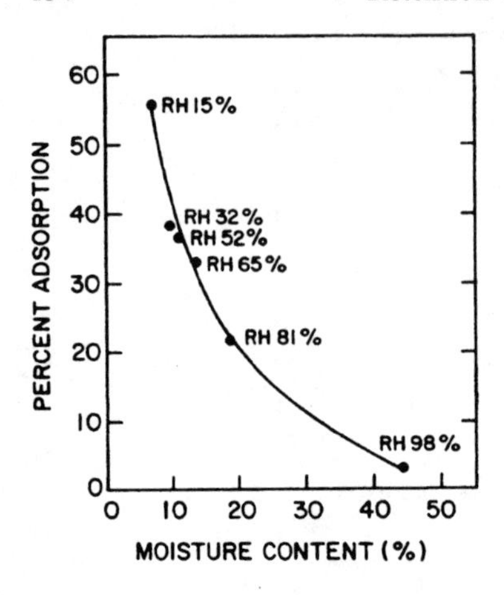

Fig. 5.12. Adsorption of parathion by attapulgite from hexane solution as affected by initial hydration status of the mineral. (Gerstl and Yaron 1981). *RH* Relative humidity

replace the strongly adsorbed water molecules, so that parathion adsorption occurs on water-free surfaces only. These phenomena give an apparent decrease in the adsorption capacity of attapulgite for parathion. It is possible that the apparent decrease is due to the time required for the parathion molecule to diffuse through the water film to the active adsorption sites. By sufficiently increasing the time of contact between the adsorbent and adsorbate, a similar adsorption capacity may be reached for dry and hydrated parathion-hexane-attapulgite systems. If this assumption were confirmed, the presence of water in the system would affect only the rate of adsorption.

Infrared studies lead to an understanding of the adsorption mechanism of the organophosphorus compounds on the clay surfaces. For example, infrared spectra indicate that on both montmorillonite and attapulgite, parathion sorbed by the clays coordinates through water molecules to the metallic cations. When the clay-parathion complexes are dehydrated, parathion becomes directly coordinated; the type of cation determines the structures of the complex. The main interaction is through the oxygen atoms of the nitro group, although interactions through the P=S group have also been observed, especially for complexes saturated with polyvalent cations (Saltzman and Yariv 1976; Prost et al. 1977).

Soil adsorption of the organophosphorus insecticide, parathion was found to be related to the organic matter content and to the soil mineralogy (Saltzman et al. 1972; Gerstl and Yaron 1978). Swoboda and Thomas (1968), studying the adsorption mechanism of parathion by leaching experiments, found that parathion was retained in soils, mainly as a water-insoluble organic constituent of the soil, by partitioning between soil organic matter and the liquid phase. Leenheer and Ahlrichs (1971) stated that parathion affinity for organic surfaces depends on the strength of the hydrophobic nature of these surfaces, rather than on the type of organic matter. They assumed parathion adsorption on organic surfaces to be a physical adsorption with formation of weak bonds between the hydrophobic portion of the adsorbent and the adsorbed molecule.

Saltzman et al. (1972), in a study on the importance of mineral and organic surfaces in the parathion adsorption-desorption process in some semiarid soils, characterized by low organic matter content and several different mineralogies, found that the parathion adsorption by soils depended on the type of association between the organic and mineral colloids. In aqueous solutions, the parathion has a greater affinity for organic than for mineral adsorptive surfaces. The distribution coefficient, K_d (μg adsorbed g^{-1} of adsorbent at an equilibrium concentration of 1 μg ml^{-1}) was 38.5 for a montmorillonitic dry rendzina (72% clay), 76.5 for a Mediterranean red soil with mixed mineralogy (63% clay), 164.0 for a kaolinitic terra rossa (64% clay), and 875.0 for a peat soil. The differences between the mineral soils cannot be explained by differences in the organic matter and clay content, which are rather similar; therefore, it is reasonable to suppose that the surface properties of the constituents and the specific interactions between organic and mineral colloids determine the nature of the adsorptive surfaces for each soil.

Following removal of organic matter by soil treatment with hydrogen peroxide, the adsorptive affinity of the mineral soils decreased (Fig. 5.13). This decrease may be due mainly to the decrease in the organic matter content, and not to other soil modifications which may occur during organic matter oxidation.

The decrease in parathion adsorptive capacity of soils, following oxidation, shows that parathion has a relatively greater affinity for organic adsorptive

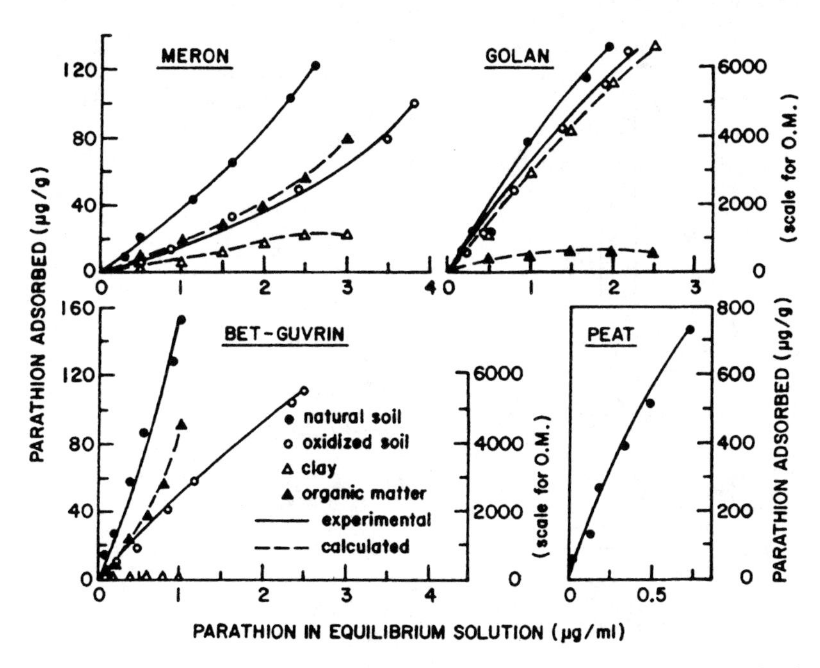

Fig. 5.13. Parathion adsorption from aqueous solutions by soils, oxidized subsamples, organic, and mineral fractions. (Saltzman et al. 1972)

surfaces than for mineral ones. In a recent experiment, Barriuso and Calvet (1992) used statistical analysis to express the relations between soil type and adsorption behavior of a cationic, an anionic, and a neutral herbicide, and showed that these relations are strongly dependent on the electrical state of the herbicide molecules. The study was done with 58 soils, covering a wide range of pH values, organic carbon contents, and mineralogical compositions. Three herbicides were used: atrazine (1-chloro-4-ethylamino-6-isopropylamino-1,3,5-triazine), terbutryn (1-ter-butylamino-4-ethylamino-60-methylthio-1,3,5-triazine), and 2,4-D (2,4-dichlorophenoxy acetic acid). These well-known chemicals were chosen as models to provide neutral, cationic, and anionic molecules. For the range of soil pH encountered, atrazine (pKa = 1.7) was always in a neutral form. 2,4-D (pKa = 2.6) in an anionic form, and terbutryn (pKa = 4.4) in either neutral or cationic form.

The distribution coefficients (K_d) were determined and analyzed in relation to soil and molecular properties. Figure 5.14 shows the results of the principal component analysis for distribution coefficients. The two axes on the graph correspond to the two principal components, which account for 84% of the total variation. This graph shows that the various soil types are distinguished quite well. Soil types such as Mollic Cambisol, Humic Cambisol, and Vertisol adsorbed atrazine and terbutryn efficiently, whereas Ferralsols and Andosols adsorbed 2,4-D more efficiently.

The vectors of the barycenter of the points corresponding to organic carbon contents, pH values, and K_d values show that:

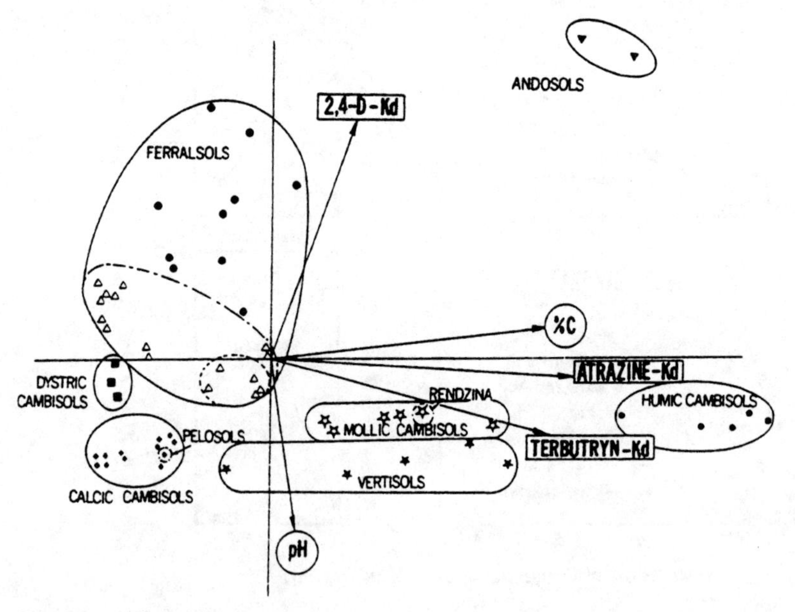

Fig. 5.14. Effect of soil type on herbicide adsorption: results of principal component analysis. (Barriuso and Calvet 1992)

- atrazine's K_d is strongly correlated with the organic carbon content, consistently with published results, but not correlated with the soil pH.
- terbutryn's K_d is less correlated with the organic carbon content and shows some correlation with the soil pH.
- 2,4-D's K_d is inversely correlated with the soil pH, in agreement with the literature, but it is not correlated with the organic carbon content.
- atrazine adsorption exhibits some correlation with that of terbutryn; adsorption of these two triazine herbicides, however, does not show any correlation with that of 2,4-D.

On the basis of adsorption characteristics (example in Table 5.7), Barriuso and Calvet (1992) pointed out that the relationship between soil type and organic molecules varies with the form in which the molecules occur, i.e., nonionized, cationic, or anionic. Soils can be classified according to their adsorption properties, using distribution coefficients if, and only if, measurements are made under conditions that lead to low equilibrium concentrations.

The organic matter content appears to be a key parameter, but its relationship with the distribution coefficient values is clear only for nonionized molecules. For ionized molecules, the mineralogical composition and the soil pH have to be taken into account, especially for anionic molecules. Thus, information from soil surveys is likely to be used for assessing general features of pesticide adsorption in soils.

Trace element retention by soils will be illustrated by the case of fluoride adsorption, as reported by Bar-Yosef et al. (1989) and oxy-anion adsorption, as summarized by Kafkafi (1989). In minor concentrations fluorine (F) is beneficial to animals and humans, but when adsorbed in excessive amounts it becomes toxic. In an experiment carried out on K montmorillonite and a series of soils with a clay content range of 4–61% and organic matter content range of 0.2–7%, the F adsorption kinetics and isotherms, as affected by the pH of the solution, were determined. The fluoride adsorption isotherms for K-montmorillonite and the six soils studied are shown in Fig. 5.15a and b. It can be seen that in all cases the F adsorption was a function of its concentration in solution at the selected pH values. Soil pH was highly and inversely correlated with the maximum number of fluoride adsorption sites, and this correlation stemmed from the effect of pH on the charge density of clay edges. No significant correlation was found between the organic matter content and the maximum fluoride adsorption. In accordance with earlier results (Fluhler and Polomski 1982), fluoride partitioning between solid and liquid phases of neutral and high-pH soils is satisfactorily described by a Langmuir adsorption model. The time needed to attain quasi-equilibrium in F sorption reactions was found to be inversely proportional to the solution-to-soil ratio of the suspensions (V/M). While at $V/M = 80$ ml g^{-1} equilibrium was attained within a few hours, the time needed at $V/M = 1$ ml g^{-1} was 10–15 days (both suspensions unstirred). The different rates probably stemmed from reduced accessibility of F to adsorption sites at $V/M = 1$, relative to $V/M = 80$, due to enhanced

Table 5.7. Absorption characteristics for two pesticides. (After Barriuso and Calvet 1992)

Soil	For terbutryn				For 2,4-D			
	Kf	nf	r^2	K_d	Kf	nf	r^2	K_d
Rendzina	36.86 ± 0.03	0.87 ± 0.02	0.990	28.3	3.09 ± 0.10	0.78 ± 0.03	0.993	2.5
Humic Cambisois	39.80 ± 0.08	1.34 ± 0.01	0.989	77 ± 6	5.30 ± 0.03	0.80 ± 0.01	0.996	3.1 ± 0.3
Mollic Cambisols	25.08 ± 0.05	1.06 ± 0.04	0.997	35 ± 13	4.99 ± 0.05	0.72 ± 0.02	0.999	2.7 ± 0.9
Calcic Cambisols	5.00 ± 0.06	0.84 ± 0.03	0.995	3.2 ± 0.8	0.54 ± 0.19	0.78 ± 0.06	0.989	0.3 ± 0.1
Dystric Cambisols	8.22 ± 0.09	0.82 ± 0.04	0.986	7.2 ± 1.4	1.19 ± 0.13	0.73 ± 0.04	0.997	0.8 ± 0.2
Gleyc Cambisols	6.77 ± 0.05	0.83 ± 0.03	0.996	4.3	1.27 ± 0.03	0.68 ± 0.01	0.991	0.8
Vertisols	32.94 ± 0.06	1.24 ± 0.07	0.980	80 ± 18	2.44 ± 0.13	0.61 ± 0.04	0.971	1.3 ± 0.5
Ferraisols	27.12 ± 0.01	0.76 ± 0.01	0.998	9 ± 4	16.81 ± 0.04	0.75 ± 0.01	0.989	9.2 ± 4.3
Andosols	44.53 ± 0.06	0.87 ± 0.03	0.987	36 ± 4	32.35 ± 0.03	0.80 ± 0.01	0.998	26.6 ± 0.8

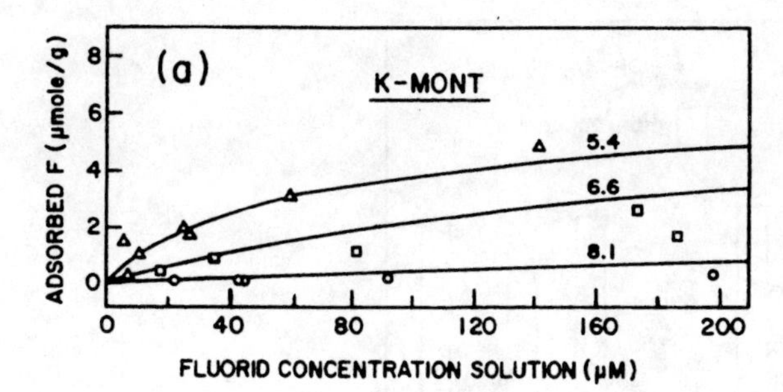

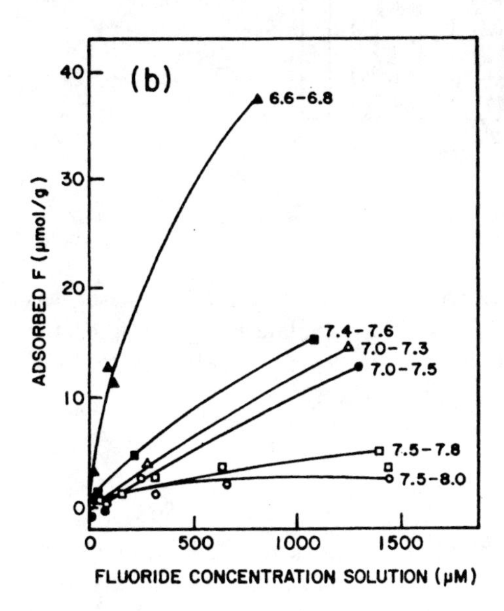

Fig. 5.15a,b. Fluoride adsorption isotherm by K-montmorillonite **a** and six soils **b** as affected by the pH of the suspensions. (After Bar-Yosef et al. 1989)

masking of binding sites on clay edges by the permanent negative electric field of soil-clay interfaces (Sposito 1984).

The formation of metal-organic complexes may increase, decrease, or have no effect on sorption, depending on the nature of both the complex and the available sorbent surface. The behavior of Cd-organo complexes will be used to illustrate the ligand effect on trace-metal adsorption on soil surfaces. Puls et al. (1991), in a study of the effects of organic acids on Cd total sorption on kaolinite, observed the reduction in Cd sorption caused by the formation of Cd complexes (Fig. 5.16). They found, however, that the decrease in adsorption is caused by the nature of ligand. If P-hydrobenzoic and O-toluic acids induced a decrease of Cd sorption on kaolinite, the complexation with 2,4-dinitrophenol brought an increase of Cd sorption. This may be due to a greater affinity of the 1:1 metal-organic complex for the kaolinite surface relative to Cd^{2+} or to a preferential adsorption of dinitrophenol, which subsequently enhanced Cd sorption. Similar behavior was reported by Lamy et al. (1991), who found that,

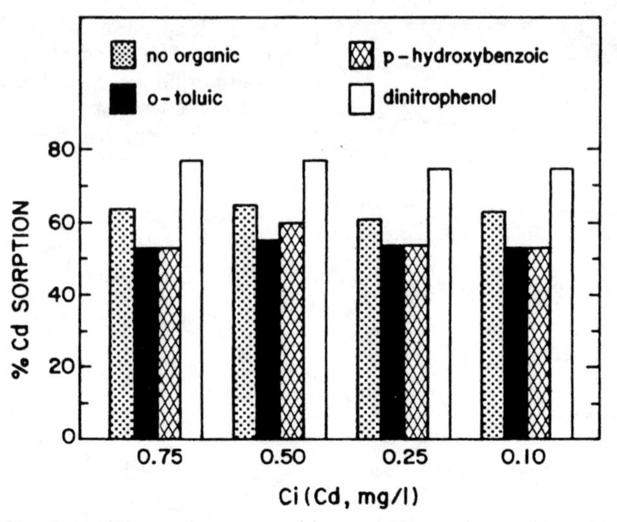

Fig. 5.16. Effects of organic acids on cadmium (Cd_{total}) sorption on kaolinite. (Puls et al. 1991). C_i Initial concentration

in the presence of oxalic acid, the adsorption of Cd on goethite mineral was enhanced due to a surface binding of Cd via an oxalate bridge between the surface and the metallic cation. This means that ligand in solution does not really compete with metallic cation for surface sites.

5.3.2 Nonadsorptive Retention

This occurs in the soil porous medium and represents a pollution hazard mainly in the case of compounds which are nonmiscible with water. This is the case with petroleum products, when the pollutants are adsorbed on particles and thereafter incorporated into the soil, following solid waste or sludge disposal, or when the pollutants are directly adsorbed by the living soil population (vegetable and animal). The following examples will illustrate such types of retention.

Polluted suspended particles originating from sludge disposal could be a source of nonadsorptive retained chemicals. Table 5.8 shows the concentration range of potential pollutant trace elements, as summarized by Chaney (1989). In this table data on maximum toxic trace element concentration are compared with median concentrations of the elements in sludges and soils. The reported data cover the situation in the USA. In a study carried out by Vinten et al. (1983), the distribution of suspended particles from a sewage effluent on various types of soils was followed. If, for example, heavy metals in the range of the values suggested in Table 5.8 with their distribution in the soil profile are adsorbed on these suspended particles, a highly concentrated conclave of polluted solid materials will be formed. According to Vinten et al. (1983), the effect of soil properties on the vertical distribution of deposited solids and on

Table 5.8. Range in concentrations of selected trace elements in dry digested sewage sludges[a]. (Chaney 1989)

Element	Reported range		Typical median sludge	Typical soil	Maximum domestic sludge
	Minimum	Maximum			
As. mg kg^{-1}	1.1	230	10	–	–
Cd. mg kg^{-1}	1	3410	10	0.1	25
Cd/Zn %	0.1	110	0.8	–	1.5
Co mg kg^{-1}	11.3	2490	30	–	200
Cu. mg kg^{-1}	84	17000	800	15	1000
Cr. mg kg^{-1}	10	99000	500	25	1000
F. mg kg^{-1}	80	33500	260	200	1000
Fe. %	0.1	15.4	1.7	2.0	4.0
Hg. mg kg^{-1}	0.6	56	6	–	10
Mn. mg kg^{-1}	32	9870	260	500	–
Mo. mg kg^{-1}	0.1	3700	20	–	35
Ni. mg kg^{-1}	2	5300	80	25	200
Pb[b]. mg kg^{-1}	13	26000	250	15	500
Sn. mg kg^{-1}	2.6	329	14	–	–
Se. mg.kg^{-1}	1.7	17.2	5	–	–
Zn. mg kg^{-1}	101	49000	1700	50	2500

[a]Composting using wood chips as a bulking agent generally produces composed sludge 50% as high in trace elements as digested sludge from the same treatment plant.
[b]Sludge Pb concentration dropped significantly during the 1980s in the USA due to reduction of Pb in gasoline.

the resulting change in flow rate may be elucidated by comparing the behavior of the silt loam with that of the coarse sand (Fig. 5.17a). Coarse sand was leached with unfiltered effluent and silt loam with filtered effluent. Passing unfiltered effluent through the coarse sand resulted in retention of only 43% of the total SS by the soil. This behavior is caused by the coarseness of the porous medium. The finer SS are completely mobile and are hardly retained by the column. In a heavier silt loamy soil, a very large relative reduction in flow rate occurred. After 400 mm of filtered effluent were added the flow rate declined to 20% of its initial value. A very small fraction of SS was present in the leachate ($\ll 0.1$) and most of the deposit was in the top 10 mm of soil.

The retention of solid particles can be can be described by a ratio characterizing the deposition as a function of the depth. This ratio is:

$$\frac{\sigma}{\sigma_i} = \frac{\text{deposit mass per unit length at a depth l (g/m)}}{\text{deposit mass per unit length a depth l}=0\text{(g/m)}} .$$

Variations of $-\text{Ln}\,(\sigma/\sigma_i)$ against depth are shown in Fig. 5.17b. It appears that the coarse solids are efficiently filtered by the coarse sand soil and the fine solids (of colloidal size) are not retained at all, as suggested above.

The silt loam by contrast shows again the characteristic accumulation at the surface associated with matt formation. This occurs even though filtered effluent was used and reflects the considerably finer pore sizes in the silt loam. Because of the smaller pores and associated very much slower flow rate, solids

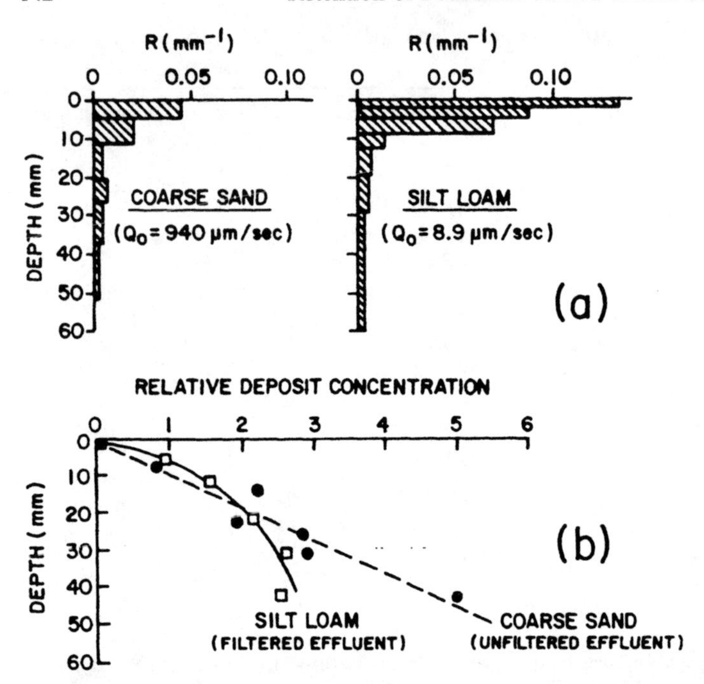

Fig. 5.17. Suspended solids retention in soils. (After Vinten et al. 1983.) **a** Deposited solids distribution in coarse sand and silt loam soil; *R* deposited solids present per unit length of section. **b** Relative deposition of SS, expressed as $-\text{Ln}\ (\sigma/\sigma_i)$, in sandy loam soil leached by a filtered and unfiltered effluent

are removed efficiently at the soil surface. As can be seen, the retention of pollutants adsorbed on solid particles is affected by the type of soil and the ways of incorporation.

Retention of chemicals on the nondecomposed soil organic residues is an additional type of retention of pollutants in the soil medium. In a laboratory experiment carried out by Tegen et al. (1991) to investigate the influence of microbial activity on the migration of Cs^+, the effect of the organic matter residues on the Cs retention was proved. This experiment was performed to understand the consequences of the accident at Chernobyl in 1986. Figure 5.18 shows the depth distribution of Chernobyl Cs at the end of the experiment on soil columns which were leached at the temperatures of 20, 30 and 40 °C, respectively. Based on these results, the authors pointed out the dominant influence of the turnover of soil nondecomposed organic matter on Cs^+ (retention) in soils. Recollecting the aspects of their study, the authors make the following hypothesis for the distribution of Cs^+ in the soil: after deposition at the soil surfaces, Cs^+ is mainly fixed by the nondecomposed organic material. Cs is redistributed in the soil profile as a result of microbial decomposition of organic matter and the dissolution of organic C compounds in the percolating water as introduced by the heating of the soil cores to 30 and 40 °C, respec-

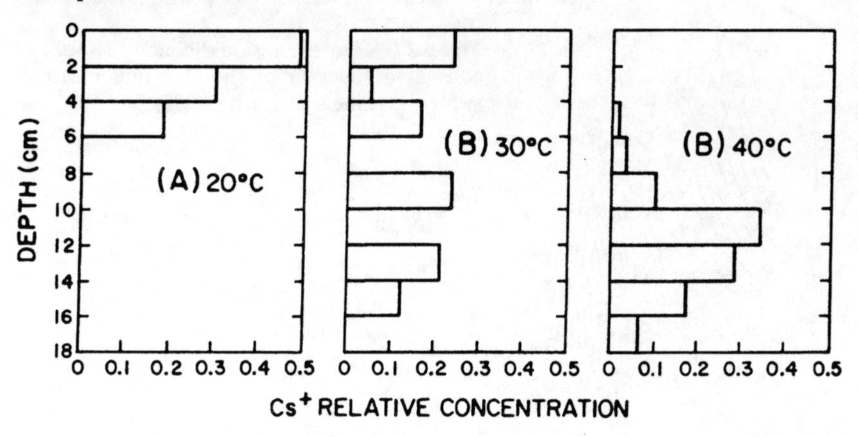

Fig. 5.18A,B. Simulated Chernobyl Cs^+ distribution in the soil ecosystem **A** in leaf pads and **B** in soil columns when the microbial decomposition of the organic material is enhanced by an increase in the ambient temperature. (After Tegen et al. 1991)

tively. As long as the organic material is not decomposed (e.g., at 20 °C), the Cs^+ is retained in the top layer of the soil. By decomposition of organic matter, the Cs^+ is complexed and transported to the depth.

Retention of immiscible with water liquids in the soil medium is done by its being trapped in the soil pores. Retention by trapping of the immiscible with water liquid occurs when soil pore geometry permits their flow at a level greater than their degree of saturation. Figure 5.19 (after Schwille 1984) exhibits in a graphic form the retention of such a liquid in a porous media after the infiltration ended. The experiment reported by Fine and Yaron (1993) shows how the soil constituents and soil moisture content affect the retention of kerosene – a petroleum product – in the soil medium. This retention is called kerosene residual content (KRC). Ten soils with a broad spectrum of clay and organic matter content, as well as four soil moisture contents corresponding to oven-dry, air-dry, 33-kPa tension (field capacity), and 100-kPa tension were studied. The KRC of the oven-dry soils ranged from 3.5 to 18.1 ml/100 g. It

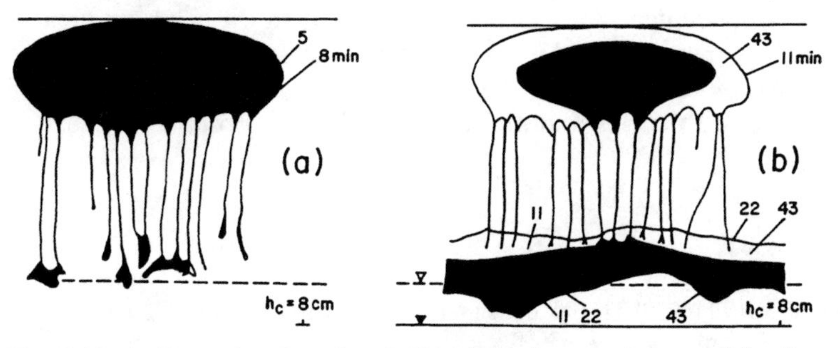

Fig. 5.19a,b. Example of an immiscible fluid (kerosene) trapped in the soil (**a**) immediately after the spill and (**b**) after infiltration ended. (After Schwille 1984)

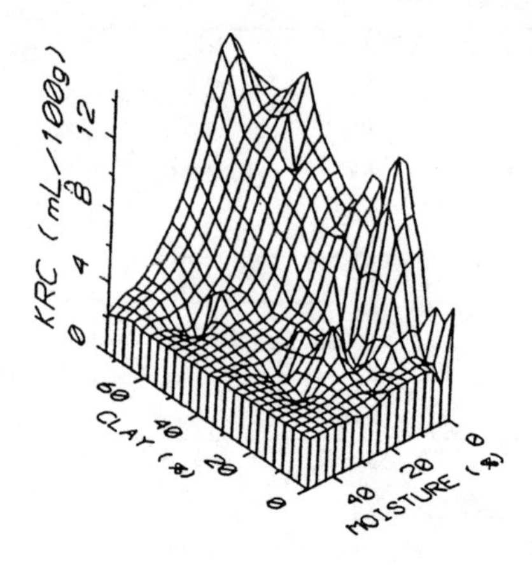

Fig. 5.20. Kerosene residual content (KRC) of soils as a function of the clay and moisture contents. (Fine and Yaron 1993)

was affected strongly by the clay content of the soil, with the more clayey soils having KRC values 1.502 times greater than the less clayey ones. The silt and organic matter (OM) contents also contributed to the KRC. The relationship between these soil components and KRC is described by the relation:

$$KRC = 0.08[\%clay] + 1.68[\%OM] + 0.06[\%silt] + 5.29 \ ,$$

with a highly significant ($p < 0.01$) coefficient of determination (R^2) of 0.91. The ranges of values of the three independent variables (X components) were: clay, 0.3–74%; OM, 0–5.2%; and silt, 1.2–59%.

KRC exhibited a linear relationship to the combination of soil texture (characterized by clay content) and OM content, and was inversely linearly decreased with the moisture content as follows:

$$KRC = 0.13\%[\%clay] + 1.48[\%OM] - 0.32[\%moisture] + 4.31 \ .$$

$R^2 = 0.833$, and was highly significant (p, 0.01). Increasing the soil moisture content caused marked reduction of the KRC of the soils to a rather uniform level. The crucial effect of moisture content on KRC can be exemplified by comparing the fine-clayey Chromoxerert with the dune sand. The respective KRCs were 14.8 and 3.2 ml/100 g for air-dry soil, and 2.17 and 1.84 ml/100 g for moist soils (at 33 pKa tension). The reduction in KRC of the soils owing to increasing moisture content was considerably larger than the ~30% reduction at "field capacity" reported by Marley and Hoag (1986) for a medium-sized sand. A graphic representation of the kerosene residual contents (KRC) of soils as a function of the clay and moisture content is shown in Fig. 5.20.

References

Ashton FM, Sheets TJ (1959) The relationship of soil adsorption of EPTC to oats injury in various soil types. Weeds 7: 88–90

Barriuso E, Calvet R (1992) Soil type and herbicides adsorption. Int J Environ Anal Chem 46: 117–128

Bar-Yosef B, Afic I, Rosenberg R (1989) Fluoride sorption and mobility in reactive porous media. In: Bar-Yosef B, Barrow NJ, Goldschmid J (eds) Inorganic contaminants in the vadose zone. Springer, Berlin Heidelberg New York, pp 75–89

Bolt GH (1955) Ion adsorption by clays. Soil Sci 79: 267–278

Bolt GH, De Boodt MF, Hayes MHB, McBride MB (1991) Interactions at the soil colloid-soil solution interface. NATO ASI Series – Series E. Applied sciences vol 190. Kluwer, Dordrecht, 603 pp

Boyd GE, Adamson AW, Mayers LS Jr (1947) The exchange adsorption of anions from aqueous solution by organic zeolites. J Am Chem Soc 69: 2836–2848

Burchill SM, Hayes MHB, Greenland DJ (1981) Adsorption. In: Greenland DJ, Hayes MHB, (eds) Chemistry of soil processes. Wiley, New York, pp 224–400

Burns IG, Hayes MHB, Stacey M (1973a) Studies of the adsorption of paraquat on soluble humic fraction by gel filtration and ultrafiltration techniques. Pestic Sci 4: 629–641

Burns IG, Hayes MHB, Stacey M (1973b) Some physico-chemical interactions of paraquat with soil organic materials and model compounds. II. Adsorption and desorption equilibria in aqueous suspensions. Weed Res 13: 79–90

Calvet R (1984) Behavior of pesticides in unsaturated zone. Adsorption and transport phenomenon. In: Yaron B, Dagan G, Goldschmid J (eds) Pollutants in porous media. Springer, Berlin Heidelberg New York, pp 143–151

Calvet R (1989) Adsorption of organic chemicals in soils. Environ Health Perspect 83: 145–177

Chaney RL (1989) Toxic element accumulation in soils and crops protecting soil fertility and agricultural food chains. In: Bar-Yosef B, Barrow NJ, Goldschmid J (eds) Inorganic contaminants in the vadose zone. Springer, Berlin Heildelberg New York, pp 140–159

Daniels SL (1980) Mechanisms involved in sorption of microorganisms to solid surfaces. In: Bitton G, Marshal KV (eds) Adsorption of microorganisms to surfaces. John Wiley, New York pp 7–59

De Boodt MF (1991) Application of the sorption theory to eliminate heavy metals from waste water and contaminated soils. In: Bolt GH, de Boodt MF, Hayes MHB, McBride MB (eds) Interactions at the soil colloid-soil solution interface. NATO ASI Series. Series E. Applied Sciences, vol 190. Kluwer, Dordrecht, pp 293–321

Fine P, Yaron B (1993) Outdoor experiments on venting enhanced volatilization of kerosene components from soils. J Cont Hydrol 12: 355–374

Fluhler HJ, Polomski BP (1982) Retention and movement of fluoride in soils. J Environ Qual 11: 461–468

Fripiat J, Chaussidon J, Jelli A (1971) Chimie-physique des phénomènes de surface. Masson, Paris, 386 pp

Gaillardon P, Calvet R, Terce M (1977) Adsorption et desorption de la terbutryne par une montmorillonite-Ca et des acides humiques seuls ou en mélanges. Weed Res 17: 41–48

Gerstl Z, Yaron B (1978) Adsorption and desorption of parathion by attapulgite as affected by the mineral structure. J Agric Food Chem 26: 569–573

Gerstl Z, Yaron B (1981) Attapulgite-pesticides interactions. Residue Rev 78: 69–99

Giles CH, MacEwan TH, Nakhwa SN, Smith D (1960) Studies in adsorption. Part XI. A system of classification of solution adsorption isotherms and its use in diagnosis and adsorptive mechanism and measurements of specific surface area of solids. Chem Soc J 3: 3973–3993

Grahame DC (1947) The electrical double layer and the theory of electro capillarity. Chem Rev 41: 441–449

Green RE (1974) Pesticide-clay-water interactions. In: Guenzi WD (ed) Pesticides in soils and water. Soil Science Society of America, Madison, pp 3–37

Greenland DJ (1965) Interactions between clays and organic compounds in soil. Soils Fert 28: 415–425

Greenland DJ, Hayes MHB (eds) (1981) The chemistry of soil processes. John Wiley, Chichester, 714 pp

Hassett IJ, Banwart WL (1989) The sorption of nonpolar organics by soils and sediments. In: Sawhney BL, Brown K (eds) Reactions and movement of organic chemicals in soils. SSSA Spec Publ 22. Soil Science Society of America, Madison, pp 31–45

Hayes AHB, Mingelgrin U (1991) Interaction between small organic chemicals and soil colloidal constitutents. In: Bolt GH, De Boodt MF, Hayes MHB, McBride MB (eds) Interactions at the soil colloid-soil solution interface. NATO ASI Series – Series F. Applied Sciences vol 190. Kluwer, Dordrecht, pp 324–401

Huang PM, Grover R, McKercher RB (1984) Components and particle size fractions involved in atrazine adsorption by soils. Soil Sci 138: 220–224

Kafkafi U (1989) Oxanion sorption on soil surfaces. In: Bar-Yosef B, Barrow NJ, Goldschmid J (eds) Inorganic contaminants in the vadose zone. Springer, Berlin Heidelberg New York, pp 33–43

Kennedy VC, Brown TC (1965) Experiments with a sodium ion electrode as a means of studying cation exchange rates. Clays Clay Minerals 13: 351–352

Lamy I, Djafer M, Terce M (1991) Influence of oxalic acid on the adsorption of cadmium at the goethite surface. Water Air Soil Pollut 57–58: 455–467

Leenheer JA, Ahlrichs JL (1971) A kinetic and equilibrium study of the adsorption of carboxyl and parathion upon soil organic matter surfaces. Soil Sci Soc Am Proc 35: 700–705

Malcom RL, Kennedy VC (1970) Variation of cation exchange capacity and rate with particle size and stream sediment. J Water Pollut Control Fed 42: 153–160

Marley MC, Hoag GE (1986) Induced soil venting for recovery restoration of gasoline hydrocarbons in vadose zone. In: Water Well Assoc Am Pot Inst, Conf on Petroleum Hydrocarbons and Organic Chemicals in Ground Water. Nat Water Well Assoc, Houston, 1984

Mills AC, Biggar JW (1969) Solubility-temperature effect on the adsorption of gamma and beta-BHC from aqueous and hexane soultions by soil materials. Soil Sci Soc Am Proc 33: 210–216

Mortland MM (1970) Clay-organic complexes and interactions. Adv Agron 22: 75–117

Muntz C, Roberts PV (1986) Effects of solute concentration and cosolvents on the aqueous activity coefficient of halophenal hydrocarbons. Environ Sci Technol 20: 830–836

Nye PH, Tinker PB (1977) Solute movement in the soil-root system. Blackwell, Oxford, 324 pp

Overbeek JTh (1952) Electrochemistry of the double layer. Colloid Sci 29: 119–123

Pignatello JJ (1989) Sorption dynamics of organic compounds in soils and sediments. In: Sawhney BL, Brown K (eds) Reactions and movement of organic chemicals in soils. Soil Science Society of America, Madison, pp 45–80

Prost R, Gerstl Z, Yaron B, Chaussidon J (1977) Infrared studies of parathion attapulgite interaction. In: Behaviour of pesticides in soils. Israel-France Symposium INRA-Versailles, pp 108–115

Puls RW, Powell RM, Clark D, Eldred CY (1991) Effect of pH solid-solution ratio, ionic strength and organic acids on Pb and Cd sorption on kaolinite. Water Air Soil Pollut 57–58: 423–430

Quirk JP, Posner AM (1975) Trace element adsorption by soil minerals. In: Nicholas DJ, Egan AR (eds) Trace elements in soil plant animal systems. Academic Press, New York, pp 95–107

Rao PSC, Hornsby AG, Kilcrease DP, Nikedi-Kizza P (1985) Sorption and transport of hydrophobic organic chemicals in aqueous and mixed solvent systems: Model development and preliminary evaluation. J Environ Qual 14: 376–383

Rao PSC, Hornsay AJ, Kilcrease DP, Nikedi-Kizza P (1989) Sorption and transport of hydrophobic organic chemicals in aqueous and mixed solvent systems. J Environ Qual 14: 376–383

Saltzman S, Yariv S (1976) Infrared and X-ray study of parathion montmorillonite sorption complexes. Soil Sci Soc Am Proc 35:700–705

Saltzman S, Yaron B (eds) (1986) Pesticides in soil. Van Nostrand Reinhold, New York, 379 pp

Saltzman S, Kliger L, Yaron B (1972) Adsorption-desorption of parathion as affected by soil organic matter. J Agric Food Chem 20: 1224–1227

Schwille F (1984) Migration of organic fluids immiscible with water in the unsaturated zone. In: Yaron B, Dadon G, Goldschmid J (eds) Pollutants in porous media. Springer, Berlin Heidelberg New York, pp 27–48

Sparks DL (ed) (1986) Soil physical chemistry. CRC Press, Boca Raton, 324 pp

Sparks DL (1989) Kinetics of soil processes. Academic Press, San Diego, 210 pp

Sparks DL, Huang PM (1985) Physical chemistry of soil potassium. In: Munson RE (ed) Potassium in agriculture. American Society for Agronomy, Madison, pp 201–276

Sposito G (1984) The surface chemistry of soils. Oxford University Press, Oxford, 234 pp

Swoboda AR, Thomas GW (1968) Movement of parathion in soil columns. J Agric Food Chem 16: 923–927

Talibuden O (1981) Cation exchange in soils. In: Greenland DJ, Hayes MHB (eds) The chemistry of soil processes. John Wiley, Chichester, pp 115–178

Tengen I, Doerr H, Muennich KO (1991) Laboratory experiments to investigate the influence of microbial activity on the migration of cesium on a forest soil. Water Air Soil Pollut 57–58: 441–449

Terce M, Calvet R (1977) Some observations on the role of Al and Fe and their hydroxides in the adsorption of herbicides by montmorillonite. Sonderdruck Z Pflanzen-kr Pflanzenschutz, Sonderheft VIII

Theng BKG (1974) The chemistry of clay-organic reactions. Adam Hilger, London, 324 pp

Thomas GW, Yaron B (1968) Adsorption of sodium from irrigation water by four Texas soils. Soil Sci 106: 213–220

US Salinity Laboratory Staff (eds) (1954) Diagnosis and improvement of saline and alkali soils. US Dept of Agric Handbook 60, Washington, 160 pp

Van Dam J (1967) The migration of hydrocarbon in a water bearing stratum. In: Hepple P (ed) The joint problems of oil and water industries. Institute of Petroleum, London, pp 55–96

Veith GA, Sposito G (1977) Reactions of aluminosilicates aluminum hydrous oxides aluminum oxides with 0-phosphate. Soil Sci Soc Am J 41: 8670–8874

Vinten AJA, Mingelgrin U, Yaron B (1983) The effect of suspended solids in waste water on soil hydraulic conductivity. Soil Sci Soc Am J 47: 402–412

Weber WJ (1984) Evolution of a technology. J Environ Eng Div (Am Soc Civ Eng) 110: 899–917

Weber JB, Miller CT (1989) Organic chemical movement over and through soil. In: Sawhney BL, Brown K (eds) Reactions and movement of organic chemicals in soils. SSSA Spec Publ 22. Soil Science Society of America, Madison, pp 305–335

Weber JB, Shea PJ, Weed SB (1986) Fluoridone retention and release in soil. Soil Sci Soc Am J 50: 582–588

Weed SB, Weber JB (1974) Pesticide organic matter interaction. In: Guenzi WD (ed) Pesticides in soil and water. Soil Science Society of America, Madison, pp 39–66
Yamane VK, Green RE (1972) Adsorption of ametryne and atrazine on an oxisol, montmorillonite and charcoal in relation to pH and solubility effects. Soil Sci Soc Am Proc 36: 58–64
Yaron B (1978a) Organophosphorus pesticides-clay interactions. Soil Sci 125: 412–417
Yaron B (1978b) Some aspects of surface interactions of clays with organophosphorus pesticides. Soil Sci 125: 210–216
Yaron B, Saltzman S (1978) Soil-parathion surface interactions. Residue Rev 69: 1–34

Part III
Pollutant Behavior in Soils

CHAPTER 6

Reversible and Irreversible Retention – Release and Bound Residues

Pollutants retained in the soil may be released into the soil solution, displaced as water-immiscible liquids, or transported into the gaseous phase. The form and rate of release are controlled by the properties of both the pollutants and the medium, as well as by environmental conditions. Nonadsorbed pollutants may be released from the soil through dissolution and volatilization, processes which have been described in Chaps. 3 and 4.

As was described in Chapter 5, several biotic and abiotic phenomena may lead to pollutant retention as the result of complex interactions with soil minerals and organic constituents. Many observations have shown that a more or less important fraction of retained molecules cannot be released or, at least, is released only slowly. The amount and rate of release are variable, depending on the nature of the medium-pollutant system, on the experimental procedure, and conditions used to study the release.

Release can be assessed on the basis of physicochemical and biological processes. In the case of the former, a change in the physicochemical characteristics of the fluid phase surrounding the retaining solid phase is the usual cause of a release. This is classically obtained, for example, by lowering the pollutant concentration in the solution previously brought in equilibrium with the solid phase. Another way to obtain a release can be through processes controlled by the pollutant bioavailability.

Theoretically, the adsorption-desorption process should be expressed by isotherm singularity. In many cases, however, the release isotherms do not coincide with the retention isotherms, in which case we are dealing with the phenomenon of hysteresis or nonsingularity. This means that not all retained molecules can be transferred back in the solution phase. Published results show a range of experimental behavior, from complete reversibility to total irreversibility.

6.1 Retention Hysteresis

Taking into account the possible role of phenomena other than adsorption and desorption, it is better to speak of retention hysteresis than of adsorption hysteresis, except for situations where reversible adsorption exists. Retention hysteresis is not yet fully understood, and several interpretations have been presented.

The rate of release and the amount of pollutant released vary according to the nature of pollutant, the nature of solid phase, and experimental procedures. They also depend strongly on experimental conditions and on the soil sample history. Whatever the situation, the key fact is that a fraction of retained pollutant cannot be released, indicating that retention is not reversible. The variety of observed behavior and of implied phenomena makes the release of pollutants a complex problem. Besides the nature of the causes of hysteresis, there are other interesting questions:

- What is the relationship between hysteresis and the factors which influence retention?
- Are nonreleasable molecules definitely retained?
- What are the mobility and the bioavailability of irreversibly retained molecules?
- Are there any hazards due to these molecules?

These questions are far from being completely answered.

Brusseau and Rao (1989) defined two types of hysteresis:

1. True hysteresis, for which it is assumed that observed data are real and that equilibrium data can be explained on the basis of well-identified phenomena, chemical, physicochemical, or biological;
2. Apparent hysteresis, which could be the result of a failure to reach retention or release equilibrium. It could also be an artifact of the experimental method and/or of measurement errors and additional phenomena such as degradation. Observed data do not necessarily correspond to retention phenomena and are often not equilibrium values.

On the basis of a large number of experimental results, obtained with various sorbates and sorbents, Brusseau and Rao (1989) discussed the possible causes of hysteresis.

6.1.1 True Hysteresis

Release has been widely explained on the basis of desorption, which was considered as the only phenomenon implicated in it. Experimental data can be interpreted in terms of true desorption if two conditions are fulfilled: the system must be at equilibrium, and released molecules must be those which were adsorbed on the surfaces of the solid phase. Molecules which were brought back into the solution as the result of dissolution, diffusion out of the solid matrix, or of biotic/abiotic chemical transformations cannot be considered as desorbed molecules. In most situations, it is obviously impossible to distinguish real desorbed molecules from the others, except with solids which can only behave as adsorbent.

Desorption isotherms may differ from adsorption isotherms in two situations. The first corresponds to systems which are not at equilibrium because the desorption rate is smaller than the adsorption rate. Theoretical treatments presented by Ponec et al. (1974) for gas adsorption, and by Giles et al. (1974) for solute adsorption, consider that the activation energy for adsorption is zero or near zero. Under this condition, they showed that the activation energy for desorption is greater then that for adsorption. Thus, the rate of desorption must be smaller than the rate of adsorption. This behavior is also true if adsorption is accompanied by the dissociation of the adsorbed molecule (Ponec et al. 1974). Attention must be drawn to the fact that equilibrium may not be reached because of the occurrence of phenomena other than adsorption/ desorption phenomena, in which case, different mechanisms are involved (see Sect. 6.1.2). The second situation occurs when adsorbed molecules undergo some modification(s) of their physicochemical state, due to chemical and/or biochemical reactions with the solid phase. Barriuso et al. (1992) assumed that adsorbed molecules may be divided into two categories: molecules retained through physical interactions and able to be desorbed; and molecules which have evolved towards states where they interact strongly with the solid matrix and thus are released slowly or not at all. For the first category, desorption may be described by a linear isotherm; for the second category, the release isotherm may be described by an exponential function of the equilibrium concentration of the solution. Figure 6.1 gives an example of such a splitting for desorption of dimefuron from a clay loam soil. However, a question remains: what changes are occurring and what are the driving forces for these changes? No satisfactory physicochemical theory has been developed until today to describe the mechanism of molecule-surface interactions. Several

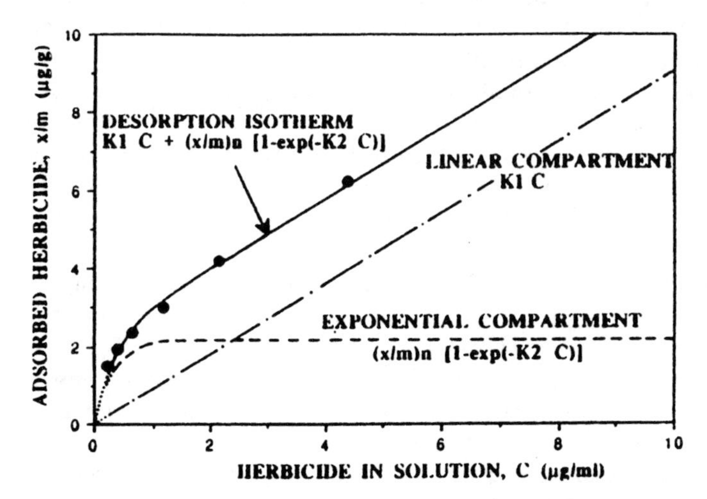

Fig. 6.1. Splitting of the desorption isotherm of dimefuron in the presence of 0.01 M $CaCl_2$ into two other isotherms corresponding to the two-compartment model of desorption isotherms. $x/m = K_1C + (x/m)_n(1\text{-}e^{K_2C})$. (After Barriuso et al. 1992a)

other explanations have been put forward to explain retention hysteresis. These are:

- surface precipitation, as proposed for metallic cations whose hydroxides, phosphates, or carbonates are sparingly soluble;
- chemical reaction with the solid surface, which is essentially possible with organic surfaces which can form complexes with metallic cations and combine with organic molecules such as pesticides;
- incorporation in the soil organic matter through chemical reactions or biochemical transformations, involving either exocellular or endocellular enzymes.

This last process probably plays a major role in the formation of what are called bound residues, and will be examined further in more detail.

6.1.2 Apparent Hysteresis

Apparent hysteresis, due to a lack of equilibrium, is one of the processes most frequently affecting pollutant release. Diffusion into the solid matrix or into micropores of aggregates has frequently been assumed to prevent the system from reaching equilibrium, and then to be one of the main causes of hysteresis. In a transitory state, sorption will still occur concurrently with adsorption, and the concentration of the liquid phase attributed to desorption will be erroneously low, since some fraction is really associated with sorption. If complete release equilibrium is not reached, excessive sorbed molecules will remain associated with the sorbent and, consequently, the calculated sorbed concentration is greater than those calculated under equilibrium conditions. These effects, associated with small deviations in measurement of the individual desorbed fractions, can lead to a significant apparent hysteresis.

Apparent hysteresis is also caused by other phenomena. During application of the consecutive desorption and dilution techniques used as a common procedure in release studies, *weathering* of the sorbent might occur, resulting in a possible increase of sorption, which will give the appearance of a decrease in the amount desorbed.

Degradation in solution can decrease the concentration, which is thus underestimated, and, when associated with the retained amount remaining in the solid phase, can give the appearance of hysteresis. Such a situation was clearly observed by Koskinen and Chang (1982) in the case of sorption of 2,4,5-T by a Palouse soil. Sterilization of the medium before release measurement decreased the hysteresis, but did not suppress it completely (Fig. 6.2). *Volatilization* can also lead to the same effect.

Soil moisture effect. Clays, particularly smectites, and humic substances are important sorbents in the soil, but it is well known that they can hydrate and swell or dehydrate and shrink when subjected to wetting and drying. This may change their surface properties as regards the adsorption of water and

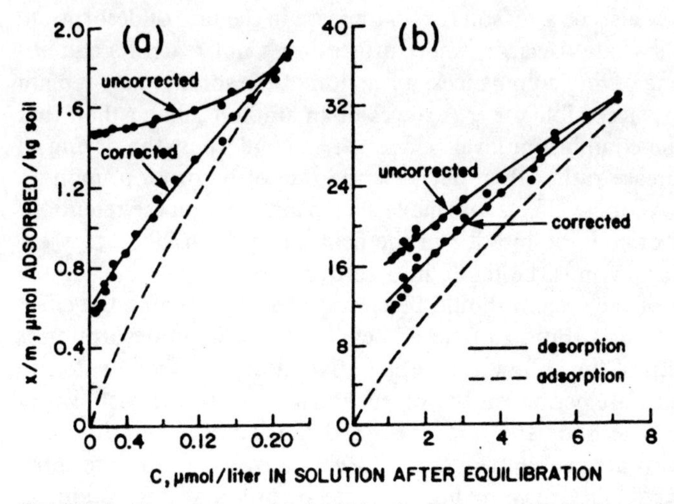

Fig. 6.2a,b. Desorption of 2,4,5-T in the Palouse soil. Impact of correction for 2,4,5-T degradation at two concentrations: **a** 2 μmol L^{-1}. Coefficient of variation for experimental data = 1.2%. **b** 39 μ mol L^{-1}. Coefficient of variation for experimental data = 3.8%. (Koskinen et al. 1979)

pollutants. The effect of drying on hydration of humic substances has been interpreted by assuming the formation of intramolecular bonds between hydrophilic functions, keeping the hydrophobic moieties on the external part of the molecular aggregates (van Dijk 1971). This interpretation was further developed by Chassin and Leberre (1979), who proposed to describe the evolution of molecular associations with the moisture content in terms of intra- and intermolecular bonding. This modifies the adsorption properties of humic substances and is responsible for the water-repelling property of dried humic soils which have been freshly rewetted. Mingelgrin and Gerstl (1993) considered that the humic anionic groups present in the soils interact intramolecularly with di- or polyvalent cations, created in the drying of a tight soil aggregate with a hydrophobic outer surface. Only after a considerable time do enough water molecules diffuse into the soils to hydrate both cationic and anionic groups to an extent that causes swelling of the soil. The higher the fraction of monovalent counter cations (excluding protons) in the organic soil, the less tightly it coils upon drying, and the higher its tendency to swell or uncoil upon rewetting.

The physicochemical state of a more or less great fraction of molecules sorbed in wet conditions may be modified upon drying, making them more difficult to desorb. The retention of sorbate molecules during drying of slow-swelling organic soil may be a reason, in many instances dominant, for the slowdown in desorption upon rewetting. When molecules are sorbed at polar surface sites, they orient their hydrophobic part towards the solution phase, creating a plug which may considerably reduce the access of water, and thus slow down the swelling and, with it, desorption (Mingelgrin and Gerstl 1993).

Drying of soil may also be a reason for an increase in the rate of desorption. If penetration of a sorbate towards inner surfaces does not reach its equilibrium level by the time drying commences, a fraction of the sorbate may remain localized at the more accessible outer surfaces in an amount greater than that corresponding to the equilibrium level. Under these conditions, the drying of the system may increase rather than decrease the rate of desorption upon an eventual rewetting. All these effects may have a deep influence on the retention/release behavior of organic pollutants, in the field, but also in laboratory experiments, in which they must be taken into account.

The past history of the system should be considered in predicting the effect of moisture on desorption. For example, the effect of drying on desorption is affected by the length of time allowed for the initial adsorption from water. In sorbing systems which are permanently wet, such as sediments, the past history of the system may determine the fate of sorbed molecules if sorption sites offering limited access are present (Pignatello 1989). Slow diffusion into inner pores with low accessibility may reduce the rate of eventual desorption, as compared with the initial sorption. A practical consequence of the above is that if a pollutant which is introduced into a water body diffuses into internal surfaces of a sediment sufficiently slowly, the rate of removal of the pollutant may depend on the elapsed time between the inception of the pollution and cleanup.

6.1.3 Method-Created Hysteresis

It has been shown that desorption characteristics may depend on the experimental procedure.

The effect of measurement techniques used for defining desorption parameters is exemplified by the study of Hodges and Johnson (1987) on the kinetics of the sulfate adsorption-desorption process. Sulfate may be retained in soils by Coulomb attraction onto positively charged sites, formation of inner-sphere complexes with Fe, or the total replacement of surface water on hydroxyl groups of Al hydroxides and precipitation of basic Al sulfates. Hodges and Johnson (1987) used various experimental techniques such as the rapidly stirred batch process, or miscible displacement with a slower flow rate, to provide additional information concerning the rate-limiting mechanism of sulfate desorption. They found that the desorption of sulfate in stirred batch experiments was much less than that obtained by the miscible displacement technique.

The amount of sulfate desorbed from a Bt_2 horizon of a Cecil soil in the miscible displacement experiment was affected by the time of leaching. Downward movement of solution of 0.2 mol m^{-3} KCl during 600 s removed only 23–34.4% of the adsorbed sulfates. Leaching of a sample at 22 °C for 14400 s (44 h) recovered $< 53\%$ of the adsorbed sulfate. The estimated time (by extrapolation) required for complete desorption was 10–20 times greater than for adsorption.

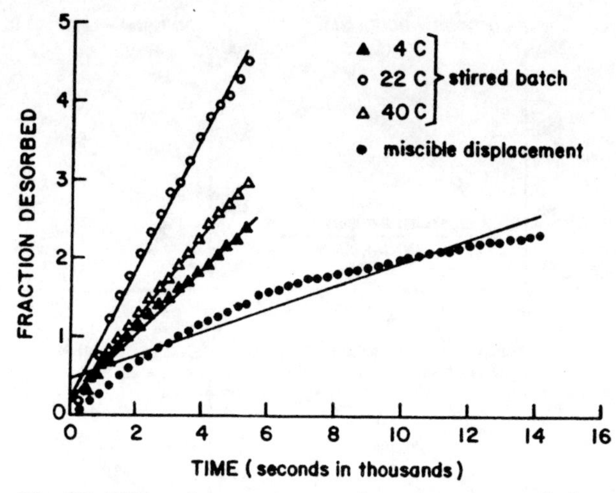

Fig. 6.3. Effect of temperature and measurement technique on sulfate desorption from soils. The *lines* represent the shell progressive particle diffusion equation plots. (Hodges and Johnson 1987)

Temperature also plays an important role in defining desorption parameters. Figure 6.3 shows the results of Hodges and Johnson (1987) for the desorption of sulfate from the Bt_2 horizon of the Cecil soil. The slope for the desorption data shows an initial increase with increasing temperature followed by a slight decrease at the highest temperature. The lowest slope value occurred at 4 °C.

The effect of desorption methodology on the desorption of selected synthetic organic pesticides will be exemplified by the research carried out by Bowman and Sans (1985). These authors compared the dilution method of measuring desorption with the consecutive desorption method. It is worth noting that dilution of suspensions may increase the accessibility of an adsorbing surface, so that the study of desorption by the dilution method is not strictly comparable with the classical method. With the latter method, the soil/solution ratio remains constant, which is not the case with the dilution method. The results of Bowman and Sans (1985) are presented in Fig. 6.4. In their opinion, the degree of desorption hysteresis exhibited by the various pesticide-soil (clay) systems was clearly related to the method used in deriving the data. In all cases, there was only a minimal amount of hysteresis in the desorption isotherms obtained using the dilution method, whereas almost all the systems investigated with the consecutive desorption method exhibited considerably more hysteresis. Differences in water solubility and partitioning tendencies of the two organophosphate compounds produced isotherms in different concentration ranges, making direct comparisons among the different systems difficult. However, the results of the Fig. 6.4 concour with the proposals of Rao and Davidson (1980), that the centrifugation-resuspension step was in some way resposible for the observed hysteresis effects. Perhaps a partially irreversible

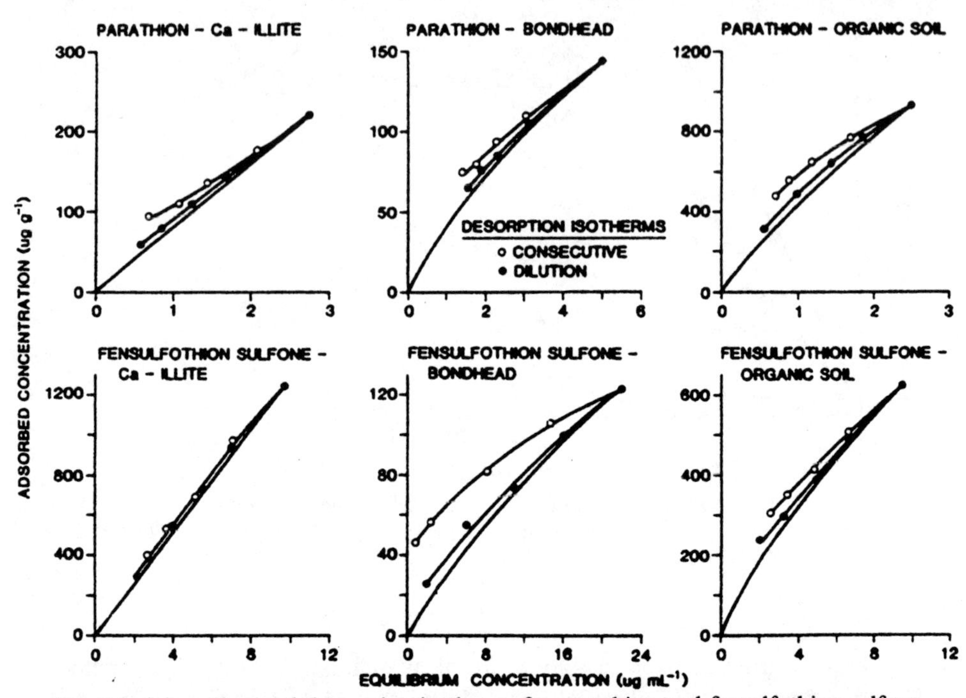

Fig. 6.4. Adsorption and desorption isotherms for parathion and fensulfothion sulfone in aqueous suspensions of three sorbent materials: Ca-illite, Bondhead loam, and an organic soil. (Bowman and Sans 1985)

compaction of the adsorbent by centrifugal forces greatly increased the time required for desorption processes to reestablish equilibrium, thereby creating the impression that partitioning had shifted in favor of the adsorbent.

Although the dilution-desorption data presented in Fig. 6.4 minimize the observed magnitude of hysteresis, it should be kept in mind that there are situations where "true" hysteresis effects (as opposed to method-created artifacts) can have an important bearing on predicting pesticide behavior in soil-water systems.

6.2 Bound Residues

6.2.1 Experimental Definition

The hysteresis we have just discussed corresponds to irreversible retention observed when release is allowed to take place in an aqueous solution. In order to extract the maximum amount of pesticide associated with the solid phase, several organic solvents are used. Many experiments made with radiolabeled

Table 6.1. Examples of bound pesticide residues in soils. (After Calderbank 1989)

Structural type	Bound residues (% of applied)	Parent detected[a]
Herbicides		
Anilides and ureas	34–90	No
Bipyridyliums	10–90	Yes
Nitroanilines	7–85	No
Phenoxy	28	No
Phosphonate (glyphosphate)	12–95	Yes
Triazines	47–57	Yes
Insecticides		
Carbamates	32–70	Yes
Organochlorines	7–25	?
Organophosphates	18–80	Yes
Pyrethroids	3–23	No
Fungicides		
Chlorophenols	45–90	Yes?
Nitroaromatic (dinocap)	60–90	Yes?

[a]Indicates where parent was positively identified. In cases with "no" or question mark, the method of extraction may have decomposed the parent molecule.

molecules show that the extraction is never complete, even with solvents in which pesticides are highly soluble. Up to 90% of the applied radioactivity can become unextractable (Table 6.1), and this has been attributed to strongly "bound residues."

Several authors have reported experimental data showing clearly that bound residues comprise both parent products and degradation products. Parent products have been found to represent various fractions of bound residues: 18% for pyridilcarbamate, more than 50% for prometryn, 10% for atrazine, and traces for several other pesticides such as picloram, triallate, and methyl-parathion (cited by Calderbank 1989). Study of unextractable residues is further complicated by the fact that the extractability of ^{14}C-labeled organic compounds depends on the size of the granulometric fraction and is also time-variable, probably as a result of the soil microflora activity (Fig. 6.5).

It has been observed that the extractability decreases with aging, probably because the phenomena responsible for hysteresis become more efficient with increasing residence time in the soil.

6.2.2 Mechanisms

Irreversible retention of pesticides is not fully understood. Several hypotheses have been put forward to explain the formation and properties of bound residues, but they have not been completely verified.

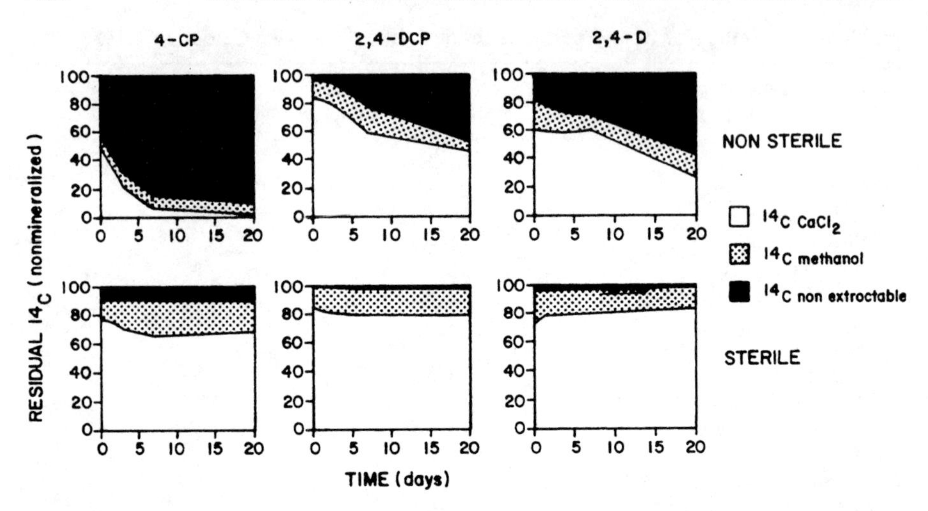

Fig. 6.5. Distribution of the nonmineralized residual radioactive carbon during the incubation in sterile and nonsterile wet straw of 4-CP, 2,4-DCP, and 2,4-D. (Benoit 1994)

Diffusion out of the solid phase may partly account for hysteresis, particularly for molecules which have diffused into the organic matter aggregates. However, some molecules would be unable to diffuse, even slowly. Entrapping in humic polymer aggregates as suggested by Khan (1982), and more recently by Wershaw (1986) and Singh et al. (1989), is a possible explanation for hysteresis, compatible with the structure of humic substances (Wershaw, 1986). Hydrophobic molecules may be trapped inside the most hydrophobic part of humic aggregates and thus prevented from release by aqueous and organic solvents. Another process, probably a major process leading to an irreversible retention, is the chemical binding of pesticide molecules to the organic matter (Bollag and Loll 1983). Fulvic and humic acids are the compounds most commonly involved in such binding. Reactions are due to microorganisms and plant enzymes and can be considered as some kind of degradation. The reactions involved appear to be the same as those responsible for humic substance formation. Phenol and aromatic amines may bind to organic matter by oxidative coupling, while substituted ureas and triazines may not. This was clearly shown, together with the part played by mineral surfaces, by Bollag et al. (1992). Bertin et al. (1991) have also shown that binding can take place during humic substance formation through polymerization processes.

Another process to be considered is biological incorporation. This retention phenomenon is due to pesticide absorption by plants and soil microorganisms. Some fraction of this absorbed pesticide may return to the soil with either crop residues or dead microorganisms, and may thus be incorporated in the soil organic matter.

It has recently been shown that microorganisms play a major part in the formation of bound residues (Benoit 1994). Figure 6.5 gives the comparison between sterilized and nonsterilized soils. Soils were amended with 4-chlorophenol (4-Cl,2,4-DCP) and 2,4-dichloro-phenoxy-acetic acid (2,4-D) and allowed to incubate at 28 °C for 20 days. The graphs represent the amounts of residual ^{14}C as water-soluble, methanol-soluble, and unextractable fractions. They clearly show that sterilization considerably reduces the amount of unextractable ^{14}C and, therefore, of bound residues, and that the distribution of residues among the three fractions is time-depenent.

Retention can be viewed as the result of two broad processes: a physicochemical process and a biological process. Distinction between the two processes has been done by comparing retention with and without sterilization. Figure 6.6 gives this comparison for two sorbents: a prehumified straw (Fig. 6.6a) and a mycellium of *Emericella nidulans* (Fig. 6.6b). It appears that

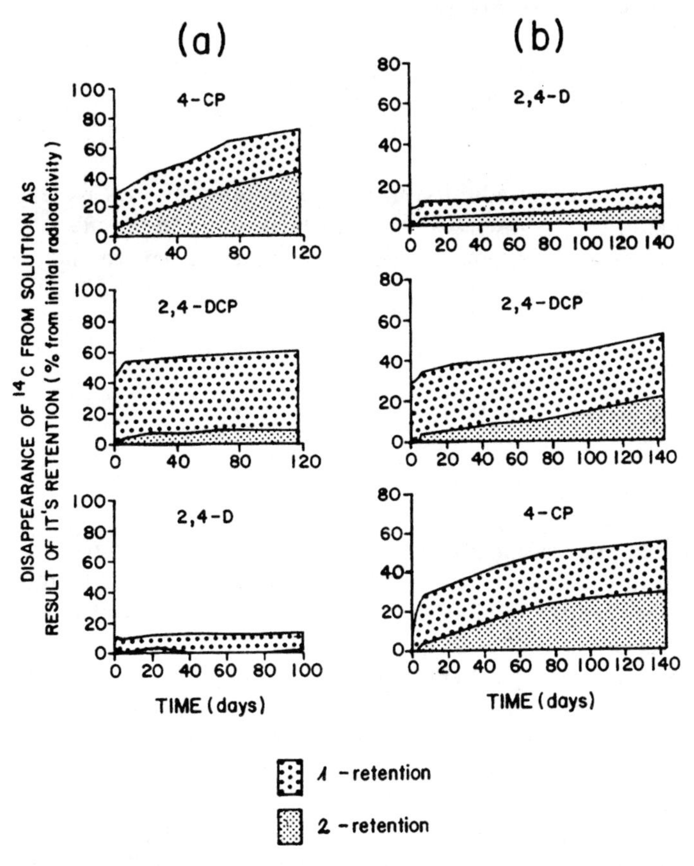

Fig. 6.6a,b. Relative proportions of physicochemical (*1*) and biological (*2*) retention of 4-chlorophenol (4-CP), 2,4-dichlorophenol (2,4-DCP), and 2,4-dichloro phenoxyacetic acid (2,4-D) on straw (**a**) and on mycelium *Emericella midulans* (**b**) These proportions were estimated from retention kinetics on sterilized and nonsterilized materials. (Benoit 1994)

the natures of both sorbate and sorbent have an influence on the relative importance of the two processes. Further, it can be seen that physicochemical process rates vary during only a short period of time (< 2 days), while the biological retention rate increases continously with time.

Physicochemical retention proceeds through three steps: adsorption, diffusion into the organic matter, and chemical and biochemical reaction with the organic matter. The release process under natural conditions, that is, a solid phase in contact with an aqueous phase, can follow only two steps: diffusion out of the sorbent (essentially the organic matter) and desorption, the latter being more rapid. If desorbed molecules were previously in a physicochemical state which is accounted for by an adsorption isotherm, then the release isotherm is a desorption isotherm. In this case, we are dealing with a fully reversible process. As long as intraparticle diffusion exists, release may be slow, and equilibrium will be reached slowly.

6.3 Examples

The examples which follow – in addition to those illustrated by Figs. 6.5 and 6.6 – were selected in order to show various release behaviors and the effects of important factors such as pollutant and soil properties and environmental physicochemical conditions. They are presented according to the nature of the pollutant.

6.3.1 Inorganic Pollutants

6.3.1.1 Major Elements

It has been shown that major elements such as N, P, and K may be irreversibly retained by soil constituents. Since we are only concerned with soil pollutants, potassium will not be further examined because no adverse effect seems to be attributable to this element.

Irreversible retention of NH_4^+, often named fixation of NH_4^+, was first reported by McBeth in 1917 and cited by Scherer (1993), who has recently reviewed this subject. It is now generally recognized that soil NH_4^+ can be divided into three fractions: in-solution, exchangeable, and nonexchangeable fractions (Nommik 1957). According to Mengel and Scherer (1981), nonexchangeable NH_4^+ is that which is nonexchangeable in neutral salt solutions and is rapidly recovered only from clay minerals treated with a 5 N hydrofluoric acid/1 N hydrochloric acid mixture. Fixation is explained on the basis of a mechanism which involves strong electrostatic interactions between NH_4^+ cations and the negative electrical charge of layer silicates, due to the penetration of the cations into the ditrigonal cavities of two adjacent layers.

According to this, the main factors in NH_4^+ fixation are the clay content, the type of clay, vermiculites being particularly efficient (Page et al. 1967), and the K content. Of total soil nitrogen, 3–14% may be in the fixed NH_4^+ form. A slow release is possible when the solution concentration is lowered by nitrification or plant absorption and is controlled by an exchange/diffusion process favored by Ca^{2+} and H^+ ions. Availability of fixed NH_4^+ appears to be very variable and quite low. From a pollution point of view, fixation may play an interesting role. As a matter of fact, many observations show that fixation may be responsible for the retention of a large fraction (up to 90%) of the NH_4^+ introduced by organic manures, preventing its nitrification and thus making it an important pollutant source.

Phosphorus can also be retained as apparently unreleasable forms. Information can be found in several reviews, e.g., those of Hingston (1981) and of van der Zee and van Riemsdijk (1991). Anion adsorption is reversible with respect to pH and temperature changes but displays several behaviors at constant pH and ionic strength, ranging from complete reversibility to almost complete irreversibilty. The low desorbability of ortho-phosphate anions is not yet clearly understood. Some insight into this problem may be given by identifying situations in which phosphates are in competition with other complexing molecules. In some cases, desorption may not be complete, and has been shown to depend on pH, ionic strength, the nature of the minerals, the time of incubation, and the concentration of phosphate in the solution. Surface complexation has been invoked to explain these observations, but another explanation based on kinetic considerations seems to be more accepted now.

The effect of time of contact with the solid phase on pollutant desorption is exemplified by the experiment of Barrow (1979) on the desorption of phosphate from soil. In this experiment, desorption occurred after both short and long periods of contact between soil and chemical. After short periods of contact, the amount of phosphate desorbed into the soil solution increased rapidly at first but, due to a readsorption process, the concentration of the phosphate in solution then decreased. Figure 6.7 shows changes through time in the amount of phosphate in the soil solution after various periods of prior incubation (from 0 to 25 days). It is observed that, independently of the soil-solution ratio, the concentration in the soil solution during 1 day of incubation exhibits a desorption and readsorption process. With extension of the time of contact, the concentration of phosphate in the solution increases only as a result of the desorption of phosphate from the soil solids surface. The extent of desorption increases with increasing solution:soil ratio. With time, however, the concentration of phosphate in the highly concentrated suspensions is reduced, compared with the low-concentration suspensions case. From the point of view of environmental concern, this behavior of phosphate is of major importance.

Considering that the same theoretical description fits the adsorption/desorption data of several anions (fluoride, phosphate, molybdate), and taking into account the kinetics of the sorption process, we may conclude that hys-

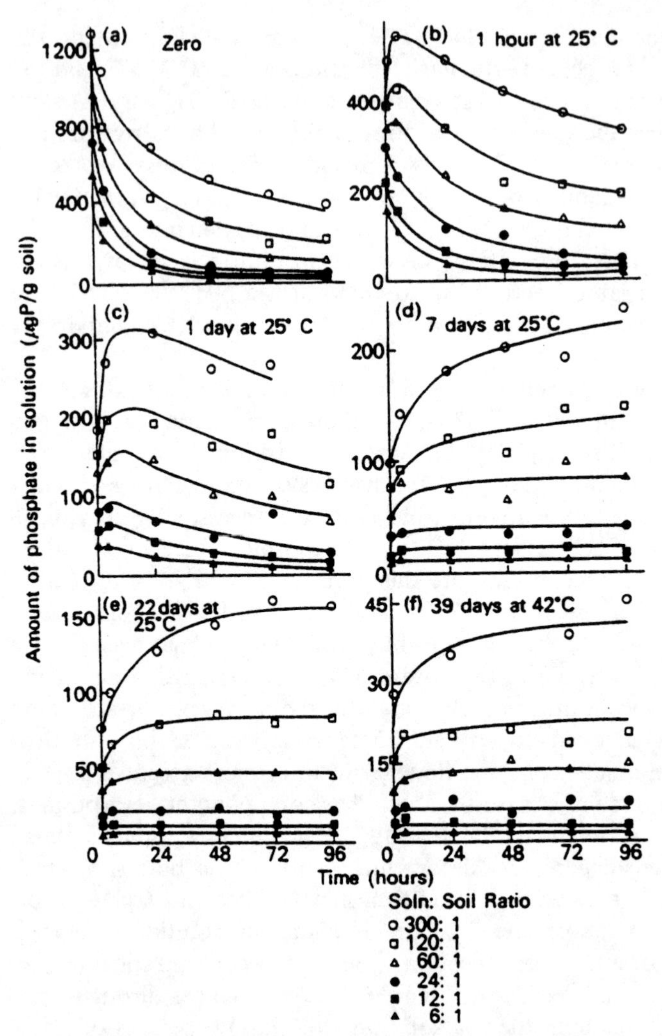

Fig. 6.7a-f. Desorption of phosphate in solution after various times of contact (0, 1 h, 1, 7, 22 and 39 days) with the soil. (After Barrow 1979)

teresis would probably be due to diffusion out of the sorbing particles (Barrow 1983, 1986). The behavior of sorbed species is a direct consequence of the sorption process, which is considered to be the result of adsorption and intraparticle diffusion, the latter being generally very slow (van der Zee and van Riemsdijk 1991). Complexing molecules modify the desorption of anions, as shown by many experimental data on the competition between phosphates on the one hand and silicate (Obihara and Russel 1972), selenite (Hingston 1981), and citrate, fluoride, and bicarbonate anions, on the other. This has long been used to extract and determine what is known as the bioavailable phosphorus.

6.3.1.2 Trace Elements, Heavy Metals

Release behaviors. Release of trace elements can be due to a decrease in the concentration in solution of the metallic cation, with no modification of the background electrolyte. In this case, true desorption is taking place. Release may also result from addition of protons and/or complexing molecules to the background electrolyte. This situation lies behind extraction methods for metallic cations, which render some strongly sorbed cations mobile. Those cations which still remain on the sorbent materials may be said to be completely irreversibly retained; then, their passage into a solution implies, at least partly, destruction of the solid phase.

Figure 6.8 gives several examples of desorption isotherms displaying a range of behaviors. Selenite is reversibly adsorbed on gibbsite at pH = 6.7 and in 0.01 M $CaCl_2$ but other experiments showed that it is almost irreversibly adsorbed on goethite, while fluoride is reversibly adsorbed on goethite at pH = 4.5 in 0.01 M NaCl. As shown Fig. 6.9, quite large amounts of sorbed metallic cations can remain on the solid phase. Radioisotopes provide an interesting way to study the release of sorbed cations. Figure 6.9 clearly shows that the amount of nonisotopically exchangeable cations increases with time, more for soil oxides than for humic acid. Such behavior has also been observed for Cu^{2+} sorbed on iron and manganese oxides.

Irreversibility has not yet been completely explained. It could be due to a lower desorption rate than adsorption rate. If such an explanation holds, it implies different activation energies for forward and reverse reactions. According to the theory of chemisorption, the desorption activation energy is always greater than the adsorption activation energy, which leads to a slower rate of desorption than of adsorption (McBride 1991). For several authors (Brummer et al. 1983; Barrow 1986; Swift and McLaren 1991), a more likely interpretation is the occurrence of a second process after the initial adsorption. This could be either a surface precipitation as a new solid phase, comprising

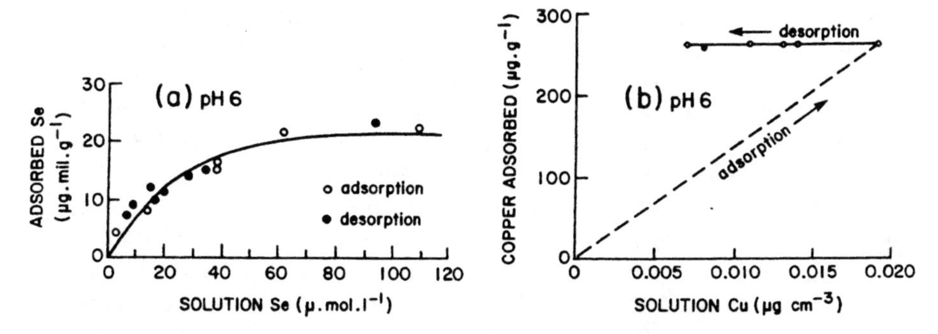

Fig. 6.8a,b. Desorption of **a** selenite from gibbsite and **b** copper from montmorillonite (**a** after Hingston et al. 1974; **b** McLaren et al. 1983). Background electrolyte: **a** 0.1 M NaCl, **b** 0.05 M $CaCl_2$

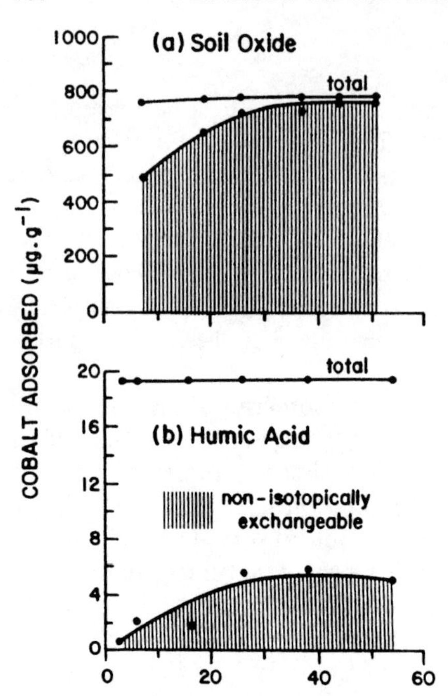

Fig. 6.9a,b. Isotopic exchangeability of Co adsorbed during 60 days by soil oxide and humic acid at pH 6.0; background electrolyte 0.01 M $CaCl_2$. (After McLaren et al. 1986)

hydroxy-polymers on layer silicates, or solid-state diffusion into the sorbing material, for retention by oxides and allophanic clays (McBride 1991).

Hysteresis due to precipitation reactions. In many cases, a pollutant retained on the soil solid surface may undergo chemical reactions which will induce the hysteresis phenomenon in the course of the release process. An example in this direction is the behavior of Cr(VI), an industrially originated pollutant, which, in the soil environment, may be subjected to the oxidation and/or reduction reactions. When an available source of electrons, such as organic matter, is present, Cr(VI) will be reducted to Cr(III) (Ross et al. 1981); the rate of this reduction reaction increases with decreasing pH. Bartlett and James (1979) found that Cr can also be oxidized from Cr(III) to Cr(VI) under conditions commonly found in soils. A study of Stollenwerk and Grove (1985) on adsorption and desorption of hexavalent chromium in the alluvial aquifer deals with these types of problems. Adsorption and desorption data for a large-column experiment are presented in Fig. 6.10a. All Cr(VI) in the first seven pore volumes of groundwater containing 960 μmol L^{-1} Cr(VI) was retained by the alluvium. A rapid increase in the effluent concentration of Cr(VI) then occurred, until the capacity of the alluvium for retained Cr(VI) was exhausted (222 pore volumes). Approximately 50% of the Cr(VI) was released from the alluvium by the first 10 pore volumes of Cr-free water used to wash the column. This rapid initial rate of release of Cr(VI) from the alluvium decreased, and an additional 50 pore volumes of groundwater removed only 34% more

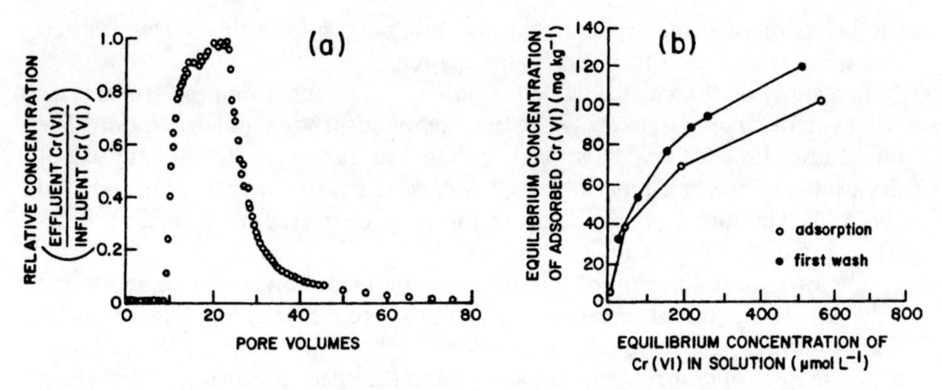

Fig. 6.10a,b. Adsorption and desorption of Cr(VI) by a telluride alluvium in **a** column experiment at a flow rate of 7.1×10^{-4} cm s^{-1}, and **b** on alluvium under batch conditions at pH 6.8 and 25 °C. (After Stollenwerk and Grove 1985)

Cr(VI) from the alluvium. Almost 16% of the adsorbed Cr(VI) was still associated with the alluvium when the experiment was stopped after 82 pore volumes (232 days). Stollenwerk and Grove (1985) attribute the difficulty in removing some of the adsorbed Cr(VI) to the presence of specific adsorption sites, and possible reduction to Cr(III) followed by precipitation.

The same authors show that the difference in the rate of adsorption and desorption of Cr(VI) by alluvium, noticed in the column experiments, was also evident in batch experiments (Fig. 6.10b). The desorption curve does not coincide with the adsorption curve, as it would if adsorption were a completely reversible process. Allowing 21 days for the desorption reaction to take place instead of only 5 days did not change the shape of the desorption curve. This indicates that reaction kinetics are not the reason for the discrepancy between the two curves (Fig. 6.10b). Specific adsorption of some Cr(VI) or Cr(III) was the most likely cause of this difference.

On the basis of the above experimental results, these authors reached the conclusion that the quantity of Cr(VI) adsorbed was a function of its concentration, as well as of the type and concentration of other anions in solution, and that the hexavalent Cr adsorbed through nonspecific processes is readily desorbed by Cr-free salt solution. Stronger bonds formed between Cr(VI) and alluvium during specific adsorption result in very slow release of this fraction. Given sufficient contact time, some of the adsorbed Cr(VI) is apparently fixed by the alluvium.

The Cr(VI) desorption from the alluvium material illustrated very well the presence of a hysteresis process resulting from a chemical transformation of the part of the pollutant retained on the soil surface.

Kinetics of desorption is one of the most crucial aspects of the process in defining the behavior of the pollutant in the soil environment. The kinetics are affected by the characteristics of both adsorbent and adsorbate, and encompass the behavior of both organic acid and inorganic chemicals. A few selected examples will be used to illustrate the matter.

In a study on the kinetics of desorption of metal ions from peat, Bunzl et al. (1976) made a comparison between the rates of adsorption and desorption. The initial rates (in μeq Me^{2+} adsorbed or desorbed during the first 10 s by 1 g of dry peat) and the half-times obtained from kinetic experiments are included in Fig. 6.11. The initial absolute adsorption rates decreased in the order Pb^{2+}> Cu^{2+} > Cd^{2+} > Zn^{2+} Ca^{2+}.

The rates of desorption of the several metal ions show a more complicated behavior, as the initial metal contents of peat are different for different metal ions. Following the selectivity series from Pb^{2+} to Ca^{2+} (Fig. 6.11), the initial absolute rates of desorption first increase and then decrease again. In desorbing lead ions, the initial absolute rate of desorption is low, since only a small amount is present in peat to be desorbed. The desorption of Cd^{2+} is comparatively rapid, since there is a fairly large amount to be desorbed, and it is, furthermore, not tightly bound. Bunzl et al. (1976) showed that relative rates of desorption, as characterized by the half-times, show a similar behavior: longer half-times for Pb^{2+}, Cu^{2+}, and Ca^{2+} but shorter ones for Cd^{2+} and Zn^{2+}.

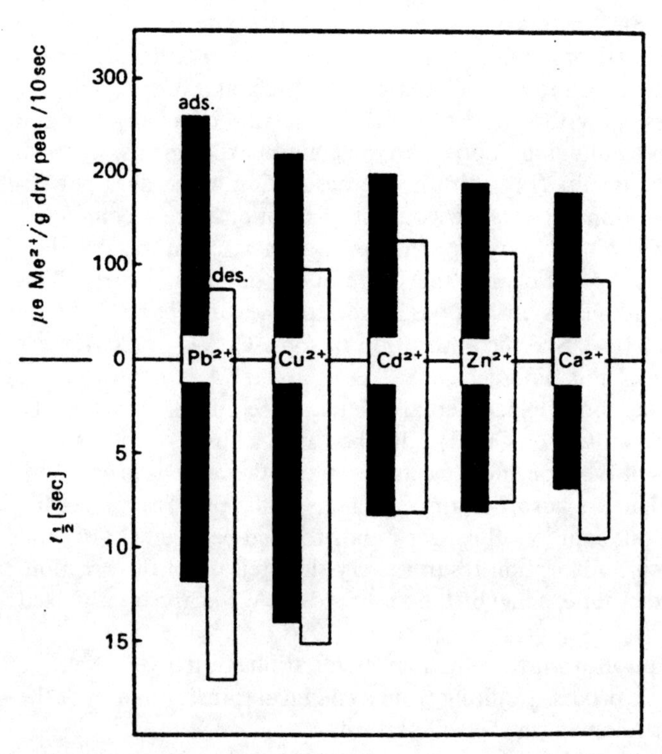

Fig. 6.11. Initial rates of adsorption and subsequent desorption during the first 10 s and corresponding half-times of Pb^{2+}, Cu^{2+},Cd^{2+}, Zn^{2+}, and Ca^{2+} in peat. (Bunzl et al. 1976)

In many cases, the time required for obtaining a completely reversed desorption is much greater than that necessary for the adsorption of chemicals on surfaces. In a study on borate adsorption-desorption on pyrophyllite, Keren et al. (1994) found that at pH = 9 the adsorption of borate on the surface was completely reversible. However, the time required to accomplish the reversal is much greater than that required for adsorption.

This is consistent with observations of Brummer et al. (1988) for Ni, Zn, and Cd sorption on goethite, which showed a slow disappearance of metal from the solution, following a fast first step. This is also to be related to the fact that the longer the period which elapses before desorption, the less complete the desorption is (Padmanabham 1983). Thus, diffusion into the solid phase would explain apparent irreversibility. This is supported by the experimental results of Barrow (1986), who has shown that there is no discontinuity in the sorbed-amount/solution-concentration relationship providing that the measurements are done on a system which has reached equilibrium (Fig. 6.12). The long time, often several months, for example, observed by Barrow (1986) as necessary to reach equilibrium for Zn retention at 25 °C, is also consistent with a diffusion-controlled release. Nevertheless, it is worthwhile to emphasize that some questions remain about the real nature of the process. As a matter of fact, surface properties of soil constituents may be modified when the soil solid phase is in contact with an electrolyte solution for a long period. This may increase the difficulty in interpreting release data, because diffusion out of the solid phase and desorption would not be the only process involved. Strictly speaking, cations which are released under such conditions are irreversibly retained because their passage into the solution is the consequence of a modification of physicochemical properties of the sorbent/solution system.

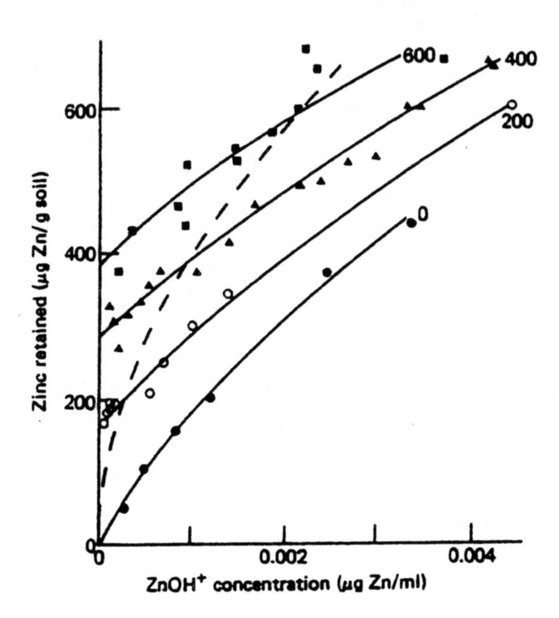

Fig. 6.12. Desorption and further sorption of zinc in 0.03 M sodium nitrate solution. Samples of the soil had been incubated with the indicated levels of zinc (μg Zn g^{-1}) for 7 days at 60 °C before desorption/sorption was measured over 24 h at 25 °C. In each case, the *broken line* joins the concentrations at which neither desorption nor further sorption occurred. (Barrow 1986)

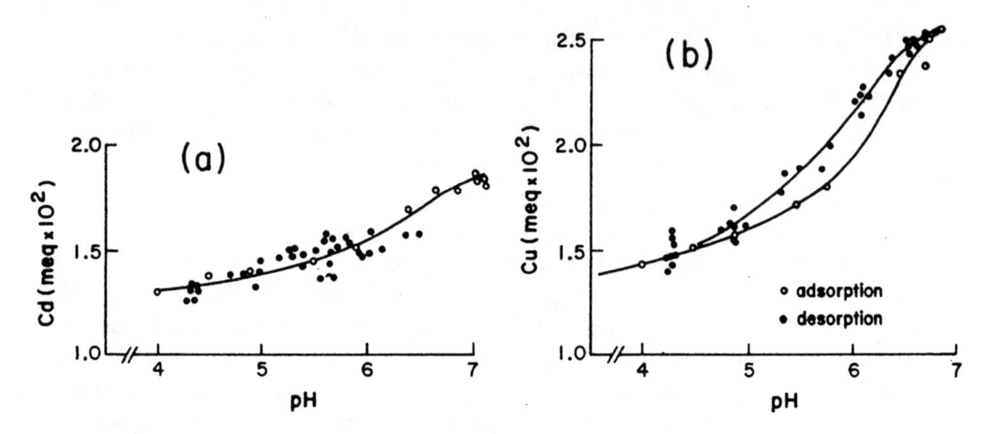

Fig. 6.13a,b. Sorption of Cd (**a**) and Cu (**b**) on montmorillonite, as function of pH. Increasing pH (●) and decreasing pH (o). Initial concentrations are 5.08 10^{-5} mol l^{-1} for Cd and 5.19 10^{-5} ml l^{-1} for Cu. Background electrolyte: 10^{-3} M $CaCl_2$. (Calvet 1989b)

pH effect. Nondesorbable metallic cations can be released, at least partly, by decreasing the pH, by addition of either complexing molecules or another metallic cation. Figure 6.13 gives an example of Cd^+ and Cu^{2+} release from a montmorillonite with decreasing pH. The difference in the behavior between the two cations can be explained by the surface complexation of Cu^{2+}.

The ligand effect on the desorption process is exemplified by an experiment on copper behavior carried out on a Paxton soil (pH = 5.6; CEC = 16.4 OM = 3.81%) by Lehman and Harter (1984). The kinetics of copper desorption was investigated, following the addition of Cu at the rate of 1.6, 3.9, 6.3, and 7.9 μmol Cu/g soil (100, 250, 400, and 500 mg soil). The Cu was allowed to react with the soil for 30 min to 4 days. At time t = 0, after the period of Cu-soil reaction, Na_2-oxalate, Na_3-citrate, or Na_4-EDTA was added in an amount stoichiometric to the initial Cu addition (i.e., a charge ratio 1:1 Cu/ligand). In each case, the suspension pH rose to 6.0 following the ligand addition. Suspension aliquots were periodically removed and filtered. The suspension was also sampled immediately prior to ligand addition, to establish the Cu concentration at time t = 0. Figure 6.14 shows the time-dependent release of soil-adsorbed Cu in solutions where three different organic ligands are present. It may be observed that the presence of EDTA ligand enhanced Cu release compared with citrate and oxalate. In the case of the Cu-oxalate complex, it appeared that it had been readsorbed onto the surface. It is also clear that each ligand in the soil will differentially affect the desorption of soil-adsorbed Cu. Since in many cases, Cu reaches the soil in sewage effluents or sludges, the presence of organic ligands will be able to alter the process of retention-release of the metal on the soil surface. A characterization of the type of organic ligands found in sewage effluents and/or sewage sludge would be useful in predicting the release of Cu adsorbed on soil surfaces.

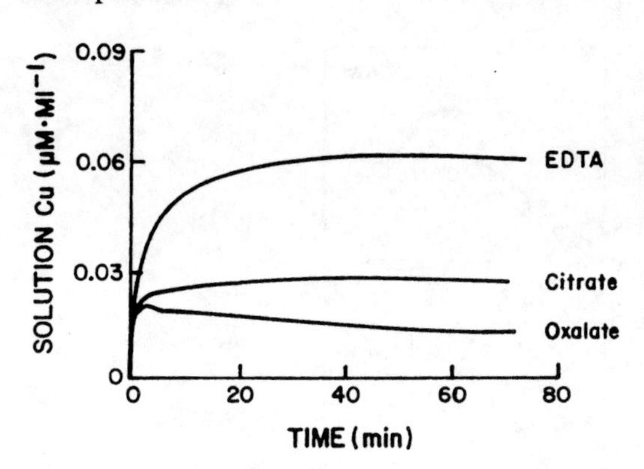

Fig. 6.14. Time-dependent release of soil-absorbed copper by EDTA, citrate, and oxalate; 0.079 μmol Cu ml^{-1} were added to the solution. At time = 0, chelator was added at a rate equivalent to a 1:1 Cu/chelator charge ratio. (Lehman and Harter 1984)

Related to this topic is the competition among metallic cations for sorption sites. Metals which can be bound through surface complexation as strongly as copper cations can displace a large proportion of these cations, as was observed with allophanes by McBride (1991). So Cu^{2+} can appear to be irreversibly bound when no other competing cations are present.

6.3.2 Organic Pollutants

As in the case of inorganic pollutants, a more or less important fraction of the retained organic molecules cannot be released from the solid phase. This has been observed in many instances with various pesticide/soil systems; experimentally, the effects include retention hysteresis, partial non-extractability of residues and time variations of the extractability.

6.3.2.1 Release Behaviors

It is well known that retention is not completely reversible, so that release isotherms differ from retention isotherms to an extent which depends on the nature of the sorbent (Fig. 6.15).

Strictly speaking, sorption is a reversible process, which includes adsorption, desorption, and diffusion in and out of the solid phase. Several explanations have been proposed to account for a hysteretic behavior, some of which relate to experimental conditions which lead to an understimate of the equilibrium concentration of the liquid phase, as we have seen.

Other potential causes for hysteresis correspond to a real incorporation into the solid phase: modifications of molecule-surface interactions, irreversible retention into the solid phase (mainly the organic matter) leading to entrapment of the molecules, and chemical reactions between the pesticide molecules and the solid phase, which is equivalent to a degradation phenomenon.

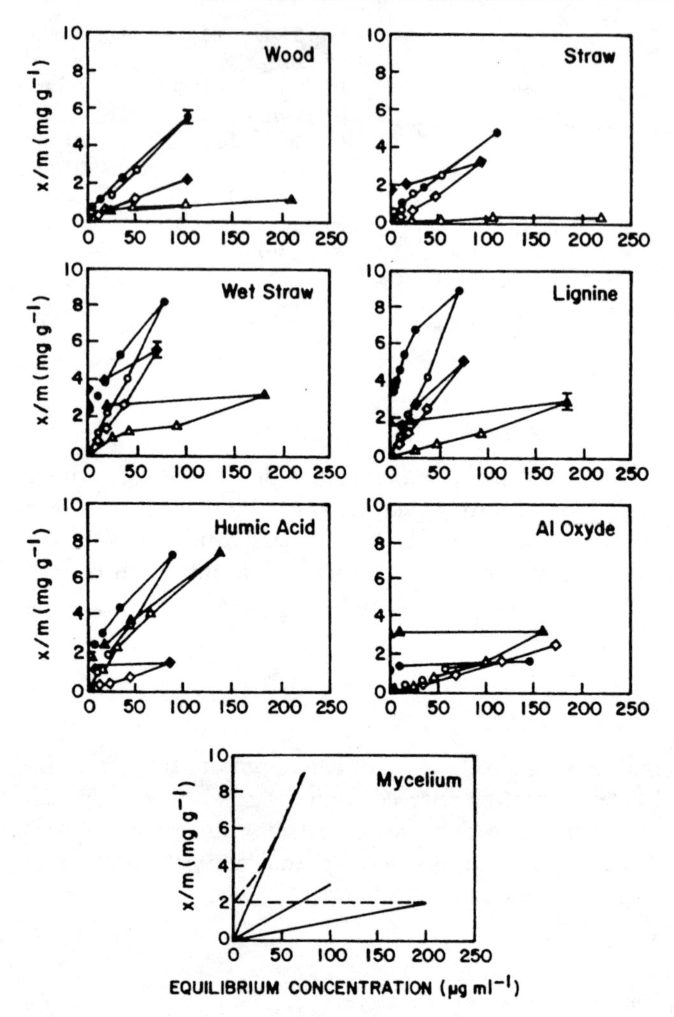

Fig. 6.15. Adsorption isotherms (*white circles*) and desorption isotherm (*black circles*) of 2,4D, 2,4-DCP, and 4-CP for various adsorbents. (Benoit 1994)

Since mineral and organic constituents are involved in the retention of organic micropollutants, it is interesting to know their respective roles. Information about this subject can be obtained from some experiments such as the two following ones.

The effect of soil constituents on the extent of hysteresis effect in pesticide adsorption-desorption was proven by Saltzman et al. (1972) for organic matter and Gaillardon et al. (1977) for clay materials. The influence of the soil organic matter on the extent of hysteresis can be observed in Fig. 6.16, which summarizes results of the determination of parathion desorption by water from natural soils and oxidized subsamples. The desorption isotherms of the natural soils do not overlap the adsorption isotherms, showing that there is a "hysteresis" in the parathion adsorption-desorption process. In contrast, smaller differences between adsorption and desorption isotherms were found for soils treated with hydrogen peroxide (oxidized subsamples).

Parathion desorption by water was also dependent on the nature of the soil adsorptive surfaces. After five consecutive desorptions, all the natural soils still contained more parathion than the treated soils. The parathion content of the natural soils was greater (8% in Golan, 24% in Meron, and 31% in Bet Guvrin soils) than in the oxidized subsamples. After the same number of desorptions, the peat soil retained two to three times more parathion than the natural soils. The stronger retention of parathion by natural soils and peat shows that parathion-organic complexes are stronger than parathion-mineral ones.

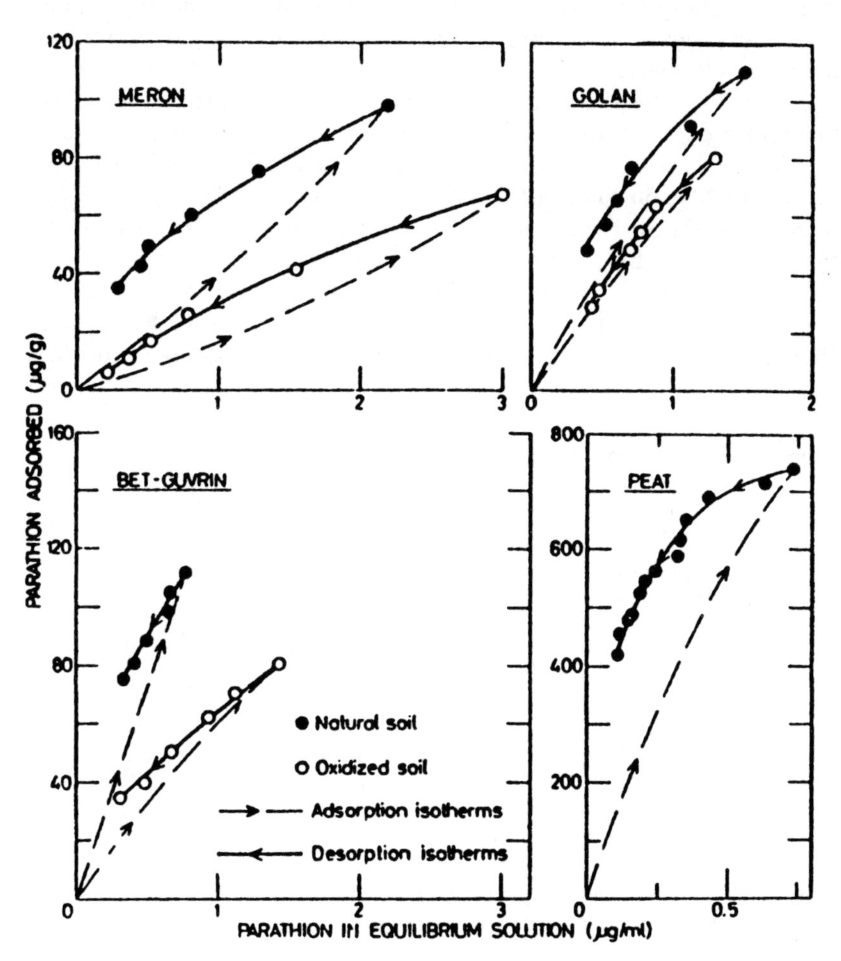

Fig. 6.16. Parathion desorption by water from soils and oxidized subsamples. (Saltzman et al. 1972)

If one assumes that only the changes in organic matter content are responsible for the changes in parathion desorption, then it is possible to estimate desorption curves for the organic matter of each soil. The adsorptive capacity for organic matter and clay was estimated as follows: (1) the differences in desorption between natural and treated soil at several concentrations of the equilibrium solution were calculated, and (2) the desorption per unit weight of organic matter was calculated as the ratio between each of these values and the amount of organic matter lost by soil oxidation. Desorption from the mineral soil fraction was obtained by subtracting the calculated adsorption on organic material from the adsorption on the natural soil. From the desorption curves obtained (Fig. 6.17), a net difference between parathion release by organic and mineral colloids may be noted. The slope of the desorption curves of the mineral fraction is rather steep (especially in Golan soil), showing that adsorption is easily and totally reversible. From the organic matter, in the range of concentrations studied, only very small amounts of parathion seem to be released.

The extent of hysteresis is controlled by the different bonding mechanisms of the pollutant on specific soil constituents, under various environmental conditions. The pH, for example, may cause changes which are reflected more often in the desorption than in the adsorption behavior. The paper of Gaillardon et al. (1977) on terbutryne adsorption-desorption on model soil constituents, as affected by the properties of constituents and the pH of soil solution, will serve as an example for the above statement. Figure 6.18 shows the adsorption-desorption isotherm of terbutryne on a clay material (montmorillonite), a humic acid, and a mixture – in various proportions – of these

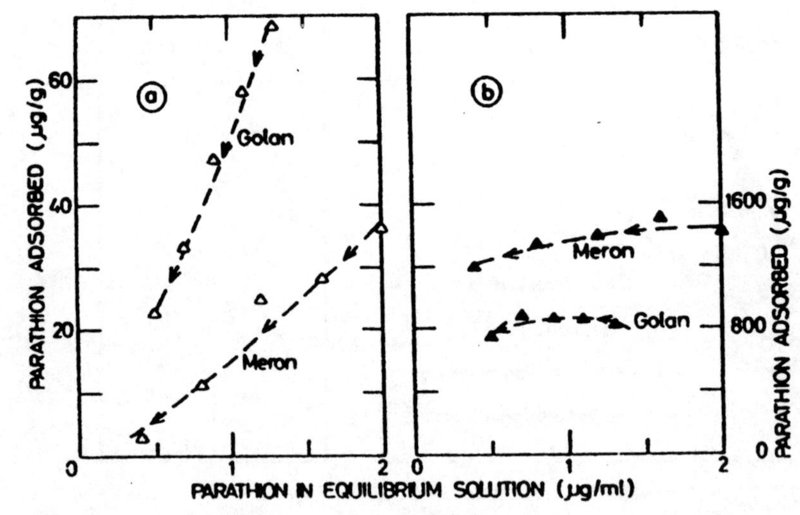

Fig. 6.17a,b. Calculated parathion desorption from the mineral (**a**) and organic (**b**) fractions of Golan and Meron soils. (After Saltzman et al. 1972)

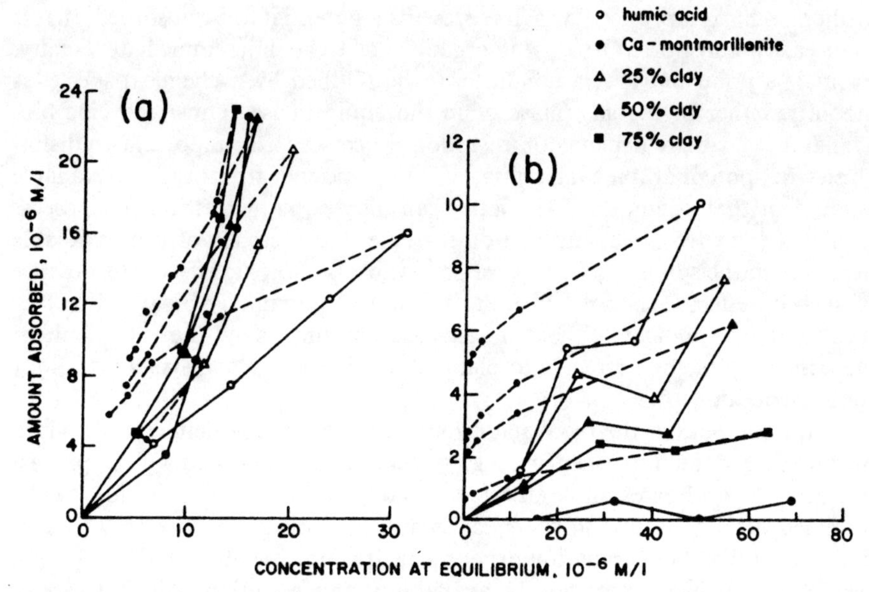

Fig. 6.18a,b. Influence of pH on adsorption isotherms (———) and desorption isotherms (- - - -) of terbutryne on and from a humic acid, a Ca-montmorillonite kand their mixtures with 25, 50 and 75% clay content. **a** pH = 3–3.2; **b** pH = 5.6–6. (Gaillardon et al. 1977)

materials in an acid and a weak acid environment. The authors reached the conclusion that, independently of the acidity of the medium, the desorption isotherm exhibited hysteresis in the case of humic acid or clay-humic acid mixtures. This phenomenon is more accentuated in an acidic medium. No evident hysteresis was observed in the case of the adsorption-desorption isotherm of the atrazine on clay surfaces (montmorillonite). The extent of the hysteresis was greater in a weakly acid medium (pH = 5.6–6) than in a strongly acid medium (pH = 3–3.2), probably due to nonionic interactions.

6.3.2.2 Observations on Bound Residues

The distribution of atrazine-bound residues in soil has been described as a function both of the localization of organic matter in granulometric soil fractions and of time (Barriuso and Koskinen 1995). Figure 6.19 illustrates the experimental results and shows that the >200-μm fraction, a less humified one, plays an important role, which is time-variable.

6.3.2.3 Bioavailability of Bound Residues

Another interesting question is the bioavailability of bound residues. A compound is bioavailable when it can be absorbed by a living organism in a given medium. This is a property which represents a potential to be absorbed, that is to be easily transported to a living organism or to be in its immediate vicinity. From this point of view, a condition to be fulfilled by a chemical species is mobility, either in the gas phase or in the liquid phase. Thus, pesticide bioavailability strongly depends on retention processes. It is important to distinguish the potential bioavailability of a compound from the bioavailable amount of that compound. The latter can only be known through a process mediated by a living organism. For example, the presence of a bioavailable herbicide in the soil is attested by an observable phytotoxic effect. The absence of such an effect does not prove the absence of potentially bioavailable herbicides; it may indicate either an adsorbed amount below the toxic limit or inaccessibility of the herbicide to plant roots (due to the soil structure and/or root distribution).

For a compound, the bioavailable amount can only be defined for a given medium, a given living organism, a given set of conditions, and a given process (Calvet 1989a). For example, the bioavailable amount of a herbicide, as determined through a phytotoxic effect, is not the same as that which would be determined from a degradation experiment. It is not possible to give a general relationship between pesticide bioavailability and retention. All that can be done is to refer to a given process for a compound or a family of compounds. Even in this limited situation, it is not a simple matter to deal with this problem because rather few experimental data are available today. Some examples will be discussed below.

Pesticide bioavailability can be assessed through:

- toxic effects against living organisms (plant, soil microflora, fauna, etc),
- biodegradation,

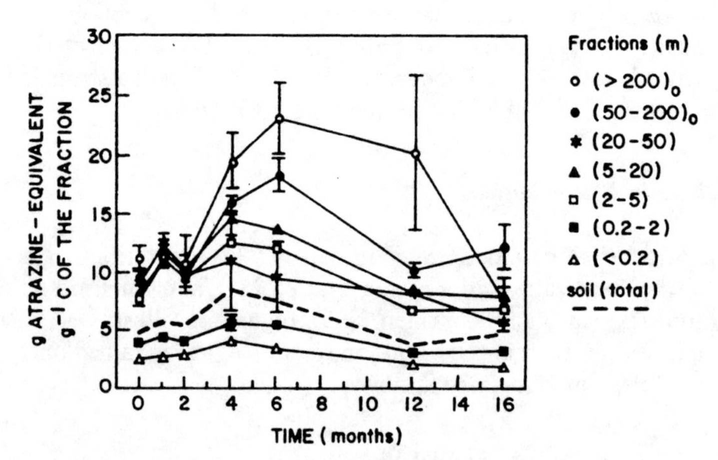

Fig. 6.19. Dynamics of atrazine-bound residues in soil fraction during incubation under field conditions. The results are expressed as mass of equivalent atrazine calculated on the basis of specific radioactivity of atrazine applied. (After Barriuso and Koskinen 1995)

- the health hazard for humans, arising from consumption of contaminated plants, animals, or water.

As we have already seen, a pesticide molecule can be in a fluid phase (soil solution and/or soil atmosphere), sorbed on colloidal materials, and retained in an unextractable form. Molecules in solution are immediately bioavailable during a short period of time (8–10 days), as shown by Stalder and Pestemer (1980) for several herbicides. In the case of long periods, bioavailable amounts probably depend on the processes of transport to the living organisms, as was earlier stated by Bailey and White (1964). This is the simplest situation, compared with those of sorbed and unextractable molecules. As we have seen, bound residues partly comprise parent molecules and partly degradation products. The question of their bioavailability is of some concern, and it has recently been discussed by Bollag et al. (1992) and Scheunert et al. (1992).

Bioavailability of pesticide residues can be analyzed through their toxic effects. A first example is found in the work of Bollag et al. (1992), who observed that phenolic compounds were not toxic for *Rhizoctonia practicola* when the culture medium contained extracellular lactase. They attributed this observation to the ability of the enzyme to transform or to polymerize the parent compound and then to decrease its toxicity. Another example is provided by the results of Scribner et al. (1992), who studied the bioavailability of aged simazine. They showed that simazine residues were less available for plant uptake and microbial degradation. Atrazine desorption seems to display the same kind of behavior, according to the results of Barriuso et al. (1992a).

As shown in Fig. 6.20, the mineralization rate of bound residues depends on the nature of the organic solid phase. (Benoit 1994).

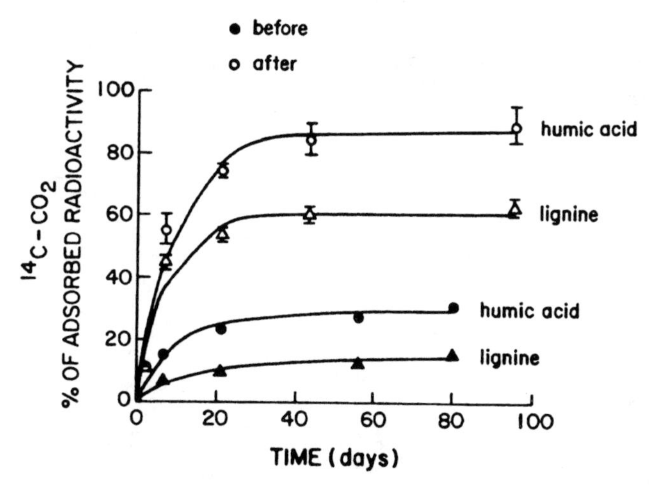

Fig. 6.20. Kinetics of the mineralization of 2.4-D sorbed on humic acid (circles) and lignin (triangles), before (●) and after (○) exhaustive extraction by 0.01 M $CaCl_2$. *Full line* corresponds to first-order kinetics. (After Benoit 1994)

From a general point of view, bound residues probably have a very low bioavailability, if any, for plants and soil organisms. They are retained through covalent linkages and by entrapment in organic matter aggregates, and they can only be released by drastic treatments such as chemical extraction or pyrolysis. It is certainly unlikely to find natural phenomena in soil with such strong effects, so that only slow release of bound residues is probable. Some doubts still remain about a possible delayed hazard. The first reason is that several observations show that bound residues can be recovered quite a long time after a pesticide application. For example, Andreux et al. (1993) were able to extract from soil a chloroaniline which represented 15% of ^{14}C-labeled buturon applied 16 years previously. The second reason is that retention by organic matter does not necessarily lead to the immobilization of the pesticide or its degradation products. When retention takes place on hydrosoluble organic polymers, pesticide water solubility can be enhanced (Chiou et al. 1986; Barriuso et al. 1992a) and transport may be favored (Ballard 1971; Bouchard 1989; Andreux et al. 1993). This may contribute to increased bioavailability of some compounds to living organisms, particularly in aquatic media.

6.3.3 Organo-Clays

The use of organo-clays in wastewater treatment as alternatives to activated C for sorbing pollutants, and in clay liners, is of great interest to soil and environmental scientists. There is a lack of information, however, on the kinetics and mechanisms of organo-clay interactions with pollutants. The time-dependent adsorption and desorption of phenol and aniline on hexadecyl-trimethyl-ammonium-montmorillonite (HDTMA-montmorillonite) from aqueous solutions were determined by Zhang and Sparks (1993). Results on kinetics of phenol and aniline adsorption-desorption are presented in Fig. 6.21a and b. Phenol could not be completely desorbed from the

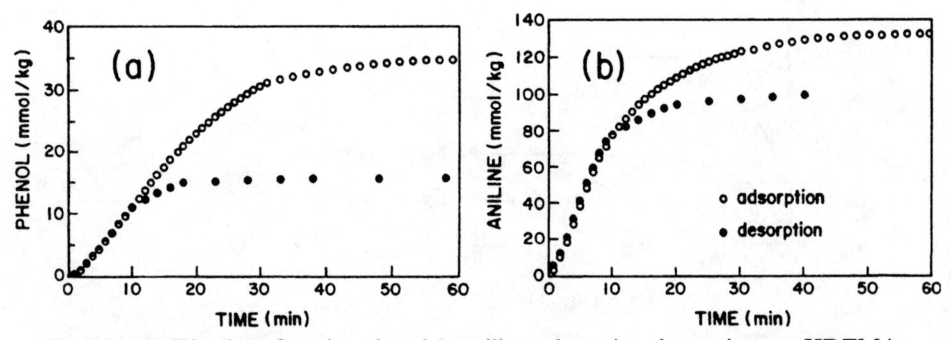

Fig. 6.21a,b. Kinetics of **a** phenol and **b** aniline adsorption-desorption on HDTMA-montmorillonite at an influent concentration of 0.01 mol l^{-1}. (After Zhang and Sparks 1993)

HDTMA-montmorillonite; about 42% of the adsorbed phenol was subsequently desorbed in 30 min. Desorption is relatively rapid, with little occurring after 30 min. These data indicate that phenol adsorption-desorption on HDTMA-montmorillonite is nonsingular, i.e., there is irreversibility. Desorption kinetics were investigated at higher aniline concentration. About 70% of adsorbed aniline was desorbed within 30 min, proving again the nonsingularity of the process.

From the data of Zhang and Sparks (1993), it may be observed that in both cases, the adsorption and desorption rates are the same during the initial stages of the process, and start to exhibit nonsingular behavior during the remaining time of contact. However, both phenol and aniline adsorption on HDTMA-montmorillonite were partially reversible. These results may indicate that adsorption does not follow a single mechanism. Modification of the clay surface by adding a large organic molecule such as HDTMA complicates the nature of the surface, which may show both hydrophilic and hydrophobic parts, as well as electrostatic properties. The properties of the modified surface vary with coverage of the surface with large molecules. Therefore, adsorbed phenol or aniline may be retained via various mechanisms. Some of the adsorbed molecules may be weakly bound to the surface and can be desorbed, while other molecules may be strongly bound to various organic parts of the adsorbent and cannot be easily removed. This behaviour, observed in a short time, is contrary to that observed by Keren et al. (1994) in borate adsorption on pyrophyllite, in which the adsorption-desorption reversibility was observed only after a long period of contact. In this last case, we are dealing with a homogeneous inorganic adsorbing surface and the mechanism of adsorption is completely different. The different mechanisms of adsorption in the two cases described control the kinetics of pollutant desorption.

References

Andreux F, Sheunert I, Adrian P, Mansour M (1993) The binding of pesticide residues to natural organic matter, their movement, and their bioavailability. In: Manour M (ed) Fate and prediction of environmental chemicals in soils, plants and aquatic systems. Lewis, Boca Raton

Bailey GW, White JL (1964) Review of adsorption and desorption of organic pesticides by soil colloids, with implications concerning pesticide bioavailability. Agric Food Chem 12: 324–382

Ballard TM (1971) Role of humic carrier substances in DDT movement through forest soil. Soil Sci Soc Am Proc 35: 145–147

Barriuso E, Koskinen WC (1995) Incorporation of non-extractable atrazine residues into soil size fractions as a function of time. Soil Sci Soc Am J (in press)

Barriuso E, Baer U, Calvet R (1992a) Dissolved organic matter and adsorption-desorption of dimefuron, atrazine and carbetamide by soils. J Environ Qual 21: 359–367

Barriuso E, Koskien W, Sorenson B (1992b) Modification of atrazine desorption during field incubation experiments. Sci Total Environ 123/124: 333–344

Barrow NJ (1979) The description of desorption of phosphate from soil. J Soil Sci 30: 259–270

Barrow NJ (1983) A mechanistic model for describing the sorption and desorption of phosphate by soil. J Soil Sci 34: 733–750

Barrow NJ (1986) Testing a mechanistic model. II. The effects of time and temperature on the reaction of zinc with soil. J Soil Sci 37: 277–286

Bartlett RJ, James BR (1979) Behavior of chromium in soils. J Environ Qual 8: 31–35

Benoit P (1994) Rôle de la nature des matières organiques dans la stabilisation des résidus de polluants organiques dans les sols. Thèse Institut National Agronomique Paris-Grignon, Paris

Bertin G, Schiavon M, Portal JM, Andreux F (1991) Contribution to the study of non-extractable pesticide residues in soils: incorporation of atrazine in model humic acids prepared from catechol. In: Berthelin J (ed) Diversity of environmental biogeochemistry. Elsevier, Amsterdam, pp 105–110

Bollag JM, Loll MJ (1983) Incorporation of xenobiotics in soil humus. Experientia 39: 1221–1231

Bollag JM, Myers CJ, Minard RD (1992) Biological and chemical interactions of pesticides with soil organic matter. Sci Total Environ 123/124: 205–217

Bouchard DC, Enfield CG, Piwoni MD (1989) Transport process involving organic chemicals. In: Sawhney BL, Brown K (eds) Reactions and movement of organic chemicals in soils. SSSA Spec Publ 22. Soil Science Society of America, Madison

Bowman BT, Sans WW (1985) Partitioning behavior of insecticide in soil-water system: desorption hysteresis effect. J Environ Qual 14: 270–273

Brummer G, Tiller KG, Herms U, Clayton PM (1983) Adsorption-desorption and/or precipitation dissolution processess of zinc in soils. Geoderma 31: 337–354

Brummer GW, Gerth J, Tiller KG (1988) Reaction kinetics of the adsorption and desorption of nickel, zinc, and cadmium by goethite. I. Adsorption and diffusion of metals. J Soil Sci 39: 37–52

Brusseau ML, Rao PSC (1989) Sorption kinetics during organic contaminant transport in porous media. Crit Rev Environ Control 19: 33–99

Brusseau ML, Jessup RO, Rao PSC (1991) Non-equilibrium sorption of organic chemicals: elucidation of rate limiting processes. Environ Sci Technol 25: 134–142

Bunzl K, Schmidt W, Sansoni B (1976) Kinetics of ions exchange in soils organic matter. IV. Adsorption and desorption of Pb^{2+}, Cu^{2+}, Cd^{2+}, Zn^{2+} and Ca^{2+} by peat. J Soil Sci 17: 32–41

Calderbank A (1989) The occurrence and significance of bound pesticide residues in soil. Rev Environ Contam Toxicol 108: 69–103

Calvet R (1989a) Analyse du concept de biodisponobilité d'une substance dans le sol. Sci Sol 26 (3): 183–201

Calvet R (1989b) Recherches sur l'influence des molecules organiques hydrosolubles sur l'adsorption de cations metalliques traces. Adsorption du cadmium et du cuivre sur une montmorillonite et une kaolinite. Rapport de recherche, Ministère de l'Environnement No 86–192, Paris-Grignon, France

Chassin P, Leberre B (1979) Influence des substances humiques sur les proprietés d'hydratation des argiles-V-deshydratation des acides humiques. Clay Minerals 14: 193–200

Chiou CT, Malcolm RL, Brinton TI, Kile DE (1986) Water solubility enhancement of some organic pollutants and pesticides by dissolved humic and fulvic acids. Environ Sci Technol 20: 502–508

Curl RL, Keoleian GA (1984) Implicit adsorbate model for apparent anomalies with organic adsorption on natural sorbents. Environ Sci Technol 18: 910–912

Gaillardon R, Calvet R, Terce M (1977) Adsorption et désorption de la terbutryne par une montmorillonite-Ca et des acides humiques seuls ou en mélange. Weed Res 17: 41–48

Giles CH, Smith D, Huitson A (1974) A general treatment and classification of the solute adsorption isotherm. J Colloid Interface Sci 47: 755–765
Hingston FJ (1981) A review of anion adsorption. In: Anderson MA, Rubin AJ (eds) Adsorption of inorganics at solid-liquid interfaces. Ann Arbor Science, Ann Arbor
Hingston FJ, Posner AM, Quirk JP (1974) Anion adsorption by goethite and gibbsite. II. Desorption of anions from hydrous oxide surfaces. J Soil Sci 25: 16–26
Hodges SC, Johnson GC (1987) Kinetics of sulfate adsorption and desorption by Cecil soil using miscible displacement. Soil Sci Soc Am J 51: 323–327
Keren R, Grossl PR, Sparks DL (1994) Equilibrium and kinetics of borate adsorption-desorption on pyrophyllite in aqueous suspension. Soil Sci Soc Am J 58: 1116–1122
Khan SU (1982) Bound pesticide residues in soil and plant. Residue Rev 84: 1–25
Koskinen WC, Chang HH (1982) Elimination of aerobic degradation during characterization of pesticide adsorption-desorption in soil. Soil Sci Soc Am J 46 (2): 256–269
Koskinen WC, Chang HH (1983) Effects of experimental variables on 2,4,5-T adsorption-desorption in soil. J Environ Qual 12: 325–330
Koskinen WC, O'Connor GA, Cheng HH (1979) Characterization of hysteresis in the desorption of 2,4,5-T from soils. Soil Sci Soc Am J 43: 871–874
Lehmann, RG Harter RD (1984) Assessment of copper-soil bond strength by desorption kinetics. Soil Sci Soc Am J 48: 769–772
McBride MB (1991) Processes of heavy and transition metal sorption by soil minerals. In: Bolt GH, De Boot MF, Hayes MHB, McBride MB (eds) Interactions at the soil colloid-soil solution interface. Kluwer, Dordrecht
McLaren RG, Williams JG, Swift RS (1983) Some observation on the desorption and distribution behavior of copper with soil components. J Soil Sci 34: 3325–3331
Mclaren RG, Lawson DM, Swift RC (1986) Sorption and desorption of cobalt by soils and soil components. J Soil Sci 37: 413–426
Mengel K, Scherer HW (1981) Release of non-exchangeable (fixed) soil ammonium under field conditions during the growing season. Soil Sci 131: 226–232
Mingelgrin U, Gerstl Z (1993) A unified approach to the interactions of small molecules with macrospecies. In: Beck AJ, Jones KC, Hayes MHB, Mingelgrin U (eds) Organic substances in soil and water. Royal Society of Chemistry, Cambridge, pp 102–128
Nommik H (1957) Fixation and defixation of ammonium in soill. Acta Agric Scand 7: 395–436
Obihara CH, Russel EW (1972) Specific adsorption of silicate and phosphate by soils. J Soil Sci 23(1): 105–117
Padmanabham M (1983) Adsorption-desorption behavior of copper (II) at the goethite solution interface. Aust J Soil Sci 21: 309–320
Page AL, Burge WD, Gange TJ, Garber MJ (1967) Potassium and ammonium fixation by vermiculitic soils. Soil Soc Am Proc 31: 337–341
Pignatello JJ (1989) Sorption dynamics of organic compounds in soils and sediments. In: Sawhney BL, Brown K (eds). Reactions and movement of organic chemicals in soils. SSSA Spec Publ 22: 45–81
Ponec V, Knor Z, Cerny S (1974) Adsorption on solids. Butterworths, London 234 pp
Rao PSC, Davidson JM (1980) Estimation of pesticide retention and transformation parameters required in nonpoint source pollution models. In: Overcash MR, Davidson JM (eds) Environmental impact of nonpoint source pollution. Ann Arbor Science, Ann Arbor, pp 23–67
Ross DS, Sjogren RE, Bartlett RJ (1981) Behavior of chromium in soils: IV. Toxicity to microorganisms. J Environ Qual 10: 145–148
Saltzman S, Kliger L, Yaron B (1972) Adsorption-desorption of parathion as affected by soil-organic matter. J Agric Food Chem 20: 1224–1226
Scherer HW (1993). Dynamics and availability of the non-exchangeable NH_4-N, a review. Eur J Agron 2 (3): 149–160

Scheunert I, Mansour M, Andreux F (1992) Binding of organic pollutants to soil organic matter. Int J Environ Anal Chem 46: 189–199
Scribner SL, Benzing TR, Shoabaiu Sun CC, Boyd SA (1992). Desorption and bioavailability of aged simazine residues in soil from a continuous corn field. J Environ Qual 21: 115–120
Singh R, Gerritse RG, Aylmore LAG (1989) Adsorption-desorption behavior of selected pesticides in some western Australian soils. Aust J Soil Res 28: 227–243
Stalder L, Pestemer W (1980) Availability to plants of herbicide residues in soil. Part I: A rapid method for estimating potentially available residues of herbicides.Weed Res 20: 341–347
Stollenwerk KC, Grove DB (1985) Adsorption and desorption of hexavalent chromium in an alluvial aquifer near Telluride, Colorado J Environ Qual 14: 150–155
Swift RS, McLaren RG (1991) Micronutrient adsorption by soils and soil colloids. In: Bolt GH, De Boot MF, Hayes MHB, McBride MB (eds) Interactions at the soil colloid-soil solution interface. Kluwer, Dordrecht
van der Zee SE, van Riemsdijk WH (1991) Model for the reaction kinetics of phosphate with oxides and soils. In: Bolt GH, De Boot MF, Hayes MHB, McBride MB (eds) Interactions at the soil colloid-soil solution interface. Kluwer, Dordrecht
van Dijk H (1971) Colloid chemical properties of organic matter. In: Soil biochemistry, 2, 16. Marcel Dekker, New York
Wershaw RL (1986) A new model for humic materials and their interactions with hydrophobic organic chemicals in soil-water or in sediment-water systems. J Contam Hydrol I: 29–45
Zhang P-C, Sparks DL (1993) Kinetics of phenol and aniline adsorption and desorption on organo-clay. Soil Sci Soc Am J 57: 340–345

CHAPTER 7

Transformation and Metabolite Formation

After reaching the soil medium, the pollutants are partitioned among the solid, liquid, and gaseous phases. Part of the chemical fraction is transported into the gaseous phase and is lost in the atmosphere. The remaining portion may be dissolved in the soil solution or retained within the soil solid phase as adsorbed molecules, entrapped ganglia, or precipitates. These chemicals are submitted to chemical, photochemical, and surface-induced reactions, leading to their transformation in the course of time. As a result of pollutant transformation, various metabolites – which may be more or less toxic than the parent material – are introduced into the soil medium.

Transformation processes in soils are the results of chemical, physical, abiotic, and/or biologically induced processes and comprise mainly:

1. those leading to chemical degradation and metabolite formation in the liquid phase and/or at the solid-liquid interface;
2. those resulting in complexation of chemicals, leading to a change in their physicochemical properties; and
3. those occurring in the liquid phase, in aqueous and/or water-immiscible mixtures which result from changes in the liquid composition during its contact with the soil medium.

From the point of view of soil pollution, we cannot restrict our consideration of the "transformation" of chemicals to their molecular changes only, but we have to include all the transformations of pollutants relevant to their behavior in soil.

7.1 Decomposition Rate

Independently of the factor inducing the decomposition of a chemical in the soil medium, the "transformation" process is controlled by a specific rate of decomposition which is affected by the environmental factors. Hamaker (1972) and Sparks (1988) showed that two basic patterns may characterize the rate of a decomposition process:

1. The rate is proportional to some power of the concentration as follows:

$$\text{Rate} = -dC/dt = kC^n \quad -dC/dt = kC \quad \text{if } n = 1, \tag{7.1}$$

where c is concentration, k is a rate constant, and n is the order of the reaction. This has been referred to as a "power rate model"; it becomes a first-order kinetic equation when $n = 1$.

2. The rate depends directly upon the concentration, and simultaneously upon the sum of concentration and additional terms. In the case where "other terms" are a single constant, (7.1) can be written:

$$\text{Rate} = \frac{dC}{dt} = \frac{k_1 C}{k_2 + C} = \frac{k_1}{k_2} C \quad \text{if } k_2 \gg C, \tag{7.2}$$

where k_1 is a maximum rate, approached with increasing concentration, and k_2 is a pseudoequilibrium constant (pseudo, because as the reaction proceeds, it constantly unbalances the equilibrium represented by the constant). This has been referred to as a "hyperbolic rate model."

The constant, k_1, depends upon the concentration of the catalytic agent. In the case of an enzyme (E), thc Michaelis-Menten approach expresses the kinetics by Eq (7.3), which is a hyperbolic one

$$\frac{dC}{dt} = \frac{V_m E C}{K_m + C}, \tag{7.3}$$

where E is the concentration of enzyme and $V_m E$ corresponds to k_1. On the other hand, hyperbolic rate models may apply to surface-catalyzed reactions as well as to enzyme kinetics.

Hamaker (1972) emphasized the fundamental difference between the mechanisms behind these two basic rate models. The power rate law arises from the situation where reaction follows activation of molecules, singly or in groups, by collision. Concentration affects the rate by determining how often collisions and, therefore, activations can occur. This type of kinetics usually applies to reactions in homogeneous solution. The hyperbolic rate law reflects an equilibrium between a reactant and a catalyst. The complex formed from the two is then capable of undergoing reaction with release of the products and regeneration of the catalyst for further activity. The rate depends not only upon the concentrations of the reactant and the catalyst, but also upon the equilibrium constant for the formation of the complex between them that is the true reactant: k_2 in Eq. (7.2).

The half-life ($t_{1/2}$) of a reaction is the time required for half of the original reactant to be conserved. A first-order reaction has a half-life that is related only to k and is independent of the concentrations of the reacting species. Sparks (1988) indicated that the half-life depends on reactant concentration for a second-order kinetic reaction equation. In this case, the corresponding equation is:

$$\frac{dy}{dt} = k(a - y)^2, \tag{7.4}$$

Table 7.1. Summary of some rate equations for abiotic transformations. (Wolfe et al. 1990)

Type of equation	Differential form ($-dC/dt=$)	Integrated form[a]	Half-life	Linear form in t
Zero order	k	$C=C(0)-kt$ (at $C>0$)	$0.5C(0)/k$	$C=C(0)-kt$
First order	kC	$C=C(0)e^{-kt}$	$-Ln(0.5)/k$	$Ln(C)=Ln[C(0)]-kt$
Simple second order	kC^2	$C=C(0)/[1+C(0)kt$	$1/[kC(0)]$	$1/C=1/C(0)+kt$
Hyperbolic	$k_1C/(k_2+C)$	$e^C C^{k_2}=e^{[C(0)-k_1t]}\times C(0)^{k_2}$	$[0.5C(0)-k_2\times Ln(0.5)]/k_1$	$A(C)=A[C(0)]-k_1t$ where $A=C+k_2Ln(C)$

[a]C, concentration, t, time, and k_1, constants.

where a is the initial concentration of the reacting species and y, the concentration of the resulting product.

Upon rearrangement and integration we obtain

$$\frac{1}{(a-y)}-\frac{1}{a}=kt. \tag{7.5}$$

Replacing $(a-y)$ by $a/2$, the following expression is obtained:

$$t_{1/2}=\frac{1}{ak}. \tag{7.6}$$

Table 7.1 summarizes the rate expressions that are in use to describe abiotic pollutant transformations in the soil system.

7.2 Abiotically Induced Transformation

Both inorganic and organic compounds can be transformed in the soil medium as a result of chemical processes. For didactic reasons, we can distinguish the following abiotic transformation pathways: (1) hydrolysis and redox reactions, (2) surface-enhanced degradation, (3) photodecomposition, (4) complex formation, and (5) mixture transformations. We will discuss the abiotically induced transformation separately for the soil-liquid phase and the soil-liquid interface. Whereas the transformation in the liquid phase encompasses both organic and inorganic toxic chemicals, at liquid-solid interfaces it is mainly the organic compounds which are affected.

7.2.1 Transformation in Soil Water

The abiotically induced transformation of toxic chemicals in the soil liquid phase is a result of hydrolysis, redox, and photolysis reactions, complexation, and mixture transformation.

7.2.1.1 Hydrolysis

A bond-making-bond-breaking process characterizes hydrolysis, which may be described by the equation.

$$RX + HO^- \xrightarrow{H_2O} ROH + X^- . \tag{7.7}$$

Wolfe (1989) shows that for abiotic hydrolysis the general expression for the observed global rate constant, k_{obs}, is given by

$$k_{obs} = k_H[H^+] + k_{OH}[OH^-] + k_w + \sum_i (k_{HA}[HA] + k_A[A^-]) , \tag{7.8}$$

where k_H and k_{OH} are the specific acid-base catalyzed second-order rate constants, respectively; k_w is the neutral hydrolysis rate constant; and k_{HA} and k_A are the general acid-base catalyzed hydrolysis rate constants, respectively. In Eq. (7.8), $[H^+]$ and $[OH^-]$ are the hydrogen and hydroxyl ion concentrations (activities), respectively, and $[HA]$, $[A^-]$ are the concentrations (activities) of the ith pair of general acids and bases in the reaction mixture. k_{obs} can include contributions from acid- or base-catalyzed hydrolysis, nucleophilic attack by water, or catalysis by buffers in the reaction medium.

It is to be emphasized that, whereas in biologically induced hydrolysis, an initial lag period generally occurs, the abiotic hydrolysis under buffered pH conditions does not exhibit a lag phase. The lag phase can be observed, however, in the abiotic degradation, where, for example, an autocatalytic reaction takes place.

The abiotic hydrolysis of pollutants in water is pH-dependent. The predominant pathways in the soil water are acid-catalyzed, base-moderated, and neutral (pH-independent) hydrolysis. The acid-catalyzed hydrolysis reaction rate increases with decrease in pH, the rate being proportional to the hydrogen ion activity. The increase of pH as a direct proportional augmentation of the hydroxyl ion activity leads to a base-mediated hydrolysis process. In the case of a neutral hydrolysis reaction, the rates are independent of pH.

Both inorganic (e.g., metals) and organic chemicals are subjected to a hydrolysis reaction in an aqueous solution. Harris (1982) showed that rate of hydrolysis of an organic compound, for example, is affected by pH, by specific acid-base effects, or by a change in compound speciation. A change in the pH can shift the equilibrium in favor of the charged or uncharged species which often have different hydrolysis rate constants.

The rate of hydrolysis is affected by temperature, the relation between temperature and the rate constant being expressed by the "Arrhenius equation".

$$k_{obs} = A\, e^{-E_a/RT}, \tag{7.9}$$

where A is a term depending on collision frequency, E_a(Jcal mol^{-1}) is the activation energy, R is the universal gas constant, and T(K) is the reaction temperature.

If, in a water phase, a number of pollutants with different properties are present, the hydrolysis of a chemical could be affected by the presence of a *cosolute*. This is the case of dissolved metals or humic acids acting on the hydrolysis of organic toxic elements present in the same water phase. This effect is, however, of little significance. Organic and inorganic acids, along with their conjugated bases, are usually common in natural waters and exhibit a wide variation in structure and concentration. Because of this, a general approach is needed to assess their contribution to hydrolysis. Perdue and Wolfe (1983) showed that by using a mathematical approach, the maximum predicted buffer-catalysis contribution will be <10% of the uncatalyzed process.

7.2.1.2 Redox Reactions

As a result of the electron transfer, there are changes in the oxidation states of reactants and products. Reductants and oxidants are defined as electron donors and proton acceptors. Because there are no free electrons, every oxidation is accompanied by a reduction and vice versa.

In aqueous solutions the electron and proton activities are defined by the pH

$$pH = -\log\{H^+\}, \tag{7.10}$$

which measures the relative tendency of a solution to accept or transfer protons. In an acid solution, this tendency is low and in an alkaline solution it is high. Similarly, we can define an equally convenient parameter for the redox intensity:

$$p\varepsilon = -\log\{e\}. \tag{7.11}$$

$p\varepsilon$ gives the (hypothetical) electron (e) activity at equilibrium and measures the relative tendency of a solution to accept or transfer electrons.

In the soil solution, only a few elements are prominent participants in redox processes (C, N, O, S, Fe, Mn). The abiotic oxidation of organic pollutants occurs in the soil solution under aerobic conditions, and abiotic reduction may occur in saturated soils and sediments (Macardy et al. 1986). In general, the redox processes in the soil-liquid phase are light-mediated on the soil surface and biologically mediated within the soil profile. This aspect will be described separately. More data on abiotic redox processes in a water system – which could be applied to a soil solution – are to be found in the book by Stumm and Morgan (1981) on aquatic chemistry.

7.2.1.3 Photolysis

The chemical environment in the top 0.5 cm of the soil is distinctly different from that in the bulk soil, and is also affected by the fact that this layer could be affected by solar radiation. This implies a possible photochemical transformation of the pollutants dissolved in the soil liquid phase.

Direct and indirect photoreactions characterize the processes at the soil-water interface. The direct photolysis process involves the direct absorption of light by the substrate.

$$P + \text{light} \quad P^* \longrightarrow \text{products}, \tag{7.12}$$

where P is a photoreactive substance and P^* is its excited state that reacts to form products. Indirect photoprocesses are initiated through light absorption by photoreactive components symbolized by "sens", such substances being referred as sensitizers.

$$\text{sens} + \text{light} \longrightarrow \text{sens}^*$$

$$\text{sens}^* + P \longrightarrow \text{products} \tag{7.13}$$

The rate of any type of photoreaction depends upon the rate of light absorption by the photoactive species I_a and the quantum efficiency of the reaction ϕ, and is expressed by:

$$\text{Rate} = I_a\,\phi. \tag{7.14}$$

Zeep and Cline (1977) showed that the light absorption rate by an object in a water body depends upon the solar spectral irradiance as it is transmitted down through the water, and the geometry of the light field.

The presence of natural products – like humus substances – in the soil liquid phase acts as a natural sensitizer. The electronically excited molecules are capable of greatly accelerating the process and determining a number of light-induced transformations of toxic organic chemicals. Senesi and Chen (1989) proposed the following key energy transfer steps for a humic substance which becomes an excited sensitizer (HS^*) and interacts with an energy acceptor

$$HS \xrightarrow{h\nu} {}^1HS^* \longrightarrow {}^3HS^*$$

$${}^3HS^* \longrightarrow HS + \text{heat}$$

$${}^3HS^* + TOC \longrightarrow TOC^* + HS$$

$$TOC^* \quad \text{photoproducts} \tag{7.14}$$

$${}^3HS^* + O_2 \longrightarrow HS + {}^1O_2$$

$${}^1O_2 + TOC \longrightarrow TOC + O_2.$$

Light absorption raises the photosensitizer molecules (HS), to their first excited states ${}^1HS^*$, which are very short-lived and are transformed in part to excited triplet states ${}^3HS^*$, which are, in turn, considerably longer-lived. Such triplets may, in part, decay to the ground state, or transfer energy to the toxic organic chemical (TOC). Thus forming its triplet state (TOC*), which then gives its photoproducts or transfers energy to ground state triplet oxygen, to produce excited singlet molecular oxygen $^1O_2$, which is a powerful oxidant and may, in turn, decay back to its ground state or react rapidly with an acceptor (TOC), thus forming its photooxidation products.

7.2.1.4 Coordinative Relationship

If the coordinative relationships are changed as a result of a change of coordinative partner or of the coordination number, a transformation occurs in the properties of the pollutant. Any combination of cations with molecules or with anions containing free pairs of electrons (bases) is called coordination (or complex formation); it can be electrostatic, covalent, or a mixture of both. The metal cation will be called the central atom, and the anions or molecules with which it forms a coordination compound will be referred to as ligands.

The actual form in which a molecule or ion is present in solution emphasizes a specific *chemical speciation*. Since the soil solution contains a broad spectrum of dissolved and colloidally dispersed natural products, it is possible that in the soil liquid phase the same molecule exists in one or more species. By complexation with the various components of the soil solution, the initial properties of a toxic molecule can be changed. These transformations involve such properties of toxic chemicals as adsorption on the soil solid phase, transport into the soil profile, and persistence. Transformation of the pollutant properties as the result of a change in their coordinative relationship encompasses both inorganic and organic toxic chemicals.

It is operationally difficult to distinguish between dissolved and colloidally dispersed substances. Colloidal metal-ion precipitates, for example, may occasionally have particle sizes smaller than 100 Å, sufficiently small to pass through a membrane filter. Organic substances can assist markedly in the formation of stable colloidal dispersions. Information on the types of species encountered under different chemical conditions (types of complexes, their stabilities, and rates of formation) is a prerequisite to a better understanding of the transformation in the properties of toxic chemicals in the soil water.

7.2.1.5 Liquid Mixture Transformations

First, we will discuss the behavior of water-immiscible ligands. Let us consider a nonaqueous pollutant liquid (NAPL) mixture spill on a land surface. Redistribution of the NAPL mixture within the soil volume will occur with time. The highly volatile fraction of the NAPL mixture will be transported in the gaseous phase of the porous medium and subsequently into the atmosphere, adsorbed onto the solid phase, or dissolved in the soil aqueous phase. The less volatile components of NAPL mixtures will be transported as a liquid into the porous medium. During their transport through the unsaturated zone, volatilization, adsorption onto the solid phase, and dissolution into the soil water will occur. Under these conditions, the soil-liquid became a multiphase system. In this system, we can recognize two distinct liquid phases: the first is a non-aqueous pollutant liquid phase (NAPL) with chemical properties different from the original NAPL; the second is a mixed aqueous solution of partially miscible organic compounds dissolved into the aqueous phase from the NAPL mixture.

The *chemodynamic properties* of the multiphase liquid are changed, with respect to the initial state of their components. In the case of NAPL mixture, both physical and chemical properties of the initial liquids are changed. A second type of liquid transformation is that of a water with a given quality when it mixes with a second water of different chemical composition. Here, we are dealing with the mixing of two electrolyte solutions – one, for example, of a high Na content and low Ca concentration (SAR = 25), and a second one with a low Na content and a high Ca concentration (SAR = 2.5). The result is an aqueous solution with a sodium adsorption ratio different from the initial values, which depends on the ratio between the volumes of water mixed together. We can say that the chemodynamic properties of the resulting water have changed and that this affects the soil quality accordingly.

7.2.2 Transformation at the Solid-Liquid Interface

Transformations at the solid-liquid interface include the abiotic transformation of pollutants in the proximity of the charged solid surface, which presents an environment different from that of the bulk solution.

7.2.2.1 Surface-Induced Transformation of Organic Compounds

The spatial distributions of ions and charge within polarizable species are strongly affected by the electric field emanating from charged surfaces. As a result, some organic compounds in direct contact with the surfaces can undergo transformations enhanced by the adsorption sites. It is accepted in the literature to refer to the enhanced transformation of toxic organic chemical on the adsorption sites as a *catalytic* process.

Catalytic reactions on clay surfaces were reported in the reviews of Chaussidon and Calvet (1974), Mingelgrin and Prost (1989), and Wolfe et al. (1990), and the reader is referred to these articles for detailed information.

The catalytic reactions on clay surfaces include the hydrolysis of the organic molecules: hydrolysis which is affected by the type of clay and the type of saturating cation of the clay involved in the reaction. Mg-montmorillonite exhibits, for example, much weaker catalytic capacity than Ca-montmorillonite. Ca-beidellite and Ca-nontronite exhibit a lower capacity than Ca-montmorillonite (Mortland and Raman 1967). Chaussidon and Calvet (1965) showed that amines adsorbed on montmorillonite undergo chemical transformation upon dehydration of the clay. Catalytic hydroxylation of an organic molecule (atrazine) on H-montmorillonite involved the substitution of a chlorine atom by a hydroxyl ion, the degradation product apparently remaining adsorbed on the clay as the keto form of the protonated hydroxy analog (Russell et al. 1968). Redox transformations of aromatic amines can occur on clay surfaces in the presence of metal ions, which are good reducing or oxidizing agents (Theng 1974; McBride 1985).

Oligomerization reactions can be enhanced or inhibited by the clay surface. Carbonium ion formation by surface protons leads to the polymerization of styrene, and it has been proved that H and H organic C-montmorillonite induce a catalytic polymerization (Solomon and Rosser 1965; Lahav et al. 1978). The organic adsorbate containing an unsaturated C can interact with the clay acidic surface to form a carbonium ion which, being unstable, acts as an intermediate in numerous transformations (Fusi et al. 1983).

Rearrangement reactions are also catalyzed by the clay surface. Mingelgrin and Saltzman (1979) described the rearrangements of an organophosphate (parathion) when adsorbed on montmorillonite or kaolinite in the absence of a liquid phase. The rate of the rearrangement reaction increased with the polarization of the water of hydration of the exchangeable cation. Table 7.2 (after Wolfe et al. 1990) summarizes a series of reactions catalyzed by the clay surfaces, as reported in various research papers.

Organic matter is also able to enhance abiotic transformations of organic compounds. Stevenson (1982) showed that organic-matter enhancement of the transformation of the organic molecule is due to the facts that (1) the organic fraction contains many reactive groups that are known to enhance chemical changes in several families of organic substances and (2) humic substances possess a strong reducing capacity (Stevenson 1982). The presence of relatively stable free radicals in the fulvic and humic acid fractions of the soil organic matter further supports the assumption that the organic matter enhances abiotic transformations of many pesticides.

The soil organic matter enhances the transformation of the toxic chemicals in soils through various pathways: hydrolysis, solubilization, and photosynthetization (Senesi and Chen 1989). The catalytic effects of organic matter on dichloro-hydroxylation of the chloro-s-triazine herbicides in soils and aqueous solutions have been extensively studied. Armstrong and Chester (1968) reported that the formation of H-bonding between the ring- or side-chain nitrogens of the s-triazine and HA-surface acid groups cause further electron withdrawal from the electron-deficient carbon atom already surrounded by electronegative nitrogen and chlorine atoms. This enables the weak nucleophile water to replace the chlorine atom, thereby increasing the rate of hydrolysis.

Perdue and Wolfe (1983) proposed a general mechanism for the effects of organic matter on the hydrolysis kinetics of hydrophobic organic pollutants.

Table 7.2. Some reactions catalyzed by clay surfaces. (After Wolfe et al. 1990)

Substrate	Reaction	Type of clay	Remarks
Ethyl acetate	Hydrolysis	Acid clays	–
Sucrose	Inversion	Acid clays	–
Alcohols, alkene	Ester formation	Al-montmorillonite	–
Organophosphate esters	Hydrolysis	Cu- and Mg-montmorillonite, Cu-Beidellite Cu-Nontronite	Cu-mont. better catalyst than other Cu-clays or Mg–mont.
Organophosphate esters	Hydrolysis and rearrangement	Na-, Ca- and Al-kaolinites and bentonites	Room and other temperatures. No liquid phase
Phosmet	Hydrolysis	Homoinic montmorillonite	In suspension
Ronnel	Hydrolysis and rearrangement	Acidified bentonite; dominant exchangeable cations: H, Ca, Mg, Al, Fe(III)	Al-catalyzed suggested reaction
s-Triazines	Hydrolysis	Montmorillonitic clay	Cl-analog degrades faster than methoxy or methoxy-thio compounds
Atrazine	Hydroxylation	H-montmorillonite	–
Atrazine	Hydrolysis	Al- and H-montmorillonite; montmorillonitic soil clay	Ca- and Cu-clays much weaker catalysts
1-(4-methoxy phenyl)–2,3-epoxy propane	Hydrolysis	Homoionic montmorillonites; Na-kaolinite	–
DDT	Transformation to DDE	Homoionic clays	Na-bentonite is a better catalyst than H-bentonite
Urea	Ammonification	Cu-montmorillonite	Air-dried clay at 20 °C
Aromatic amines	Redox	Clays (review)	In presence of metal ions that are good redox agents
3,3′,5,5′ Tetra methyl benzidine	Redox	Hectorite	O_2 oxidizing agent
Pyridine derivatives, olefines, dienes	Oligomerization	Clays (review)	–
Glycine	Oligomerization	Na-kaolinite, Na-bentonite	–
Styrene	Polymerization	Palygorskite, kaolinite, and montmorillonite	Weakest catalyst is montmorillonite
Fenarimol	Dialdehyde formation	Homoionic montmorillonite	–

The overall model for organic matter is derived by a combination of equations that, separately, describe partitioning equilibria and, together, acid-base and micellar catalysis. The resulting model predicts that the overall effect of organic matter in modifying hydrolysis reaction rates of toxic organic chemicals may be almost totally attributed to partitioning equilibria and micellar catalysis. The hydrolysis of chloro-s-triazines (the substitution of the chlorine by a hydroxyl) in soils occurs through the interaction of the pesticides with carboxyl groups of the soil organic matter, as reported by Armstrong and Konrad (1974). Stevenson (1982) suggested that after the removal of the side chain of phenoxyalkanoates, possibly by enzymatic action, the phenolic metabolites transform into humic-like substances by interaction with the amino groups of the soil organic matter. The above author also suggested that amines produced in the biological degradation of phenylcarbamates and other related pesticides may undergo an analogous transformation through interaction with carbonyl groups in the soil organic matter. The last two reactions are examples of abiotic transformations of intermediates produced through biotic degradation of pesticides.

Among the abiotic transformations of organic toxic chemicals enhanced by soil organic matter, Khan (1978) includes hydrolysis of organophosphorus esters, the dehydrochlorination of chlorinated hydrocarbons such as DDT and lindane, the degradation of 3-aminotriazole, the S-oxidation of phorate, and the conversion of aldrin to dieldrin.

Besides that by clay and organic matter surfaces, the adsorption of organic molecules by other solid components of the soil can lead to their enhanced transformation. Metal oxides, for examples, enhanced the redox, hydrolytic, and rearrangement processes of specific organic molecules adsorbed on their surfaces.

Nonspecific enhanced surface transformations can be observed on charged surfaces which are found in the vicinity of the surface but are not specific adsorption sites. Wolfe et al. (1990) pointed out the strong effect of the electric field, originating from the charged surface, on the properties of that part of the liquid phase that is under the field's influence. Probably the most outstanding phenomenon in the interfacial region near the surface of charged solids is the strong dependence of the concentration of charged solutes on the distance from the surface. The concentration in the interfacial region of charged inductors, catalysts, reactants, and products can, therefore, be different from their concentration in the bulk solution. Another factor that can affect organic transformations in the interfacial region near charged particles is the aforementioned influence of the electric field on the polarization and dissociation of both solutes and solvents.

The organic chemical transformation at the solid-liquid interface is affected by physicochemical *environmental factors* such as surface properties and exchangeable cations, temperature, and hydration status.

The properties of the surfaces, including their saturating cations, are a major factor in defining the transformation mechanisms of organic molecules at the solid-liquid interface or at the solid-air interface in the case of a dry soil. Adsorbed molecules can be transformed directly by a component of the solid phase or by a third adsorbed species. Properties of the solid phase, such as surface area and cation exchange capacity, will control the adsorption and consequently affect the extent of the transformation. The properties of the interacting cations and their hydration shells will affect the intrinsic properties of the solid, the surface acidity and, consequently, the adsorption catalysis reaction.

The *exchangeable cation* saturating the surface will affect the transformation of organic molecules, for the following reasons:

1. The high localized charge and the relatively accessible portion result in a strong interaction with anionic and polar organic compounds, directly or through their water of hydration.
2. The dissociability of the water molecules in the first hydration shell of the exchangeable cations is higher than that of water molecules adsorbed by the uncharged surfaces, and these cations become preferential sites for surface-enhanced transformations.
3. The radius of the exchangeable cation and the arrangement of ligand water may have an important effect on the catalytic efficiency of the cation.

Temperature and *moisture* content affect both the rate and the extent of the transformation process.

The temperature dependence of surface catalysis is influenced by the temperature of adsorption. An exothermic adsorption is negatively correlated with temperature, whereas the surface catalysis rate is positively correlated. Since an adsorbate can simultaneously interact with several surfaces of a heterogeneous adsorbent, but will do so differently according to the variations in the properties of the surface, the temperature effect will also be differentiated.

Soil moisture content has a strong effect on the rate of biotic and abiotic transformation. The hydration level of the exchangeable cation may determine whether an adsorbate will interact with the cation directly or through a water bridge. The mode of interaction with the cation will, in turn, determine the tendency of the adsorbed pesticide to undergo any transformation enhanced by the cation or by its water of hydration. The dissociability of hydration water molecules will be lower in the presence of a free aqueous phase than in dry soils, where the surface (and in particular the exchangeable cations) is only partially hydrated. In the presence of hydration water, the mobility of the reactants and of the products of surface reactions is often considerably higher than in the absence of a liquid phase. In the latter case, the slow surface diffusion might dominate the transport. Under saturated conditions, when expandable clays swell, bulky solute molecules may interact with the interlayer surface sites that, under dry conditions, were inaccessible to the molecules.

Wolfe et al. (1990) emphasized the fact that many of the abiotic transformations that organic molecules undergo in soils are acido-basic-catalyzed. In the absence of a liquid phase, the capacity of the soil to enhance acido-basic reactions is determined by the number, dissociability and extent of dissociation of the acido-basic groups exposed at the surfaces of the soil solid phase. Surface acidity is the parameter that is most frequently used to define the efficiency of a surface as an acido-basic catalyst. When a liquid phase is present, its pH rather than the surface acidity may become the most important parameter defining the tendency of acido-basic catalysis to occur in the soil. Because of the equilibration between the solid surfaces and the liquid phase, the extent of dissociation of the acido-basic groups on the surfaces varies with the pH of the liquid phase. Thus, the pH of the liquid phase controls both the kinetics of acido-basic-catalyzed reactions in the soil solution and the catalytic efficiency of the soil for many acido-basic surface reactions.

7.3 Biologically Mediated Transformations

The soil microbial population is a fascinating array of organisms, with diverse enzyme systems capable of deriving energy from metabolism of organic and inorganic compounds (Alexander 1980). The toxic chemicals reaching the land surface, as a result of agricultural activity, as disposal products, or accidentally, are subjected to many biological reactions which can affect their persistence and mobility in the soil profile.

7.3.1 Microbial Reactions

Microbial life is adapted to the available energy and nutrient supply under varying environmental conditions. The chemicals applied to the soil, on purpose or by accident, may become a source of energy for the soil biomass. When a microbial population is offered an energy source, a specific population rapidly increases in numbers and activity to utilize the energy (Keeney 1983). Energy sources and environmental conditions interact to determine the microbial ecology during any microbial processes which induce the transformation of a given compound or element in the soil. Table 7.3 (after Delwich 1967; Alexander 1977) shows the classification of microbial reactions in relation to energy.

Aerobic respiration is one of the major processes mediating chemical transformations, mainly through the activity of heterotrophic microorganisms. Carbon turnover in soils, for example, involves chemicals such as N, S, and P. The large polymeric molecules which compose the bulk of biological residues are converted to smaller units by exoenzymes and used as a source of energy which ultimately becomes the soil humus, which affects the behavior and properties of the toxic chemicals reaching the soil environment. Aerobic respiration is also the cause of change in the molecular structure of organic agrochemicals. Autotrophic or chemoautotrophic reactions are among the most important aerobic processes in nitrification. Autotrophic transformation

Table 7.3. Classification of the main microbial reactions in relation to (top) energy sources and (*bottom*) which influence mobility of elements and compounds in soils. (After Delwich 1967; Alexander 1977)

Class	Electron		Products
	Donor	Acceptor	
Photoautotrophic	H_2O H_2S, H_2R	CO_2	$(HCHO)_n$ and other reduced compounds
Respiration			
Aerobic	Organic compounds	O_2	CO_2+H_2O
Anaerobic	Organic compounds	Many ranging from the same molecule, another molecule, CO_2 or inorganic compound	Many ranging from reduced compounds oxidized compounds CO_2 and H_2O
Chemoautotrophic	Inorganic compound	O_2 or another inorganic compound	Oxidized inorganic compound

Reaction

1. Mineralization of inorganic ions during organic matter decomposition when amount of element present is in excess of microbial demand
2. Immobilization of inorganic ions to organic forms to satisfy needs of microbial growth
3. Oxidation of inorganic ions, particularly as energy sources by autotrophs although numerous heterotrophs also oxidize compounds but do not obtain energy from the process
4. Reduction of oxidized elements, particularly when an alternate electron acceptor to O_2 is required. However, some reduction reactions occur in which the oxidized species is not needed as an electron acceptor
5. Indirect transformations leading to changes in the microenvironment (pH, depletion of O_2 to lower Eh, addition of CO_2)
6. Production or degradation of organic ligands
7. Methylation to produce more mobile and/or volatile compounds

controls the oxidation of S in SO_4^{2-} or $S_2O_3^{2-}$, organic Fe compounds in Fe precipitates, or arsenite (mobile and toxic) in arsenate (nontoxic). Aerobic respiration also has indirect effects on the transformation of chemicals. The release of CO_2, which affects redox potential or soil pH, leads to the transformation in the properties of metal ions by chelation.

Anaerobic metabolism occurs under concentrations where the O_2 diffusion rate is insufficient to meet the microbial demands and alternative electron acceptors are used. The type of anaerobic microbial reaction affects processes such as redox potential (Eh), denitrification, reduction of Mn^{4+} and SO_4^{2-}, transformation of selenium and arsenate, etc. Keeney (1983) emphasized that denitrification is the most significant anaerobic reaction occurring in soils, from an agronomic standpoint. However, since nitrification and NO_3^- reduction to NH_4^+ also produce gaseous N oxides, Firestone (1982) suggested that denitrification be defined as the process in which N oxides serve as terminal electron acceptors for respiratory electron transport. In this case, a reduced electron-donating substrate enhances the formation of a more oxidized N oxide through numerous electron carriers.

Anaerobic conditions also lead to the transformation of organic toxic compounds (e.g., DDT) and in many cases, these transformations are more rapid than under aerobic conditions.

Microbial methylation, is a reaction which mainly affects the change in the properties of toxic inorganic trace elements, and it can occur under aerobic or anaerobic conditions. Mercury methylation, for example, occurs under both conditions. Hg dimethylation leads to the input of this element into the atmosphere.

7.3.2 Transformation of Inorganic Compounds

The biologically mediated transformation of inorganic pollutants encompasses a broad spectrum of compounds. In the present discussion, we will refer only to those having a major impact on soil pollution: nitrates, phosphates, and selected metals. A more comprehensive approach to this topic can be found in the book of Paul and Clark (1989) and in various review articles.

Nitrification-denitrification. The conversion of NH_4^+ to NO_2^- and NO_3^- and the reduction of NO_3^- to NO_2^- and then to gaseous N_2O and N_2 are the microbially mediated processes involved in the nitrification-denitrification phenomenon.

Nitrification is associated with chemoautotrophic bacteria which, under aerobic conditions, derive their energy from the oxidation of NH_4^+ to NO_2^-. *Nitrobacter* is the soil bacterium which oxidizes nitrite NO_2^- to NO_3^-. In most habitats, this bacterium is found together with *Nitrosomonas, Nitrospira,* or *Nitrosovibrio,* which oxidize ammonia (NH_3) to the nitrite required for NO_3^- formation. Nitrification is affected by the soil pH (optimum value varying between 6.6 and 8.0) and by the soil water-air ratio. Once NO_3^- is formed in the soil, it becomes subject to transformation. In most soils, NH_4^+ released during soil organic matter (SOM) decomposition and not immediately reused by organisms is rapidly transformed to NO_3^-. Once NO_3^- is formed in soil, it may undergo denitrification by microorganisms to gaseous oxides of nitrogen and to N_2. NO_3^- may be taken up by organisms and used in synthesis of amino acids (assimilatory reduction) or, in the absence of O_2, it may be used by microorganisms as an electron acceptor and become reduced to NH_4^+ (Paul and Clark 1989). The denitrification process – called enzymatic denitrification – is the result of assimilatory reduction of NO_3^- by microorganisms and also of the dissimilatory reduction of nitrate to ammonium, which is accomplished by

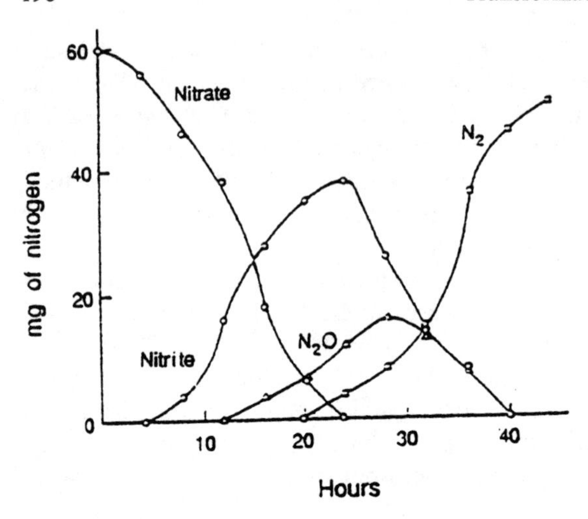

Fig. 7.1. Products formed during denitrification in Melville loam, pH = 7.8. (After Cooper and Smith 1963)

specific microorganisms in the absence of O_2. Figure 7.1 shows the following sequence of identifiable products formed during denitrification:

$$(NO_3^-) \quad (NO_2^-) \quad (N_2O) \quad (N_2).$$

Under field conditions, however, not all of the intermediate products are converted to N_2 by each of the specific reductase enzymes involved. Nitrate (NO_3^-) reductase, for example, causes a decrease in the enzymatic activity. Denitrification in the absence of O_2 is caused by a large number of bacteria and a general list of bacteria capable of denitrification is given in Table 7.4.

Denitrification occurs only in the presence of oxidized nitrogen and a limited presence of O_2, and in an environment favorable for the growth of denitrifying organisms. Since denitrification is enzyme-mediated, the substrate concentration functions as a rate determinant. The dominant denitrifying

Table 7.4. Bacteria capable of denitrification. (After Firestone 1982)

Genus	Interesting characteristics of some species
Alcaligenes	Commonly isolated from soils
Agrobacterium	Some species plant pathogens
Azospirillum	Capable of N_2 fixation, commonly associated with grasses
Bacillus	Thermophilic denitrifiers reported
Flavobacterium	Denitrifying species isolated
Halobacterium	Requires high salt concentrations for growth
Hyphomicrobium	Grows on one-carbon substrates
Paracoccus	Capable of both lithotrophic and heterotrophic growth
Propionibacterium	Fermentors capable of denitrification
Pseudomonas	Commonly isolated from soils
Rhizobium	Capable of N_2 fixation in symbiosis with legumes
Rhodopseudomonas	Photosynthetic
Thiobacillus	Generally grow as chemoautotrophs

bacteria being heterotrophic, the process is dependent on carbon aerobility. The best environment for the activity of denitrifying bacteria includes a neutral pH (pH 6–8), a favorable water/air (oxygen) ratio and a soil temperature between 20 and 30 °C.

Phosphorus transformations. Soil phosphorus may be a deleterious element when it reaches a water source where the bonding of inorganic phosphorus by organic ligands can lead to changed properties. Organic phosphorus can be found in soils in such forms as phytalis, nucleic acids and their derivatives, and phospholipids. Microbial phosphates are produced by most organotrophic members of the soil microorganism population, and microbial mineralization of organic phosphorus is strongly influenced by environmental parameters such as pH, temperature, and soil moisture content. The products produced in the various stages of microbially mediated transformation will have different properties, their transport in the soil media being affected by the type of speciation involved. Figure 7.2 shows the various forms of phosphorus in soils, which result from the microbially mediated transformation.

Metal transformations include two main processes: oxidation-reduction of inorganic forms and conversion of metals to organic complex species, and the reverse conversion of organic to inorganic forms.

Microbially mediated oxidation and reduction is the most typical pathway for metal transformation. Under acidic conditions, metallic iron (Fe^0) readily

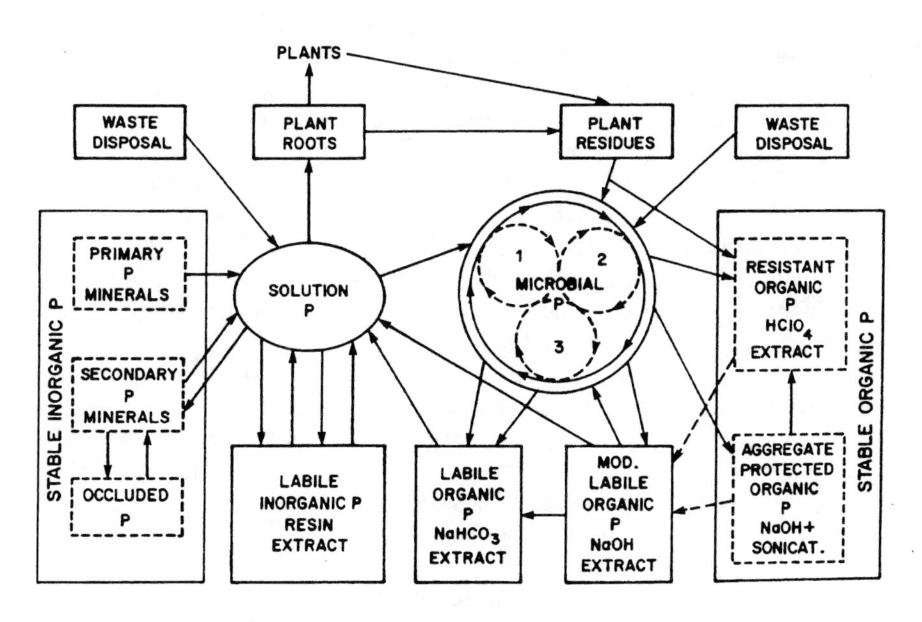

Fig. 7.2. Partitioning and flows of phosphorus. (After Stewart and McKercher 1982)

oxidizes to the ferrous state (Fe^{2+}) but at pH values greater than 5 it is chemically oxidized to Fe^{3+}. Under acidic conditions the Fe^{3+} is readily reduced. *Thiobacillus ferroxidant* mediates this reaction – in an acidic environment – and derives both energy and reducing power from the reaction. Paul and Clark (1989) showed that before Fe^{2+} is microbiologically oxidized it is chelated by organic compounds. During the oxidation, the electrons are moved through the electron transport chain, with cytochrome c being the point of entrance into the transport chain. Both oxidation and reduction of iron are controlled by a series of processes. Oxidation can be caused by direct involvement of enzymes or by microorganisms that raise the redox potential or the pH. Iron reduction occurs when ferric iron serves as a respiratory electron acceptor or reduces by reactions with microbial end products such as formol or H_2S.

Bacterial oxidation of Mn^{2+} is microbiologically mediated only in neutral and acidic environments, and a number of soil bacteria and fungi can oxidize manganese ions.

Mercury – the most toxic metal – does not remain in a metallic form in an anaerobic soil environment. Microorganisms transform metallic mercury to methylmercury (CH_3-Hg^+) and dimethylmercury (CH_3-Hg-CH_3), which are volatile and adsorbable on soil organic matter and soil organisms.

7.3.3 Transformation of Organic Toxic Chemicals

The microbial metabolic process is the major mechanism for the transformation of toxic organic chemicals in the soil environment. Bollag and Liu (1990) defined five basic processes which are involved in the microbially mediated transformation of toxic organic molecules:

Biodegradation, a process in which the toxic organic molecule serves as substrates for microbial growth. In this case, the organic molecules are used by one or more interacting microorganisms and metabolized into CO_2 and inorganic molecules. In this way, the microorganisms obtain their requirements for growth and the toxic organic molecules are completely decomposed without losing metabolities in the soil environment.

Cometabolic transformations include the degradation of toxic organic molecules by microorganisms, which grow at the expense of a substrate other than the toxic organic one, with the latter not being used as an energy source. This process, in which enzymes involved in catalyzing the initial reaction are lacking in substrate specificity, may lead to the accumulation of intermediate products which in many cases are more toxic than the parent organic molecules.

Polymerization or conjugation is the process in which the toxic organic molecules undergo microbially mediated transformation by oxidative coupling reactions. Microbially mediated polymerization leads to the incorporation of xenobiotics into soil organic matter.

Microbial accumulation is an additional microorganism-mediated transformation process which may affect the behavior of toxic organic chemicals in the soil medium. The rate of accumulation depends on the properties of the microorganisms and of the toxic organics.

Nonenzymatic transformation of toxic organics as a result of microbial activity is an indirect process, which occurs as a result of microbially induced changes in environmental parameters such as pH and redox potential. Certain microbial products act as photosensitizers enhancing the photochemical decomposition of some toxic chemicals.

We will follow the presentation of Bollag and Liu (1990), for the principal processes involved in microbial pesticide metabolism to describe the major geochemical reactions.

Oxidative reactions are the most important of the metabolic processes involving a group of oxidative enzymes such as peroxidases, lactases, mixed-function oxidases, etc. The major oxidation reactions are presented in Table 7.5.

Hydroxylation, which can occur at the aromatic ring, at aliphatic groups and at alkyl side chains, makes the compound more polar, so that its solubility in water increases. *N-dealkylation* consists of the alkyl substitutions on an aromatic molecule where the catalytic transformation of the molecule is microorganism-mediated. *β-oxidation* proceeds by the stepwise cleavage of two-carbon fragments from a fatty acid; for the normal functioning of β-oxidation, two protons are required on each α- and β-carbons. *Decarboxylation* occurs

Table 7.5. Oxidation reactions in microbial pesticide metabolism. (After Bollag and Liu 1990)

Hydroxylation
$RCH \rightarrow RCOH$
$ArH \rightarrow ArOH$
N-Dealkylation
$RNCH_2CH_3 \rightarrow RNH + CH_3CHO$
$ArNRR' \rightarrow ArNH_2$
β-Oxidation
$ArO(CH_2)_nCH_2CH_2COOH \rightarrow ArO(CH_2)_nCOOH$
Decarboxylation
$RCOOH \rightarrow RH + CO_2$
$ArCOOH \rightarrow ArH + CO_2$
$Ar_2CH_2COOH \rightarrow Ar_2CH_2 + CO_2$
Ether cleavage
$ROCH_2R' \rightarrow ROH + R'CHO$
$ArOCH_2R' \rightarrow ArOH + R'CHO$
Epoxidation
$RCH{=}CHR' \longrightarrow RCH{-}CHR'$ (with O bridging the two carbons)
Oxidative coupling
$ArOH \rightarrow (Ar)_2(OH)_2$
Sulfoxidation
$RSR' \rightarrow RS(O)R''$ or $RS(O_2)R'$

R = organic moiety, Ar = aromatic moiety.

when a carboxyl group is taken off as a result of microbial activity. It is a widespread microbially catalyzed reaction for naturally occurring compounds as well as for synthetic organic chemicals. *Ether cleavage* is the result of the separation of a hydrocarbon from an oxygen atom that functions to link it with the other moiety of a molecule. A suggested mechanism for the reaction is that cleavage is catalyzed by mixed-function oxidases, in the presence of reduced pyridine nucleotides and molecular oxygen. *Oxidative coupling* is a condensation reaction, catalyzed by phenol oxidases. In the oxidative coupling of phenol, aryloxy or phenolate radicals are formed by removal of an electron and a proton from the hydroxyl group. The resulting phenolate radicals then couple with phenolic or other compounds to yield dimerized or polymerized products (Brown 1967). *Aromatic ring cleavage* is a microorganism-mediated catabolic process. The type of linkage, the specific substitutent(s), their position, and their number determine the susceptibility of an aromatic ring to fission. Usually, the substituents have to be modified or removed and hydroxyl groups inserted in appropriate positions before oxygenase enzymes can cause ring cleavage. Dehydroxylation is usually essential for enzymatic cleavage of the benzene ring under aerobic conditions. The hydroxyl groups must be placed either ortho or para to each other, probably in order to facilitate the shifts of electrons involved in the ring-fission reaction. Dioxygenases are the enzymes responsible for ring cleavage, and they can cause ortho (intra-diol) or meta (extra-diol) fission of a catechol, forming *cis, cis*-muconic acid, or 2–hydroxymuconic semialdehyde, respectively (Bollag and Liu 1990). *Heterocyclic ring cleavage* is performed by microorganisms. In molecules with heterocyclic rings, the path followed by the degradation process is complicated by the hetero atoms, usually N, O, and S, which contribute to the decomposition reactions, through their individual characteristics. *Sulfoxidation* reaction implies the enzymatic conversion of a divalent compound to a sulfoxide such as sulfone. The oxidation of organic sulfides and sulfites to the corresponding sulfoxides and sulfates can be catalyzed by minerals. However, it is often difficult to distinguish between biologically or chemically induced reactions. Microbially mediated *reduction reactions* may include the reduction of nitro groups and of double or triple bonds, sulfoxide reduction and reductive dehalogenation (Table 7.6). The reduction of the nitro group to amine involves the intermediate formation of nitrase and hydroxyamino groups. Other reductive reactions involve the saturation of double bonds, and reduction of aldehydes to alcohols, of ketones to secondary alcohols, and of certain metals, etc.

Hydrolytic reactions could involve organic toxic molecules which have ether, ester, or amide linkages. In the case of hydrolytic dehalogenation, a halogen is exchanged with a hydroxyl group, this reaction being microbially mediated by the hydrolytic enzymes excreted outside the cell by many microorganisms.

In general, enzymes involved in the hydrolytic reactions include esterase, acrylamidase, phosphatase, hydrolase, and lyase. Bollag and Liu (1990) emphasized that it is often difficult to determine the original catalyst of the re-

Table 7.6. Reduction and hydrolysis microbial mediated reactions of synthetic pesticides. (After Bollag and Liu 1990)

Reduction of nitro group
$RNO_2 \rightarrow ROH$
$RNO_2 \rightarrow RNH_2$
Reduction of double bond or triple bond
$Ar_2C=CH_2 \rightarrow Ar_2CHCH_3$
$RC \equiv CH \rightarrow RCH=CH_2$
Sulfoxide reduction
$RS(O)R' \rightarrow RSR'$
Reductive dehalogenation
$Ar_2CHCCl_3 \rightarrow Ar_2CHCHCl_2$
Ether hydrolysis
$ROR' + H_2O \rightarrow ROH + R'OH$
Ester hydrolysis
$RC(O)OR' + H_2O \rightarrow RC(O)OH + R'OH$
Phosphor-ester hydrolysis
$(RO)_2P(O)OR' + H_2O \rightarrow (RO)_2P(O)OH + R'OH$
Amide hydrolysis
$RC(O)NR'R'' + H_2O \rightarrow RC(O)OH + HNR'R''$
Hydrolytic dehalogenation
$RCl + H_2O \rightarrow ROH + HCl$

R = organic moiety, Ar = aromatic moiety.

action since specific environmental conditions or secondary effects of microbial metabolism create opportunities conducive to hydrolysis.

Synthetic reactions occur under the influence of microbial activity, when toxic organic chemicals are linked together or linked in the natural organic compounds found in the soil medium. Bollag and Liu (1990) divided the synthetic reactions into: conjugation reactions, which involve the union of two substances, and condensation reactions, which yield oligomeric or polymeric compounds. Conjugation reactions such as methylation and acetylation commonly occur during microbial metabolism of xenobiotics. Microbial phenoloxidases and peroxidases catalyze the transformations of anilic or phenolic compounds to polymerized products.

7.4 Examples

The following examples will illustrate both abiotically and biologically mediated transformation of toxic chemicals occurring in the soil liquid phase as well as at the soil solid/liquid interface. As in the previous chapters, the examples to be included were randomly selected and are used to illustrate the basic processes of chemical transformation in the soil medium.

7.4.1 Hydrolysis

As a process leading to the transformation of a pollutant both in the liquid phase and at the solid/liquid interface, hydrolysis will be illustrated by the behavior of *parathion*, an organophosphorus compound used for many years both in plant-protection practices and for domestic use. Evidence of parathion hydrolysis in an aquatic environment was obtained during the 1960s and summarized by Gomaa and Faust (1971). Since most organophosphorus pesticides hydrolyze, their persistence in natural waters may be controlled by chemical forces rather than by biological activity. An order of magnitude of persistence and/or appearance of hydrolysis products of organophosphorus pesticides in natural waters may be obtained from kinetic studies. The rates of hydrolysis of parathion were found to be pH-dependent, i.e., the reaction was catalyzed by hydronium or hydroxide ions:

$$O_2N\text{-}C_6H_4\text{-}OP(S(O))(OC_2H_5)_2 + H_2O \xrightarrow[\bar{O}H]{H_3\overset{+}{O}} O_2N\text{-}C_6H_4\text{-}OH + HOP(S(O))(OC_2H_5)_2 \quad (7.15)$$

Table 7.7 shows the hydrolysis rate constants and half-lives of parathion at different pH at a temperature of 20 °C.

An order of magnitude of persistence and/or appearance of hydrolysis products of parathion and paraoxon in natural water is obtained from the kinetic data. Gomaa and Faust (1971) showed that, in general, at 20°C parathion hydrolysis proceeds slightly faster than that of paraoxon under acidic or neutral conditions. The reverse is true under alkaline conditions, where paraoxon hydrolysis is approximately seven times faster than that of parathion at pH = 9 and five times faster at pH = 10.4.

The surface-enhanced hydrolysis of parathion was proved by Saltzman et al. (1974) using kaolinite, a model material for the soil solid phase. The degradation of parathion was found to proceed via the hydrolysis of the phos-

Table 7.7. Hydrolysis of parathion at 20 °C. (After Gomaa and Faust 1971).

pH	Parathion[a] K_{ob} h^{-1}	$t_{1/2}$ h
3.1	1.65×10^{-4}	4182
5.0	1.88×10^{-4}	3670
7.4	2.66×10^{-4}	2594
9.0	1.32×10^{-3}	523
10.4	2.08×10^{-2}	33.2

K_{ob} = observed rate constant; $t_{1/2}$ = half-life.
[a]Initial parathion concentration = 3.948×10^{-5} M.

phate ester to p-nitrophenol and diethyl thiophosphate. The kaolinites were found to catalyze the hydrolysis. The nature of the exchangeable cation greatly affected the rate of hydrolysis. Ca-kaolinite was the most active in inducing the degradation of parathion, the hydrolysis being twice as fast as in a water solution at pH = 8.5 at the same temperature. The presence of a large excess of water decreased the catalytic effect of the kaolinite. The mechanism of the hydrolysis is via adsorption of the parathion upon the exchangeable cation or its hydration shell. The temperature dependence of the process is consistent with this proposed mechanism.

Figure 7.3 shows the persistence of parathion on various monoionic kaolinites. It may be clearly observed that parathion degradation was controlled by the clay surface. Saltzman et al. (1976) emphasized that the degradation rate of parathion on kaolinite was dependent on the nature of the saturating cation, and the process was hindered by the presence of free water. Since the degradation rate on oven-dried clays was much slower with Na- than with Ca-kaolinite, it was suggested that the determining factor in the degradation could be the presence of cation hydration water on clays, that is, the capacity of the various cations to retain their hydration shell when dried at 105 °C. Additional sorbed water on Na-kaolinite enhanced the catalytic effect of the clay, as compared with that of the oven-dried clay. A slight increase in the moisture content above that corresponding to sorbed water results in a steep decrease in the degradation rate. This statement is illustrated by Fig. 7.4, which emphasizes the relationship between the hydrolysis of parathion and the kaolinite water content during a 15-day incubation period. The differences observed between the Ca, Na, and Al-kaolinites are explained by the different hydration capacities of the above cations. In the case of bentonite (Ca-saturated), reduction of the moisture content increases the relative rate of the rearrangement reactions while decreasing the direct hydrolysis rate.

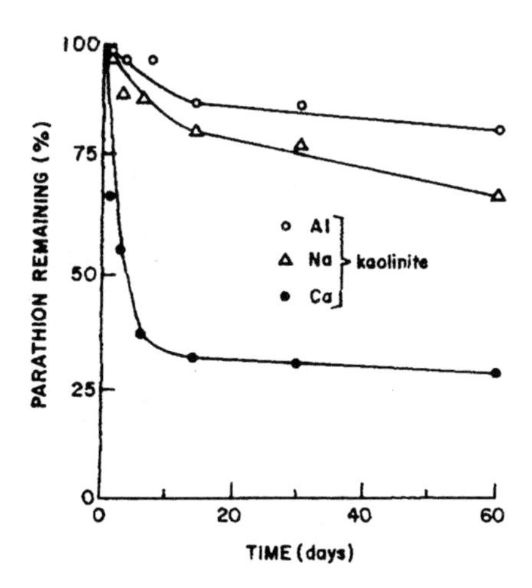

Fig. 7.3. Hydrolytic degradation kinetics of parathion on kaolinite surfaces. (After Saltzman et al. 1974)

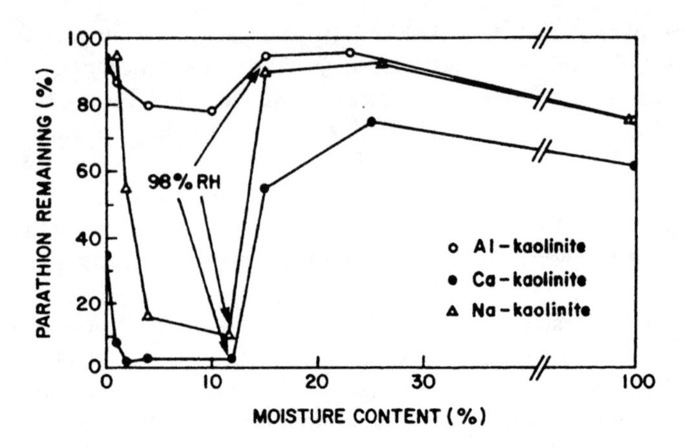

Fig. 7.4. Surface degradation of parathion on kaolinite after 15 days of incubation, as affected by the clay water content, in the presence of different exchangeable cations. (After Saltzman et al. 1976)

The hydrolysis of parathion is not only an abiotic process but can also be enzymatically mediated. Heuer et al. (1976), in laboratory experiments in which phosphatase was applied to parathion in water solution, adsorbed on soil solid surfaces and in soil extract (all previously sterilized), found an obvious effect of phosphatases on parathion in solution but no effect on parathion adsorbed on solid surfaces.

Assuming that the enzymes do not act when adsorbed on soil surfaces, the authors suggested that the interaction between them and the parathion occurred only when both components were not adsorbed. Nelson (1981) succeeded in isolating from soil a bacterium able to hydrolyze parathion (*Arthrobacter* sp.), but the bacterially induced hydrolysis ceased when the p-nitrophenol – the parathion hydrolysis product – exceeded an amount of 1 mM, and became toxic to bacteria.

7.4.2 The Redox Process

Leading to the transformation of metals in soils, the redox process can be chemically or biologically mediated. Iron, for example, exists in oxidation states Fe^0, Fe^{2+}, and Fe^{3+}. Under acidic conditions, metallic iron Fe^0 readily oxidizes to the ferrous state (Fe^{2+}). At pH values less than 5.0, the Fe^{2+} state is stable, but at pH values greater than 5.0 it is chemically oxidized to Fe^{3+}. In turn, Fe^{3+} is readily reduced under acidic conditions but is precipitated in alkaline solution. Since the oxidation of Fe^{2+} is spontaneous at pH values above 5.0, it is difficult to prove that the reaction is enzyme-controlled at higher pH values. Just as iron can be oxidized by the direct involvement of enzymes, or nonenzymatically by microorganisms that raise the redox potential or the soil pH, iron reduction can occur via a number of processes. Ferric iron can serve as a respiratory electron acceptor. It can also be reduced by reaction with microbial end products such as formate or H_2S .Fungi, such as *Alternaria* and *Fusarium*, and bacteria, such as *Bacillus*, have been found capable of Fe^{3+} reduction. The enzyme, nitrate reductase may be involved in these reactions, and the presence of NO_3^- usually provides a high enough redox potential, $p\varepsilon$, to inhibit iron reduction (Paul and Clark 1989).

As an example of oxidation kinetics, we selected the behavior of arsenic in soils. In the soil environment, there can also be found manganese (III/IV) and iron (III) oxides, which can oxidize metals such as As(III) to As(V) (Oscarson et al. 1981). This is a beneficial effect since As (III) is extremely toxic, soluble in water, and more easily transportable than As(V). When one arsenic compound such As(III) ($HASO_2$) is added to a soil containing MnO_2, it can be either oxidized to As(V) (H_3AsO_4) or sorbed by soil solids. Oscarson et al. (1983) proposed the following scheme of oxidation:

$$HAsO_2 + MnO_2 \longrightarrow (MnO_2) . HAsO_2$$

$$(MnO_2) \cdot HAsO_2 + H_2O \longrightarrow H_3AsO_4 + MnO$$

$$H_3AsO_4 \rightleftharpoons H_2AsO_4^- + H^+$$

$$H_2AsO_4^- \rightleftharpoons HAsO_4^{2-} + H^+$$

$$(MnO_2) \cdot HAsO_2 + 2H^+ \longrightarrow H_3AsO_4 + Mn^{2+}.$$

The first step is formation of an adsorbed layer. With oxygen transfer, $HAsO_2$ is oxidized to H_3AsO_4. At pH ≤ 7, the major As(III) species is arsenious acid ($HAsO_2$), but the oxidation product H_3AsO_4 should dissociate to form equal quantities of $H_2AsO_4^-$ and $HAsO_4^{2-}$ with little H_3AsO_4 present at equilibrium. Thus each mole of As(III) oxidized releases about 1.5 mol H^+ into the system. Assuming that no other reaction takes place, pH will be lowered but will remain at about 7.0. The H^+ produced after dissociation of H_3AsO_4 reacts with the $HAsO_2$ adsorbed on MnO_2, forming H_3AsO_4 and leading to the reduction and dissolution of Mn. Therefore, every mole of As(II) that is oxidized to As(V) results in a mole of Mn(IV) in the solid phase being reduced to Mn(II) and partially dissolved in solution.

The kinetics of chemical oxidation can be illustrated by the behavior of two pesticides: diquat and paraquat. Gomaa and Faust (1971) showed that the rates of oxidation of these two pesticides by potassium permanganate was dependent upon the hydronium ion concentration, pH and temperature (Table 7.8). The authors observed that the activation energy for the diquat-$KMnO_4$ reaction at pH $= 9.1$ is approximately 1 kcal smaller than that for the paraquat-$KMnO_4$ oxidation under the same conditions. The activation energy was increased by about 4 kcal as a result of decreasing the pH from 9.1 to 5.1. This was observed also for the paraquat-$KMnO_4$ system. The rate of reaction of diquat-$KMnO_4$ at pH 9.1 is about 100 times faster than the rate at pH 5.1. The same is true with respect to paraquat oxidation. This means that potassium permanganate is a more effective oxidant of dipyridylium cations in an

Table 7.8. Effect of pH and temperature on diquat and paraquat oxidation by potassium permanganate. (After Gomaa and Faust 1971)

Temp. (°C)	K_{ob} l mol^{-1} min^{-1}			
	Diquat - $KMnO_4$		Paraquat - $KMnO_4$	
	pH 5.1	pH 9.1	pH 5.1	pH 9.1
10	0.053	6.71	0.040	4.73
20	0.126	14.54	0.111	9.64
30	0.275	23.46	0.232	17.99
40	0.501	41.28	0.425	33.61

Initial concentrations were: diquat $= 4.16 \times 10^{-5}$ M, paraquat $= 3.65 \times 10^{-5}$ M, and $KMnO_4 = 10^{-3}$ M.

alkaline medium than in acid medium. This may be due to the decrease in activation energy of the oxidation as the pH value is increased.

The oxidative reactions in microbial pesticide metabolism have already been discussed in this chapter. The degradation of the herbicide 2,4-D is presented as an example. 2,4-D is degraded to 2,4-dichlorophenol, which can be oxidatively coupled by phenol oxidases. Minard et al. (1981) showed that a fungal lactase isolated from the soil fungus *Rhizoctonia praticola* polymerized 2,4-dichlorophenol to dimeric [Eq.(7.16)] and higher oligomeric products. In this investigation, it was established that the herbicidal phenol intermediate formed cross-coupling products with phenolic humus constituents such as orcinol, syringic acid, vanillic acid, and vanillin.

OCH_2COOH, Cl, Cl → OH, Cl, Cl → OH OH, Cl Cl, Cl Cl (7.16)

Reductive dissolution of oxide/hydroxide mineral can be enhanced by both organic and inorganic reductant. Stone (1986, 1987a,b), studying the reductive dissolution of manganese oxides by organic reductants, proposed two general mechanisms for electron transfer between metal ion complexes and organic compounds: inner-sphere and outer-sphere. Both mechanisms involve the formation of a precursor complex, electron transfer with the complex, and subsequent breakdown of the successor complex. In the inner-sphere mechanisms, the reductant enters the inner coordination sphere by way of ligand substitution and bonds directly to the metal center before electron transfer. In the outer-sphere mechanism, the inner coordination sphere is left intact and electron transfer is aided by an outer-sphere precursor complex. The two mechanisms can occur in parallel, and the overall reaction is dominated by the faster pathway (Fig. 7.5).

A specific example is the reactive dissolution of Mn by two microbial metabolites: oxalate and pyruvate (Stone 1987). The rate of dissolution was di-

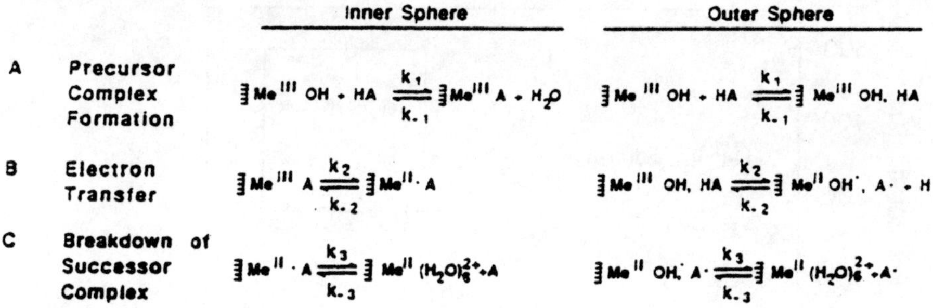

		Inner Sphere	Outer Sphere
A	Precursor Complex Formation	$\equiv Me^{III}OH + HA \underset{k_{-1}}{\overset{k_1}{\rightleftharpoons}} \equiv Me^{III}A + H_2O$	$\equiv Me^{III}OH + HA \underset{k_{-1}}{\overset{k_1}{\rightleftharpoons}} \equiv Me^{III}OH, HA$
B	Electron Transfer	$\equiv Me^{III}A \underset{k_{-2}}{\overset{k_2}{\rightleftharpoons}} \equiv Me^{II} \cdot A$	$\equiv Me^{III}OH, HA \underset{k_{-2}}{\overset{k_2}{\rightleftharpoons}} \equiv Me^{II}OH^{\cdot}, A\cdot + H$
C	Breakdown of Successor Complex	$\equiv Me^{II} \cdot A \underset{k_{-3}}{\overset{k_3}{\rightleftharpoons}} \equiv Me^{II}(H_2O)_6^{2+} + A$	$\equiv Me^{II}OH, A\cdot \underset{k_{-3}}{\overset{k_3}{\rightleftharpoons}} \equiv Me^{II}(H_2O)_6^{2+} + A\cdot$

Fig. 7.5. Reduction of trivalent metal oxide surface sites by phenol (*HA*) showing inner-sphere and outer-sphere mechanisms. (Stone 1986)

rectly proportional to organic reductant concentration and increased as pH decreased. For oxalate, the rate of reductive dissolution at pH 5.0 was 27 times greater than at pH 6.0. For pyruvate, a similar change in pH increased the rate by a factor of 3 only.

A similar effect of pH on dissolution rates of Mn(III/IV) oxides was observed by Stone (1987) with substituted phenols. In this study, phenols with alkyl, alkoxy, or other electron-donating substituents were more slowly degraded. He found that p-nitrophenol, the most resistant phenol studied, reacted slowly with Mn(III/IV) oxides.

The implications of these results are extremely important, as they show that abiotic oxidation by Mn (III/IV) oxides can be a degradation mechanism for substituted phenols.

7.4.3 Photolytic Degradation

Photolytic degradation of pesticides can be used as an example of the photolysis process.

Direct photolysis, in liquids, is controlled by the absorption of sunlight by pesticides. Depending on their molecular configurations, the pesticides exhibit different absorption spectra (Fig. 7.6). Trifluralin and methylparathion are examples of pesticides that absorb sunlight strongly. The pesticide 3,4-dichloramine absorbs at shorter wavelengths and methoxychlor absorbs sunlight only weakly (Wolfe et al. 1990). The photolytic degradation of paraquat as described by Bollag and Liu (1990), shows that paraquat is transformed to N-methyl-isonicotinate and methylamine. A Gram-positive bacterium was isolated that is capable of using N-methyl-isonicotinate as the primary C source. Whole cells and cell-free extracts of this bacterium degraded the substrate by hydroxylation and N-demethylation, followed by further hydroxylation and subsequent cleavage of the pyridine ring, resulting in the formation of malamic

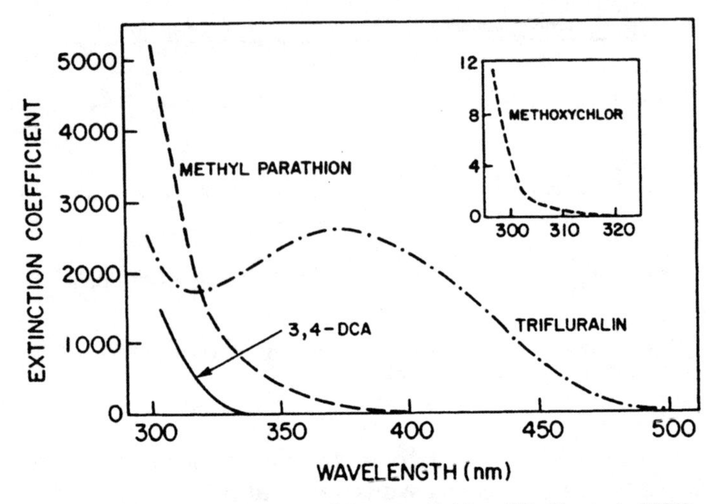

Fig. 7.6. Absorption spectra of four pesticides. (Wolfe et al. 1990)

acid. The degradation of paraquat and cleavage of the pyridine ring by bacteria is shown in Fig. 7.7.

Indirect, sensitized, photolysis of uracil derivatives in water is reported by Acher and Saltzman (1989). Bromacil solutions exposed to sunlight in the presence of dye sensitizers and dissolved oxygen undergo fast chemical transformations. The rate of the reaction is pH-dependent, being greater at alkaline pH. Under optimal reaction conditions, the reaction was completed after about 1 h. The major product formed, comprising 83% of the yield, has been identified as a mixture of diastereoisomers of 3-sec-butyl-5-acetyl-5-hydroxyhydantoin. The mechanism proposed is based on singlet oxygen reactions with the double bond of the uracil ring, followed by a fast OH^--catalyzed rearrangement, with concomitant loss of bromine.

The first step of the reaction is the formation of the intermediate compounds A and B. In the second step, A and B are transformed by an OH^--catalyzed

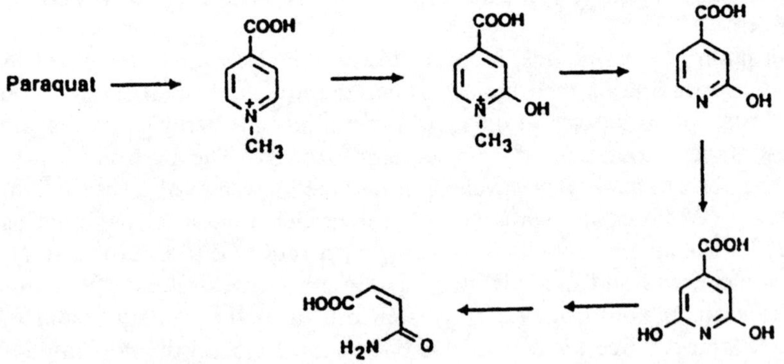

Fig. 7.7. Degradation of paraquat, photodegradation product, and cleavage of pyridine ring by a bacterium. (Bollag and Liu 1990)

rearrangement to the intermediate C, which undergoes ring opening (D) and further ring closure, to give the major photodegradation product. Further irradiation of the reaction mixture resulted in the decomposition of this product and the formation of unidentified polar compounds. The phytotoxicity of the photoreaction mixtures, and of the main photodegradation products of bromacil isolated, was tested with sorghum, which is very sensitive to uracil derivatives (Saltzman et al. 1982). At concentrations equal to or higher than those expected to result from the regular use of bromacil, neither the main photooxidation products, nor the photoreaction mixtures were phytotoxic. The dye sensitizer, methylene blue, at concentrations 20 times higher than those generally used in photoreactions, inhibited the development of the radicles of the sorghum seedlings. However, its leuco-form, resulting from irradiation, was nonphytotoxic.

A new catalytic system composed of an inorganic n-type semiconductor (TiO_2) and an organic polymer (polyaniline), both immobilized in the same PVC membrane, was used for photodegradation of atrazine, propoxur, and terbuthylazine in aqueous solutions (Militerno et al. 1993). From Fig. 7.8 it may be seen that the use of immobilized TiO_2 and polyaniline on an organic membrane consistently increased the degradation rate of the pesticides at the end of the process. Future studies will show whether the inclusion of semiconductors and organic polymers on clay membranes or on soil solid surfaces will have the same degradative effect on toxic organic molecules.

Because of the light-attenuating effect, the rates of direct photolysis of pesticides on soil surfaces are much slower than the rates of photolysis in water and on foliage surfaces. Nilles and Zabik (1975) found that more than 80% of fluchloralin had degraded after 60 h of irradiation in water, while more than

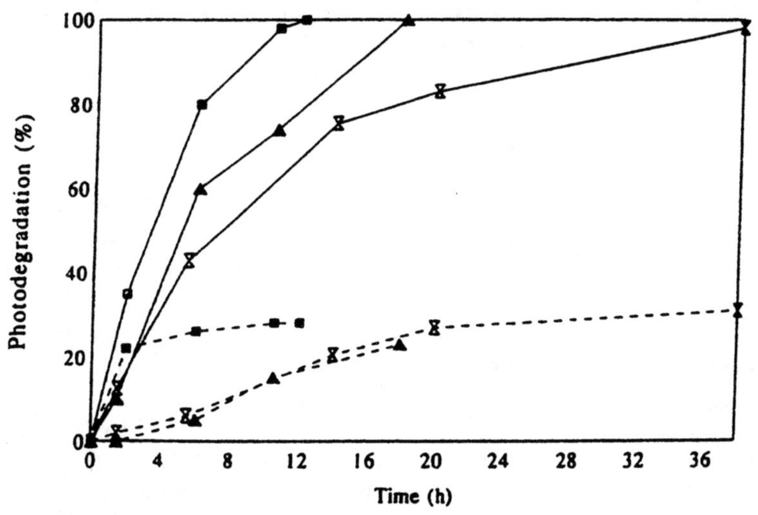

Fig. 7.8. Catalyzed (- -) and noncatalyzed (–) photodegradation of Atrazine (⧖), Terbutylazine (■), and Propaxur (▲). (Militerno et al. 1993)

80% of Basalin, a formulation of fluchloralin, remained after 80 h of irradiation on a 0.5-mm-deep sandy loam soil. Similarly, they observed more than 80% loss of bentazone when it was irradiated in distilled water for 24 h while more than 50% of Basagran, a formulation of bentazone, remained after 120 h of irradiation on the soil solid surfaces (Nilles and Zabik 1975).

A reduction in photolysis rate was observed when the herbicide flumetralin was irradiated on soil, compared with the rate on leaf surfaces, on silica gel, on glass plates, and in solution. The photolysis rate was substantially lower on soils and leaf surfaces than in the other systems (Miller and Zepp 1983). These differences were probably due to differences in light attenuation.

Light attenuation by soils is probably responsible for the relatively low rate of xenobiotic photolysis on soils. It is unclear, however, whether this light reduction is because of bulk attenuation or due to an internal filtering phenomenon. Evidence for this internal filtering effect was also provided by Miller and Zepp (1979), who demonstrated that the photolysis rate of DDE on suspended sediments is dependent on the time allowed for sorption on the sediments. The DDE equilibrated with suspensions of two different sediments for 4 days underwent photolysis more rapidly than DDE equilibrated with the sediments for 60 days. In both cases, photolysis did not follow simple first-order kinetics, and the rate constants decreased throughout the irradiation period. For the longer equilibration period, the decrease in rate was more rapid and approached zero when about 50% of the DDE had been degraded. For the shorter equilibration period, as much as 90% of the DDE was photodegraded. It has been suggested that approximately 50% of the DDE migrated to a microenvironment where the chemical may not been exposed to radiation (Wolfe et al. 1990).

The photolysis of pesticides in soils is still poorly understood, and additional studies are required to provide a basis for estimating photolytic degradation of organic molecules in the soil environment.

7.4.4 Coordinative Transformations

Coordinative transformations leading to changes in pollutant behavior in soil as a result of the changes in their physicochemical properties, can be exemplified by the effect of complexation on heavy metals and/or organic molecules. Cadmium adsorption and mobility in soils was affected by its complexation with sewage sludge components. Lamy et al. (1993) reported on an experiment in which anaerobically digested liquid sewage sludge was applied as a single treatment to a loamy hydromorphic drained soil, characteristic of agricultural soils in the north of France, at a rate of 11 Mg of dry solids ha^{-1}. Total Cd concentrations in the drainage waters, from both amended and unamended plots, were monitored at selected times to follow the mobility of Cd after sludge disposal. The drained groundwater of the plot that exhibited the highest discharge of sludge soluble organic matter (SSOM) exhibited about

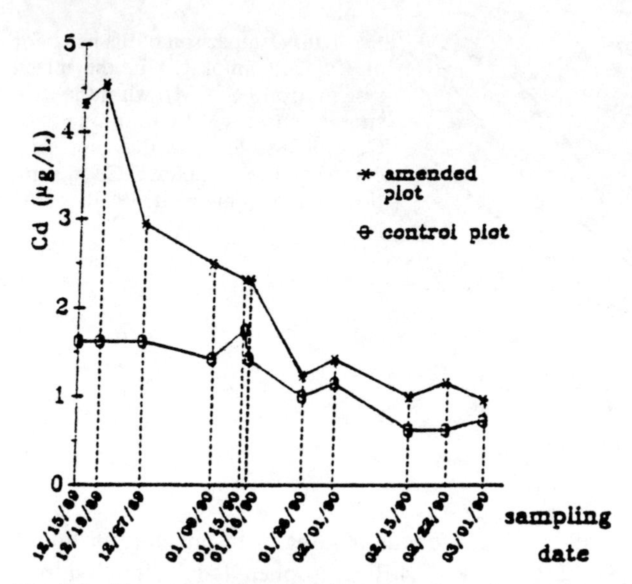

Fig. 7.9. Cd concentration in drainage waters of treated and untreated plots at selected times after sludge application. (After Lamy et al. 1993)

twice the Cd levels of the control plot during the first few weeks following sludge disposal (average of 3 and 1.5 $\mu g\ l^{-1}$, respectively). The Cd concentration in drainage waters of the treated and untreated plots at selected times after sludge application are shown in Fig. 7.9. Laboratory experiments were conducted to provide estimates of the relative binding strengths of Cd-soil interactions and Cd-SSOM interactions. The adsorption data for Cd in the presence of a mixture of soil (13.3 g l^{-1}) and SSOM (0.87 mmol H^{+} K^{-1}) are shown in Fig. 7.10, where the points are the experimental data and the dotted lines correspond to the behavior of Cd in the presence of soil alone or of SSOM alone (in this case, the data correspond to complexed Cd in solution). The behavior of Cd in a mixed sludge-soil system showed that the addition of sludge soluble organic matter to the soil leads to a decrease in the sorption of this cation across the pH range between 5.0 and 7.0.

This associated field and laboratory experiment highlighted the role of soluble organic matter from the sludge in complexing Cd in solution, and its role in diminishing Cd retention in the investigated soil. It helps in the understanding of chemical speciation in the aqueous phase of soil when sewage sludge has been added; that is, speciation which leads to the transformation of Cd properties with regard to its behavior in soil.

Toxic organic molecules can also be bound by the water-soluble organic materials which occur naturally in soils or originate from organic amendments applied to the land surface. Madhun et al. (1986) reported on the binding of herbicides by water-soluble organic substances from soils. Evidence obtained

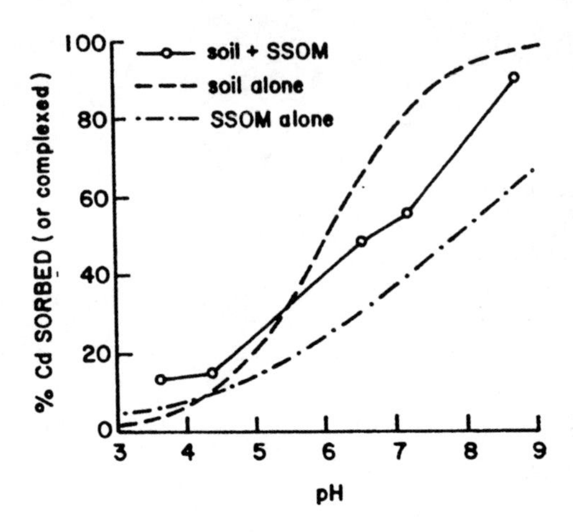

Fig. 7.10. Comparison of the behavior of Cd (200 μmol l^{-1}) in the mixed system (soil + SSOM) with the two simple ones: Cd + soil alone and Cd + SSOM alone. In this case, data correspond to complexed Cd in solution. (After Lamy et al. 1993)

by the gel filtration method is offered for the binding of bromacil (5-bromo-3-sec-butyl-6-methyluracil), diuron [3-(3,4-dichlorophenyl)-1,1-dimethylurea], chlorotoluron [3-(3-chloro-4-methylphenyl)-1,1-dimethylurea], and simazine (2-chloro-4,6-bis-ethylamino-s-triazine) by water-soluble soil organic materials (WSSOM). Infrared and gel filtration data showed that the WSSOM consist of compounds with molecular masses in the range of 700 to 5000 daltons, and very closely resemble the fulvic acids present in soil and surface waters. The tendency of these herbicides to complex with WSSOM is: diuron < chlortoluron < bromacil < simazine.

In Fig. 7.11 a representative elution diagram demonstrating the binding of herbicides with WSSOM is shown. The appearance of a herbicide-deficient zone (trough) in the elution profile (Fig. 7.11a) provides a criterion of binding of the herbicide by WSSOM, as does the presence of excess herbicide in the WSSOM peak. Figure 7.11a shows the coincidence of the bromacil-WSSOM trough with the elution volume of the herbicide. The three peaks in the WSSOM-bromacil elution diagram coincide almost perfectly with WSSOM peaks (Fig. 7.11b). This elution curve also shows that bromacil association was proportional to WSSOM binding. The amount of herbicide bound by WSSOM was calculated from the elution diagram of the herbicide-WSSOM complex. The amount of diuron, bromacil, and chlorotoluron complexed by WSSOM was about 70 times the amount sorbed by the soil from which the WSSOM was extracted. The binding by the water-soluble organic materials may determine the importance of transformations in the mobility and transport of pesticides and pollutants in the environment.

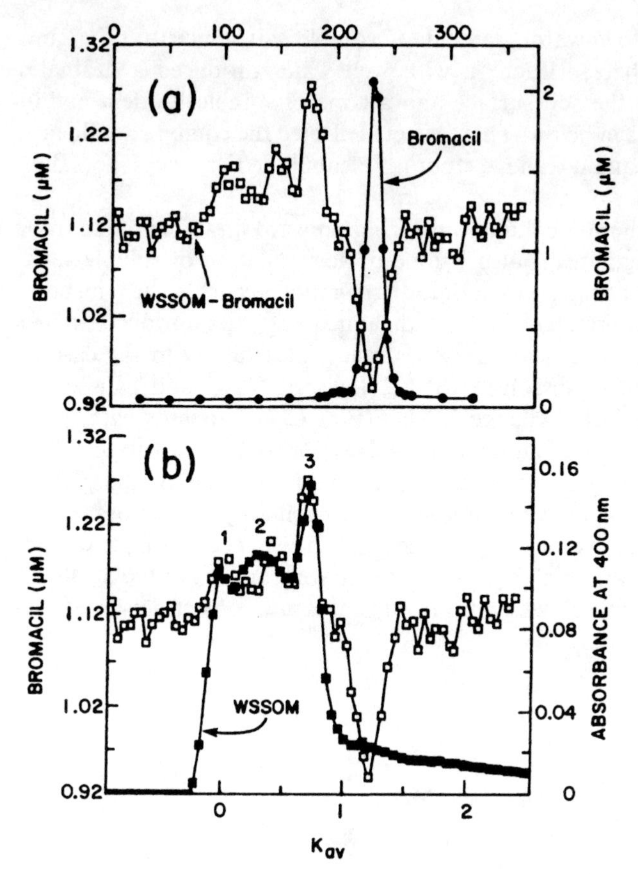

Fig. 7.11a,b. Elution curve of WSSOM-bromacil superimposed on **a** the elution profile of bromacil and **b** the elution pattern of WSSOM. (After Madhun et al. 1986)

7.4.5 Liquid Mixture Transformations

Occurring both in aqueous solution and in water-miscible liquids, these lead to changes in their interactions with the soil solid phase.

The field and laboratory studies of Thellier et al. (1990a, b) illustrate the changes in soil properties as a result of mixing irrigation water of several different compositions with saline drainage water. The soil column experiment was designed to simulate physicochemical conditions in a field experiment conducted in the western San Joaquin Valley, where an Entisol above a shallow, saline aquifer was irrigated with waters of varying quality. Columns 0.46 m long containing the Entisol were leached with "California Aqueduct water" (EC = 0.72 dS m^{-1}, SAR = 4 $mol_c^{1/2}m^{-3/2}$) or with saline "well water" (EC = 8 dS m^{-1}, SAR = 13 $mol_c^{1/2}m^{-3/2}$) for periods up to 1 year. When a simulated "aquifer" was 0.43 m below the soil surface, leaching with aqueduct water produced a positive downward gradient of soluble salt concentrations and exchangeable Na^+, whereas leaching with well water produced a dramatic increase of sodicity at the soil surface and a zone of soluble bivalent cation accumulation about 0.2 m below. These effects reflected the combined influence of the applied water quality and evaporative capillary rise from the saline "aquifer".

Evaporation from the soil columns creates an upward flow. This zone from where evaporation originates is also a zone of accumulation of soluble salts, which can be depleted, however, each time an irrigation occurs, to be reinstated by subsequent evaporation. Leaching with aqueduct water produced a net depletion of soluble cations near the soil surface, particularly in the case of Na^+. This dilution effect dissolved calcite, lowered SAR, and caused exchangeable Ca^{2+} to replace exchangeable Na (Fig. 7.12). Leaching with saline well water produced a net accumulation of Na^+ near the soil surface. This difference in the depth of accumulation is probably a result of the greater mobility of Na^+ in the water moving upward by capillarity to evaporate. The combination of evaporative flux and saline applied water led to high salinity above the 0.2-m depth and high sodicity near the soil surface (ESP≈20–40%). After the simulated "aquifer" was withdrawn, soil saturation extracts indicated

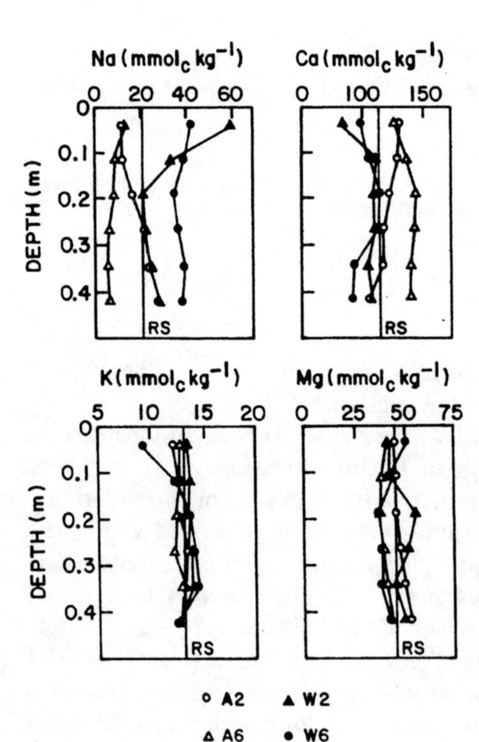

Fig. 7.12. Exchangeable cation profiles in soil columns leached with aqueduct (A2, A6) or well (W2, W6) water. *Solid vertical lines* (*RS*) denote values in the unleached soil. (Thellier et al. 1990a)

equilibration with the applied waters after 0.5–1 year, the rate being greater under leaching.

Thellier et al. (1990b) reported that the results of the accompanying field experiment were similar to those obtained in the column experiment. Both salinity and sodicity in the soil samples increased with decreasing applied water quality, and all of the samples were saline and sodic and contained more calcite at the end than at the initiation of the field experiment. The samples from plots receiving well water had developed values of electrical conductivity and sodium adsorption ratio (SAR), as well as an exchangeable sodium percentage-SAR_e relationship, identical to those observed after 1 year in a companion soil column experiment. The samples from plots receiving aqueduct water were more saline and sodic than those in the corresponding soil column experiment, however, probably because of less effective leaching in the field experiment. The similarity of the field soils and soil columns irrigated with well water indicates that the principal mechanisms governing the effects of saline irrigation water on the Twisselman soil evolved within a relatively short period. The fact that the ESP-SAR relationship for field and column soils were indistinguishable is also evidence that the column behavior simulated the natural equilibration process when saline water was applied.

The authors emphasize that the most dramatic deterioration of soil chemical properties was observed when the field soil was subjected only to saline groundwater evaporation, whereas irrigation with water of any quality tended to diminish the deleterious effect of evaporative capillary rise.

The above combined column-field experiment provides a classical example of the transformation caused in the soil solution by the mixing of waters of different qualities and the effect of this transformation on the soil properties.

When a nonaqueous liquid mixture (NAPL) has already been applied to a soil surface, the transformation in the quality of the mixture caused by the differing behavior of the several components of the mixture affects its behavior in soil. Galin et al. (1990) reported on the effect of volatilization on the physical properties of kerosene – a petroleum product, characterized by a mixture of hydrocarbons with different vapour pressures. As a result of volatilization of the light fractions of kerosene, the physical properties of the remaining liquid change progressively. Table 7.9 summarizes the effect of differential volatili-

Table 7.9. Effect of volatilization on physical properties of the remaining kerosene. (After Galin et al. 1990)

Amount volatilized (%)	Viscosity (Pa s$\times10^{-3}$)	Surface tension (N m^{-1})	Density (g cm^{-3})
0	1.32	2.75	0.805
20	1.48	2.78	0.810
40	1.78	2.80	0.818
60	1.96	2.78	0.819

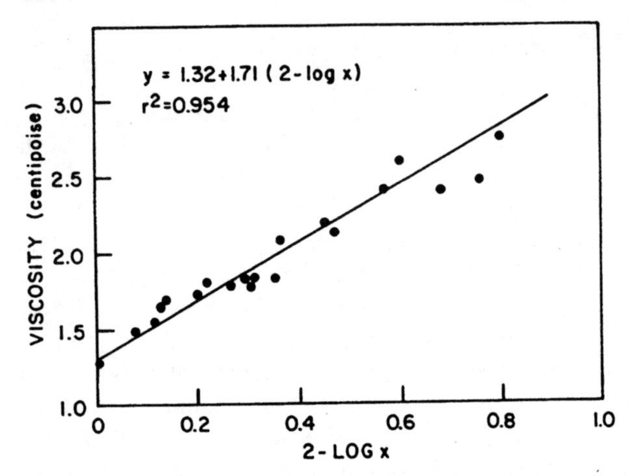

Fig. 7.13. Relationship due to losses occurring by volatilization between viscosity, Y, and the amount of remaining kerosene (%), χ. (Galin et al. 1990)

zation on the kerosene viscosity, surface tension and density. When 20, 40, and 60% of the initial amount had been lost by transfer of the light hydrocarbon fractions to the atmosphere, the viscosity of the remaining kerosene was the property most affected. Negligible effects were observed on the liquid density and no effect on the surface tension.

Since the kerosene viscosity was the property most affected by the volatilization, more attention was given to this. Figure.7.13 shows the relationship between the viscosity of the liquid and the percentage of kerosene remaining after volatilization. Viscosity and interfacial tension are the liquid properties that control the transfer of NAPL through an inert porous medium. Although surface tension measurements showed that the initial value of 2.75 N m^{-1} was not significantly changed during the volatilization process, the evaporation of the volatile fraction of the kerosene led to an increase in its viscosity during its incubation in the porous medium. Galin et al. (1990) emphasized that volatilization was the major physicochemical process affecting the transport of kerosene in the inert porous medium. For example, the infiltration rate in sands has been reported to decrease by about 20% when the viscosity increased form 1.3×10^{-3} to 2.0×10^{-3} Pa s^{-1}.

References

Acher A, Saltzman S (1989) Photochemical inactivation of organic pollutants from water. In: Gerstl Z, Chen Y, Mingelgrin U, Yaron B (eds) Toxic organic chemicals in porous media. Ecological studies 73. Springer, Berlin Heidelberg New York, pp 302–320

Alexander M (1977) Introduction to soil microbiology, 2nd edn. John Wiley, New York 467 p

Alexander M (1980) Biodegradation of chemicals of environmental concern. Science 211: 132–138

Armstrong DE, Chester G (1968) Adsorption catalyzed chemical hydrolysis of atrazine. Environ Sci Technol 2: 683–689
Armstrong DE, Konrad YG (1974) Nonbiological degradation of pesticides. In: Guenzi WD (ed) Pesticides in soil and water. Soil Science Society of America, Madison
Bollag JM, Liu SY (1990) Biological transformation processes of pesticides. In: Cheng HH (ed) Pesticides in soil environment. SSSA Book Series No 2. Soil Science Society of America, Madison
Brown BR (1967) Biochemical aspects of oxidative coupling of phenols. In: Taylor WI, Battersby AR (eds) Oxidative coupling of phenols. Marcel Dekker, New York, p 167
Chaussidon J, Calvet R (1965) Evolution of amine cations adsorbed on montmorillonite with dehydration of the mineral. J Phys Chem 69: 2265–2268
Chaussidon J, Calvet R (1974) Catalytic reactions on clay surfaces. 3rd Int Congr of Pesticides Chemistry (IUPAC) Helsinki. In: Coulton F, Albaky NY, Konle F (eds) Environmental quality and safety Vol III Gerg Thiem, Stuttgart
Cooper GS, Smith RL (1963) Sequence of products formed during denitrification in some diverse western soils. Soil Sci Am Proc 27: 659–662
Delwich CC (1967) Energy relationships in soil biochemistry. In: McLaren AD, Skujins JJ (eds) Soil biochemistry, Vol 2. Marcel Dekker, New York, pp 173–193
Firestone MK (1982) Biological denitrification. In: Stevenson FJ (ed) Nitrogen in agricultural soils. Agronomy 22: 289–326
Fusi P, Ristori GG, Cecconi S, Franci M (1983) Adsorption and degradation of fenarimol on montmorillonite. Clays Clay Minerals 31: 312–314
Galin TS, McDowell C, Yaron B (1990) Effect of volatilization on mass flow of a nonaqueous pollutant liquid mixture in an inert porous media: experiments with "kerosene" J Soil Sci 41: 631–641
Gomaa HM, Faust D (1971) Thermodynamic stability of selected organic pesticides in aquatic environment. In: Faust SD, Hunter JV (eds) Organic compounds in aquatic environment. Marcel Dekker, New York, pp 341–387
Hamaker JW (1972) Decomposition: quantitative aspects. In: Goring AJ, Hamaker JW (eds) Organic chemicals in the soil environment. Marcel Dekker, New York, pp 253–340
Harris JC (1982) Rate of hydrolysis. In: Lyman WJ et al. (eds) Handbook of chemical properties, estimation methods, environmental behavior of organic compounds. McGraw-Hill, New York
Heuer B, Birk Y, Yaron B (1976) Effect of phosphatases on the persistence of organophosphorus insecticides in soil and water. J Agric Food Chem 24: 611–614
Keeney DR (1983) Principles of microbial processes of chemical degradation, assimilation and accumulation. In: Nielsen DR (ed) Chemical mobility and activity in soil system. ASA Soil Science Society of America, Madison, pp 153–164
Khan SU (1978) The interaction of organic matter with pesticides. In: Schnitzer M, Khan SU (eds) Soil organic matter. Elsevier, Amsterdam, pp 137–173
Lahav N, White D, Chang S (1978) Peptide formation in the prebiotic area: thermal condensation of glycine in fluctuating clay environment. Science 201: 67–69
Lamy I, Bourgeois S, Bermond A (1993) Soil cadmium mobility as a consequence of sewage sludge disposal. J Environ Qual 22: 731–737
Macardy DL, Trainek PG, Grundy TJ (1986) Abiotic reduction reactions of anthrophogenic organic chemicals in anaerobic systems. A critical review. J Cont Hydrol 1: 1–28
Madhun YA, Young JL, Freed VH (1986) Binding of herbicides by water-soluble organic materials from soil. J Environ Qual 15: 64–68
McBride MB (1985) Surface reactions of 3,6′,6,6′-tetramethyl benzidine on hectorite. Clays Clay Minerals 33: 510–516

Militerno S, Campanella L, Crescenti G (1993) Photocatalytic degradation of selected chlorophenols and pesticides by immobilized polyaniline and TiO_2. In: Proc IX Symp Pesticide chemistry, Piacenzo, 1992, pp 331–339
Miller GC, Zepp RG (1979) Photoactivity of aquatic pollutants sorbed on suspended sediments. Environ Sci Technol 13: 860–863
Miller GC, Zepp RG (1983) Extrapolating photolysis rates from the laboratory to the environment. Residue Rev 85: 89–110
Minard RD, Liu SY, Bollag JM (1981) Oligomers and quinones from 2.4-dichlorophenol. J Agric Food Chem 29: 250–253
Mingelgrin U, Prost R (1989) Surface interaction of toxic organic chemicals with minerals. In: Gerstl Z, Chen Y, Mingelgrin U, Yaron B (eds) Toxic organic chemicals in porous media. Springer, Berlin Heidelberg New York, pp 91–136
Mingelgrin U, Saltzman S (1979) A possible model for the surface-induced hydrolysis of organophosphorus pesticides on kaolinite clays. Soil Sci Soc Am J 41: 519–523
Mortland MH, Raman KV (1967) Catalytic hydrolysis of some organophosphate pesticides by copper (II). J Agric Food Chem 15: 163–167
Nelson LM (1981) Biologically-induced hydrolysis of parathion in soil: isolation of hydrolysing bacteria. Soil Biol Biochem 14: 219–222
Nilles GP, Zabik MJ (1975) Photochemistry of bioactive compounds. Multiphase photodegradation of basalin. J Agric Food Chem 22: 684–688
Oscarson DW, Huang PM, Defoss C, Herbillon A (1981) Oxidative power of Mn(IV) and Fe(III) oxides with respect to As(III) in terrestrial and aquatic environments. Nature 291: 50–51
Oscarson DW, Huang PM, Hammer UT (1983) Kinetics of oxidation of arsenite by various manganese dioxides. Soil Sci Soc Am J 47: 644–648
Paul EA, Clark FE (1989) Soil microbiology and biochemistry. Academic Press, New York, 275 pp
Perdue EM, Wolfe NL (1983) Prediction of buffer catalysis in field and laboratory studies. Environ Sci Technol 17: 635–642
Russell JD, Cruz M, White JL (1968) Mode of chemical degradation of s-triazines by montmorillonite. Science 160: 1340–1342
Saltzman S, Yaron B, Mingelgrin U (1974) The surface-catalyzed hydrolysis of parathion on kaolinite. Soil Sci Soc Am Proc 38: 231–234
Saltzman S, Mingelgrin U, Yaron B (1976) The role of water in the surface-catalyzed hydrolysis of phosphate esters on kaolinite J Agric Food Chem 24: 739–742
Saltzman S, Acher A, Brates N, Horowitz M (1982) Removal of phytotoxicity of uracil herbicides in water by photodecomposition. Pestic Sci 13: 211–215
Senesi M, Chen Y (1989) Interaction of toxic organic chemicals with humic substances. In: Gerstl Z, Chen Y, Mingelgrin U, Yaron B (eds) Toxic organic chemicals in porous media. Springer, Berlin Heidelberg New York, pp 37–91
Solomon DH, Rosser MJ (1965) Reactions catalyzed by minerals. I. Polymerization of styrene. J Appl Polymer Sci 9: 1261–1271
Sparks DL (1988) Kinetics of soil chemical processes. Academic Press, San Diego, 210 pp
Stevenson FJ (1982) Humus chemistry. Wiley, New York
Stewart JWB, McKercher RB (1982) Phosphorus cycle. In: Burns RG, Slater JH (eds) Experimental microbial ecology. Blackwell, Oxford, pp 221–238
Stone AT (1986) Adsorption of organic reductants and subsequent electron transfer on metal oxide surfaces. ACS Symp Ser 323: 446–461
Stone AT (1987a) Microbial metabolites and the reductive dissolution of manganese oxides: oxalate and pyruvate. Geochim Cosmochim Acta 51: 919–925
Stone AT (1987b) Reductive dissolution of manganese (III/IV) oxides by substituted phenols. Environ Sci Technol 21: 979–988
Stumm W, Morgan JJ (1981) Aquatic chemistry. John Wiley, New York, 780 pp

Thellier C, Sposito G, Holtzclaw KM (1990a) Chemical effects of saline irrigation water on a San Joaquin Valley soil. I. Column studies samples. J Environ Qual 19: 50–55

Thellier C, Holtzclaw KM, Rhoades JD, Sposito G (1990b) Chemical effects of saline irrigation water on a San Joaquin Valley soil: II. Field soil samples. J Environ Qual 19: 56–60

Theng BKG (1974) The chemistry of clay – organic reactions. Adam Holger, London, 374 pp

Wolfe NL (1989) Abiotic transformation of toxic organic chemicals in the liquid phase and sediments. In: Gerstl Z, Chen Y, Mingelgrin U, Yaron B (eds) Toxic organic chemicals in porous media. Springer, Berlin Heidelberg New York, pp 136–148

Wolfe NL, Mingelgrin U, Miller GC (1990) Abiotic transformation in water, sediments and soil. In: Cheng HH (ed) Pesticides in soil environment. SSSA Book Series No 2. Soil Science Society of America, Madison

Zeep RG, Cline MD (1977) Rates of direct photolysis in aquatic environment. Environ Sci Technol 11: 359–366

CHAPTER 8

Pollutants Transport in the Soil Medium

After reaching the soil-solid surfaces, before, while, or after being subjected to biologically or chemically induced transformation, or retained by the soil constituents as sorbed or bound residues, the pollutants are redistributed in the soil profile as solutes and/or water-liquids immiscible, in gaseous form or adsorbed on colloids and other fine particles. The extent of this redistribution and the kinetics of redistribution are controlled by both soil and pollutant properties, by the environmental conditions, and by the management of the polluted lands.

The interphase transfer of the pollutants can also occur during the redistribution of the chemicals along the soil profile. The kinetics and extension of the processes involved, such as dissolution, volatilization, and retention, should be considered in analyzing the transport of chemicals in soils. The above processes were broadly presented in Part II of this book and will not be discussed in this chapter. The fates of the chemicals during this transport through the soil medium are also affected by the reactions of chemicals in soils. These reactions, which were discussed in the previous chapters are: speciation-dissolution-precipitation (Chap. 3), and biologically and chemically induced degradation (Chap. 7). All these reactions should be considered when dealing with the redistribution of the contaminants in the soil profile.

A tremendous amount of research has been devoted to the modeling of transport processes. The validation of the proposed models was done mainly by using parameters obtained in laboratory experiments. The little information obtained from field experiments shows, however, fundamental differences between the transport characteristics derived from the laboratory and the field experiments. The present chapter will include the description of the pollutant transport patterns – mainly in the liquid phase – exemplified by selected examples from the literature. Transport modeling will be presented in Chapter 9, together with the other models for predicting the fate of pollutants in soils.

8.1 Solute Transport

Solutes are all transported by molecular diffusion and mass flow, whatever their properties. For pollutants with low solubility and for soil with high adsorption capacity, the rate-limiting parameters are dissolution and/or release and/or diffusion. Hydrodynamic dispersion – including both diffusion and mass flow – determines solute spreading in soil and other porous media in accordance with water velocity variations. At low average water fluxes in uniform soil, this process is relatively unimportant; it becomes dominant over diffusion at high water fluxes or in structured sols where there are substantial variations in water velocities. Pollutants being adsorbed on the soil solid phase do not move freely with water through the soil. For example, Jury (1983) showed that the travel time for an adsorbed chemical, t_A, is related to the travel time for a non-adsorbed or mobile chemical, t_M, by

$$t_A = (1 + \rho_b K/\Theta) t_M \equiv R t_M \quad , \tag{8.1}$$

where ρ_b is the soil dry bulk density, K is the solid/liquid partition or distribution coefficient of the interacting chemical, Θ is the volumetric water content, and R is a generalized sink term modified for specific reactions (mass of solute/volume/time).

Under field conditions, the movement of pollutants often does not follow the anticipated general pattern. Soils with a high content of clay may shrink and crack when subjected to wetting and drying cycles. Then, when a pollutant is applied, followed by a rainfall or irrigation, the pollutant will partially leach into the soil through cracks and large pores. During transport through these large pores or cracks, only a portion of the solid phase – the external surface – comes in contact with the solute and the amount retained on the soil is relatively small.

8.1.1 Transport Mechanisms

At the soil-pore scale, there are two transport mechanisms that can move solutes through the medium: diffusion and convection.

Molecular diffusion is the process whereby material moves from a higher to a lower concentration region as a result of random molecular motion. The rate of transfer of material is proportional to the concentration gradient normal to the direction of movement. Since solutes are not distributed uniformly through the soil, solution concentration gradients will exist and solutes will tend to diffuse from where the concentration is higher to where it is lower (Nye 1979). In bulk water the rate of diffusion J_d is related to the concentration gradient by Fick's first law:

$$J_d = -D_0 \, dC/dx \quad , \tag{8.2}$$

where D_0 is the diffusion coefficient for diffusion in bulk water and dC/dx is the concentration gradient.

Letey and Farmer (1974) discussed conditions in the soil system that may affect solute diffusion in soil as follows: (1) the diffusion coefficient cannot be assumed to be independent of concentration; (2) diffusion is confined to certain segments of the system (a molecule will not diffuse through a solid particle); (3) sorption of the diffusing substance by soil particles often occurs; and (4) the diffusion coefficient is dependent upon temperature and a number of soil properties such as mineral composition, bulk density, and water content. If it is assumed that pollutant movement by volatilization is negligible, that no degradation occurs during the diffusion process, and that the adsorption is linear and single-valued, then the Olsen and Kemper (1968) relationship,

$$D = \frac{D_0 f \Theta}{(\rho_b K + \Theta)} \tag{8.3}$$

expresses the apparent diffusion coefficient (D), where f is an impedance factor that takes into account the tortuous pathway followed by the solute through the soil pores.

Gerstl et al. (1979b) showed that a mobile fraction of the adsorbed phase of an organic molecule may also contribute to diffusion. The adsorbed molecules may be divided into an immobile fraction (a) and a mobile fraction (1-a), which contributes to diffusion and has the same apparent mobility as molecules in solution. If a fraction (a) of the adsorbed molecules is mobile, then Eq. (8.3) becomes

$$D = \frac{DK_0 f(\Theta + a\rho_b K)}{(\rho_b K + \Theta)} \quad . \tag{8.4}$$

Since the diffusion coefficient of organic molecules in soil remains low even over a large range of moisture content, biological degradation may occur during the diffusion process. Gerstl et al. (1979a) proposed including microbial activity in the description of diffusion. They considered the rate of decomposition at any distance as a function of time (t) to be dependent on the local concentration of the chemical and on microbial activity.

Calvet (1984) showed that the variation in the soil water content can influence the effective molecular diffusion of organic pollutants in two ways:

1. By determining the air-filled porosity, thus controlling the gas diffusion/liquid diffusion ratio for a given organopollutant (for given solubility and vapor tension values).
2. By modifying pollutant adsorption. The general trend is for adsorption to decrease as water content increases, and this effect is probably greater at low water contents, because of the more pronounced competition between pollutant and water molecules.

The relationships between pesticide diffusion and water content, for example, are complex, since they result from combinations of properties of the chemicals and soil. According to published results, several cases can be described (Fig. 8.1).

Case A. Low vapor tension. The pesticide is entirely in the soil solution, so that diffusion occurs only in the liquid phase. Increasing the water content increases the effective diffusion coefficient, essentially through the effect on

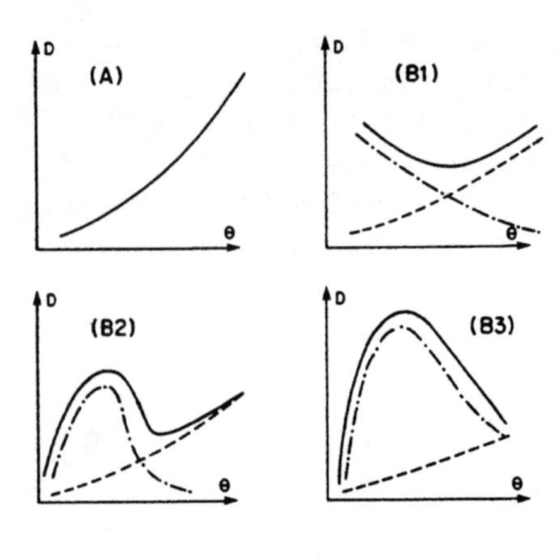

Fig. 8.1. Possible types of effects of the soil water content (Θ) on the apparent diffusion coefficient of pesticides (*D*), in the following cases: *A* low vapor tension; *B* vapor tension high enough to allow the diffusion of some organic pollutants; *B*1 low adsorption from the vapor to the solid phase; *B*2, *B*3 high adsorption from the vapor phase; (After Calvet 1984)
- - - - diffusion in the liquid phase,
–·–·– diffusion in the vapor phase,
—— apparent diffusion.

tortuosity (Nye 1979). The value of this geometric characteristic is reduced when pores are filled with water, because diffusion pathways become more rectilinear. Due to modifications of the water thermodynamic activity, an influence of adsorption on diffusion is also possible, although there is no experimental information about this. Examples of Case A are diffusion in soil of dimethoate (Graham-Bryce 1969), metribuzin (Scott and Paetzold 1978), prometone, atrazine, simazine, prometryne, chloropropham, and diphenamid (Scott and Philips 1972).

Case B. The vapor tension is high enough to allow some pesticide to diffuse in the vapor phase. Since gas diffusion in soil is closely related to the geometric characteristics of the pore space, bulk density is a factor to be taken into account (Ehlers et al. 1969; Bode et al. 1973a,b). Concerning adsorption-diffusion relationships, two situations can be distinguished, according to the extent of adsorption from the vapor.

B1. Low adsorption from the vapor phase. As water content decreases, effective diffusion in solution decreases (increasing tortuosity) and effective diffusion in the vapor phase increases, because the gas-accessible pore volume increases. If adsorption is greater, this increase can be limited, but not completely ruled out. As a result, the apparent diffusion coefficient of the pesticide goes through a minimum value as water content changes. An example of such behavior was reported by Graham-Bryce (1969) for the diffusion of disulfoton.

B2, B3. High adsorption from the vapor phase. In this case, pesticide adsorption increases as soil water content decreases. Thus, the apparent (effective) diffusion coefficient in the vapor phase increases to a maximum value, and then decreases. Under these conditions, the resulting variation of the apparent diffusion coefficient depends on the relative magnitudes of the diffusion coefficient in solution and in the vapor phase as is shown in Fig. 8.1. For lindane, the increase of solution-phase diffusion corresponds to an increase of the apparent diffusion coefficient, while for trifluraline it does not fully compensate for the decrease in the vapor diffusion coefficient (Bode et al. 1973a,b; Letey and Farmer 1974).

Convection refers to the transport of a dissolved chemical by virtue of a bulk movement of the host water phase. In general, the convective transport of solute is described by considering the two components of convective flow: mean pore water velocity and deviations from the mean, induced by variations in the flow regime within the soil. The mass flow of soil water which carries with it a convective flux, J_c, of solutes, proportional to their concentration, C, is described for a monodimensional transport in the z direction by the equation:

$$J_c = qC = -C[K(\Theta)dH(\Theta)/dz] \ , \tag{8.5}$$

where the flux density, $q = -K(\Theta)dH(\Theta)/dz$ follows Darcy's law, expressed as a linear transport equation. $K(\Theta)$ and $H(\Theta)$ are the hydraulic conductivity and hydraulic head, respectively, as functions of volumetric moisture content, Θ. Since q is usually expressed as volume of liquid flowing through unit area per unit time and C is the mass of solute per unit volume of solution, J_c is given in terms of mass of solute passing through unit cross sectional area of a soil body in unit time. The mass transport of water and solute in soils takes place through a complex three-dimensional network of pores characterized by an infinite variety of forms, sizes, and shapes. During their flow through soil pores, the pollutants are subject to a series of interactions such as adsorption, degradation, volatilization, and plant uptake. Under conditions of unsteady or transient water flow, the contact time between the solute and the soil solid phase is often insufficient to attain an equilibrium. This condition is characteristic of the simultaneous flow of water and solute during the infiltration/redistribution process which results from rain or irrigation. The convective-diffusive, dispersive transport of nonconservative pollutants during transient monodimensional water flow in the z direction is described by the equation

$$\frac{\partial}{\partial t}\left(\Theta C + \frac{\rho S}{b}\right) = \frac{\partial}{\partial z} D_h \Theta \frac{\partial C}{\partial z} - \frac{\partial}{\partial z}\left(\nu \Theta C\right) - \sum_{i=1}^{n} Q_i \ , \tag{8.6}$$

where ρ is soil bulk density, D_h is the hydrodynamic dispersion coefficient, ν is the average pore-water velocity, Q_i are various sink terms to account for pollutant losses (degradation and plant uptake), S and C, respectively, adsorbed phase and solution phase concentrations, Θ is the volumetric soil water content, and z is the soil depth (Rao and Jessup 1983).

Hydrodynamic dispersion refers to the motion of solute due to small-scale convective fluctuations about the mean motion (Bear 1972). The hydrodynamic dispersion process results from the microscopic nonuniformity of flow velocity in the soil conductive phase and includes both diffusive and convective transport.

Bear (1972) expresses the vector of the local convective solute flux v_{sc} as

$$v_{sc} = v_w C_l \ , \tag{8.7}$$

where v_w is the local water flux vector and C_l is the corresponding local fluid concentration. The local diffusion flux vector, v_{sd}, within the water phase is described by Fick's law of diffusion, expressed in Eq. (8.2).

Jury and Fluhler (1992) emphasized that the expression for the local flux of dissolved chemical $v_e = v_{sc} + v_{sd}$, describing the movement of solute within the water phase, is exact, but inapplicable in practice, because the local water flux comprises three-dimensional flow around the solid and gaseous portions of the medium and is not measurable. Instead, the local quantities are volume-averaged to produce a larger-scale representation of the system properties. The averaging volume must be large enough so that the statistical distribution of geometric obstacles is the same from place to place. If the porous medium contains the same material and density throughout, then the mean value produced by this averaging is macroscopically homogeneous over the new transport volume containing the averaged elements. If the soil matrix does not have the same properties at different locations, then a representative elementary volume may not exist for that quantity, because the value for its average will then change continually as the size of the volume is altered (Baveye and Sposito 1985).

The convective-dispersive representation of the dissolved phase (J_e) in one dimension is given by the Eq. (8.8) developed by Nielsen and Biggar (1962)

$$J_e = J_w C - D_e \partial C/\partial z \ , \tag{8.8}$$

where D_e is the effective dispersion coefficient, which combines the influence of diffusion in the dissolved phase with the long-term dispersion representation of the effect of small-scale convection on the mixing of solute around the center of motion, C is the solute concentration (mass/volume of water), and J_w is the volume-average water flux. The dispersion contribution to the solute motion is random or diffusion-like only after sufficient time has elapsed for the solute to mix over the range of solute velocities that are correlated. This mixing time is scale-dependent, so that the model depicted by Eq. (8.8) is more likely to be correct, after a given time, in applications bounded by small transverse distance scales, such as soil columns, than by large ones, such as the field regime (Jury and Fluhler 1992). The Eq. (8.8) was again extended by Jury et al. (1990), by including a situation in which a soil water regime changes over time. For a detailed description of the mathematical development, the reader is referred to the Jury and Fluhler (1992) review on transport of chemicals through soil.

8.1.2 Preferential Flow of Solute

Soils with large cracks and well aggregated soils present physical situations that are much more difficult to describe than those covered by the classical convection-dispersion approach. In the convection-dispersion approach it is considered that the pore-size distribution is rather narrow and is described by an average macroscopic pore-water velocity and dispersion coefficient. This assumption is no longer valid in the case of aggregated and cracked soils, where preferred flow pathways are developed in the macropores. In such a case, a large fraction of the flow occurs via interaggregate porosity, cracks, and

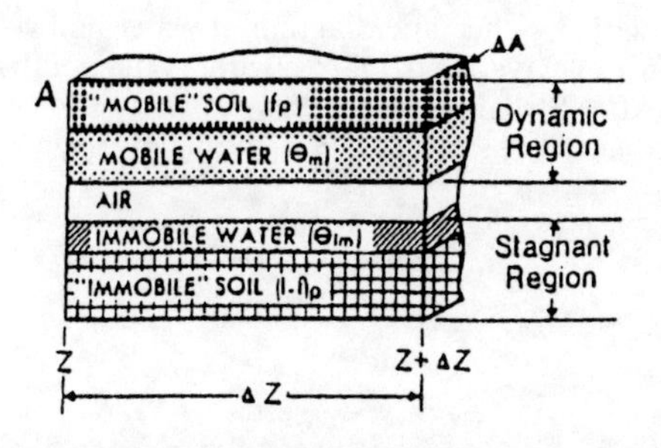

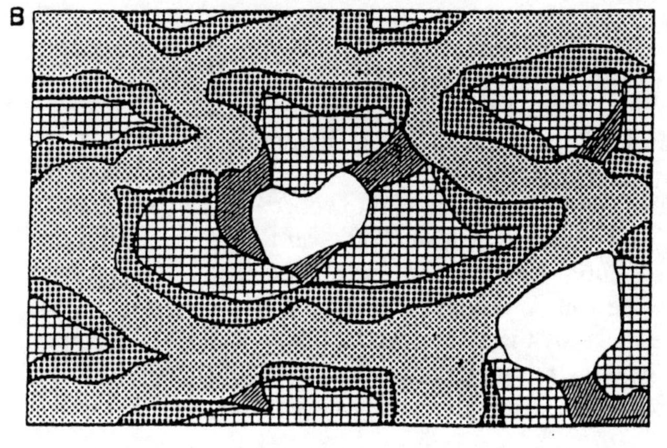

Fig. 8.2A,B. Schematic representation of mobile-immobile soil-water regions; **A** simplified model; **B** actual model. (After Van Genuchten and Wierenga 1976)

channels, while the soil solution in the pores within the aggregate is less mobile, acting as a distributed source or sink for the solute. Figure 8.2, after Van Genuchten and Wierenga (1976), presents a conceptualization of the water and solute transport through preferred paths when a "mobile" and an "immobile" region are defined. In their approach, the macropore contains mobile water through which the bulk of solute transport occurs, while the water in the micropores is relatively immobile or nonmoving. Transfer between these two regions occurs by diffusion. The mobile-immobile water approach can be used successfully in predicting the solute transport in aggregated media with specified geometric configurations.

Preferential flow may occur, not only as a result of "preferred paths", but also because of true fluid instabilities created by density or viscosity differences between the resident and invading fluids (Hillel and Baker 1988). It can be induced by structural voids with high permeability when these are filled with water (Beven and Germann 1982), or may be developed at the interface between two soil layers of differing permeability (Hill and Parlange 1972). The application of the solute may affect both the magnitude and the spatial distribution of the macropore flow (Quinsberry et al. 1994). A simple approach for predicting the solute concentration in preferential flow was developed by

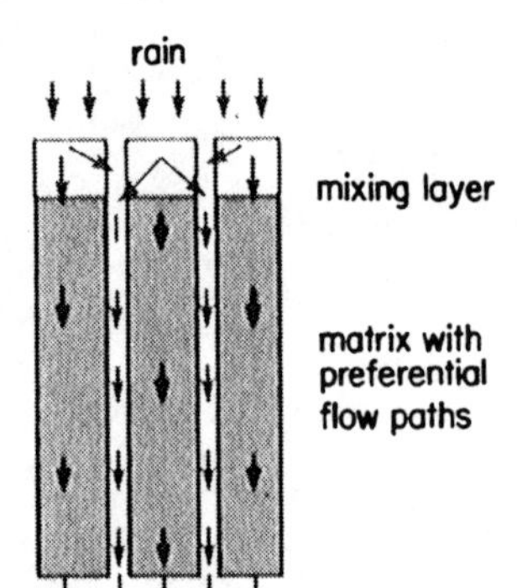

Fig. 8.3. Schematic diagram of water and solute flow in mixing layer and matrix with preferential flow paths. (After Steenhuis et al. 1994)

considering a simplified two-layer model in which the processes of adsorption and desorption were separated (Steenhuis et al. 1994). For the layer near the surface, considered as the mixing layer, the solute concentration is similar to that in the percolating water, since in the lower soil layer the flow is partitioned between matrix and preferential flow. The solute concentration in the matrix flow is characterized by the soil condition near the outlet point, whereas the preferential flow is represented by the solute concentration in the mixing layer. A schematic diagram, depicting the preferential flow pattern considered in the two-layered model proposed by Steenhuis et al. (1994) is presented in Fig. 8.3.

8.2 Transport of Nonaqueous-Phase Liquid

Synthetic organic solvents in common use are, in general, insoluble or only slightly soluble in water. Consequently, these compounds could be transported in the porous media from the land surface to the groundwater as a distinct nonaqueous-phase liquid community denoted as NAPL. After encountering the land surface, the nonvolatile fraction of NAPL could migrate downward through into the unsaturated zone and into the groundwater. The migration of a water-immiscible organic liquid into the unsaturated zone is governed largely by its density and viscosity. A schematic representation of NAPL in the unsaturated zone after Pinder and Abriola (1986) is presented in Fig. 8.4. It may be observed that the density of NAPL controls the fluid behavior of the junction between the NAPL and the water-saturated region. In the case when NAPL is less dense than water (Fig. 8.4a), it will tend to remain above the water table; when it is denser (Fig. 8.4b), it will continue its downward migration through the saturated zone. The rate of vertical flow of NAPL is determined by the properties of the liquid itself and those of the porous medium with its hydration status. The sinking phenomenon has been observed in physical model experiments (Schwille 1984) and near landfills (Mackay et al. 1985).

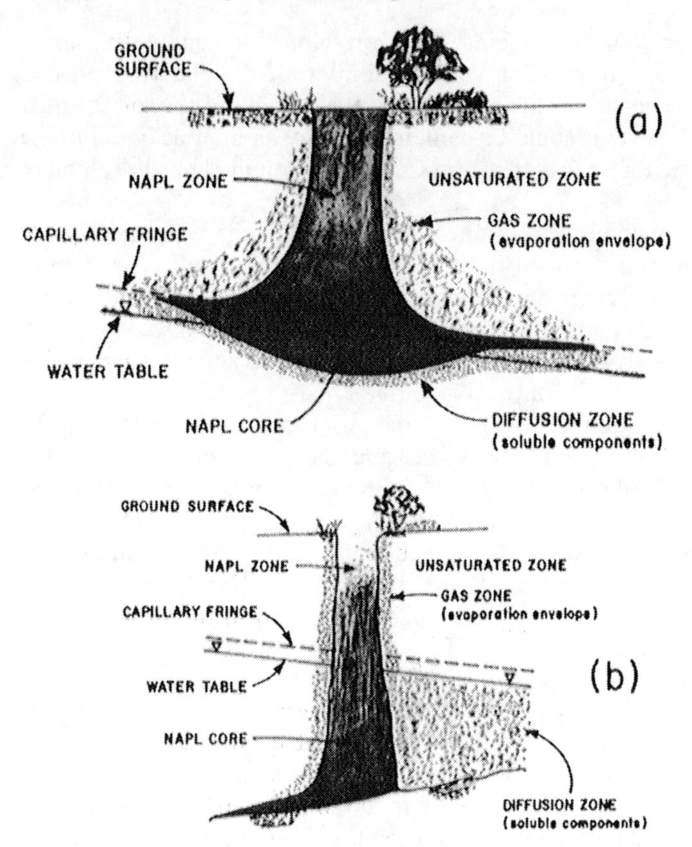

Fig. 8.4a,b. Schematic representation of lighter than water (**a**) and heavier than water (**b**) NAPL movement through the unsaturated zone and the saturated zone. (After Abriola and Pinder 1985)

The transport of a water-immiscible organic liquid phase could also be influenced by its surface-wetting properties compared with those of water. Schwille (1984) showed that, following the passage of the organic liquid, halogenated aliphatics tend to spread by capillary action into the aquifer media, where they tend to be retained in amounts of about 0.3–5% by volume. This points to the possibility of storage of large quantities of water-immiscible organic liquid contaminants as droplets dispersed within the pores of aquifer media, even if the bulk of the migrating mass of liquid is removed. The organic liquid droplets retained in the aquifer may then dissolve over time into the groundwater flowing past them. Mackay et al. (1985) reported that an organic liquid of moderately low solubility, which could be considered almost immiscible with water, can contaminate as much as 10 000 times its own volume. Organic compounds are only rarely found in groundwater in concentrations approaching their solubility limits, even when organic liquid phases are known or suspected to be present. The observed concentrations are usually more than a factor of 10 lower than the solubility limit, presumably owing to diffusional limitations of very low dissolution and the dilution of the dissolved organic contaminants by dispersion. This implies that the volume of the unsaturated zone and groundwater that could be contaminated by an organic liquid phase is much larger than that calculated by assuming dissolution to the solubility limit.

NAPL flux. As previously mentioned, the density of the water-immiscible fluid will govern its flux through the soil. When the fluid is of lower density than the water, its flux occurs only in unsaturated soils. When the fluid density is greater than that of the water, theoretically it will penetrate into a saturated soil, but on a completely different time scale. In our discussion we will examine the case of low-density fluids only.

The depth of infiltration of NAPL into the unsaturated soil depends on the spilled volume, the infiltration mechanism, and the retention capacity of the porous medium (Schwille 1984). The water-immiscible fluid flow depends on the relative permeability of the medium to NAPL mass, which is regulated by the presence of the water phase, as well as the solid matrix. Abriola and Pinder (1985), Lenhard and Parker (1987), and Parker et al. (1987) suggested the following flux equation for one-dimensional NAPL flow in the z direction:

$$J_p = -k_p K_{sp} (\partial h_p/\partial z + \rho_p/p*) , \tag{8.9}$$

where J_p is the flux of phase p (water, NAPL or air), k_p is the relative permeability of the medium to phase p, K_{sp} is the saturated hydraulic conductivity for phase p in the medium, ρ_p is the density of the phase used as the reference, and $h_p$is the fluid pressure head of phase p in the soil when p* is a scaled phase p. The fluid pressure in each phase depends on the configurations that the fluid forms at the interfaces within the soil pores. Jury and Fluhler (1992) noticed that the three-phase equilibrium pressures can be characterized as a function of the relative saturation of each phase in controlled laboratory experiments. However, the relative permeability, k_p, must be developed from an idealized geometric model of the porous medium (Parker et al. 1987) or else related empirically to the two-phase relative permeabilities of water and NAPL (Stone 1973). The flux equations for the three phases in soil are combined with mass conservation equations for each phase, to produce a transport model (Abriola and Pinder 1985).

The relative permeability is obtained experimentally as a function of void fraction or saturation. In the case of three-phase flow through an unsaturated soil, each relative permeability will depend upon the saturation of gas, aqueous phase and NAPL. Figure 8.5 (after Faust 1985) shows the relative permeability of the nonaqueous phase as a function of phase saturation. The diagram in Fig. 8.5 – which has not been experimentally verified – should be considered only on a general level.

As was previously shown (Chap. 5), as an immiscible fluid migrates through the unsaturated zone, a fraction of the fluid, termed residual saturation, remains behind as a coating of the pore walls of the porous medium, or as

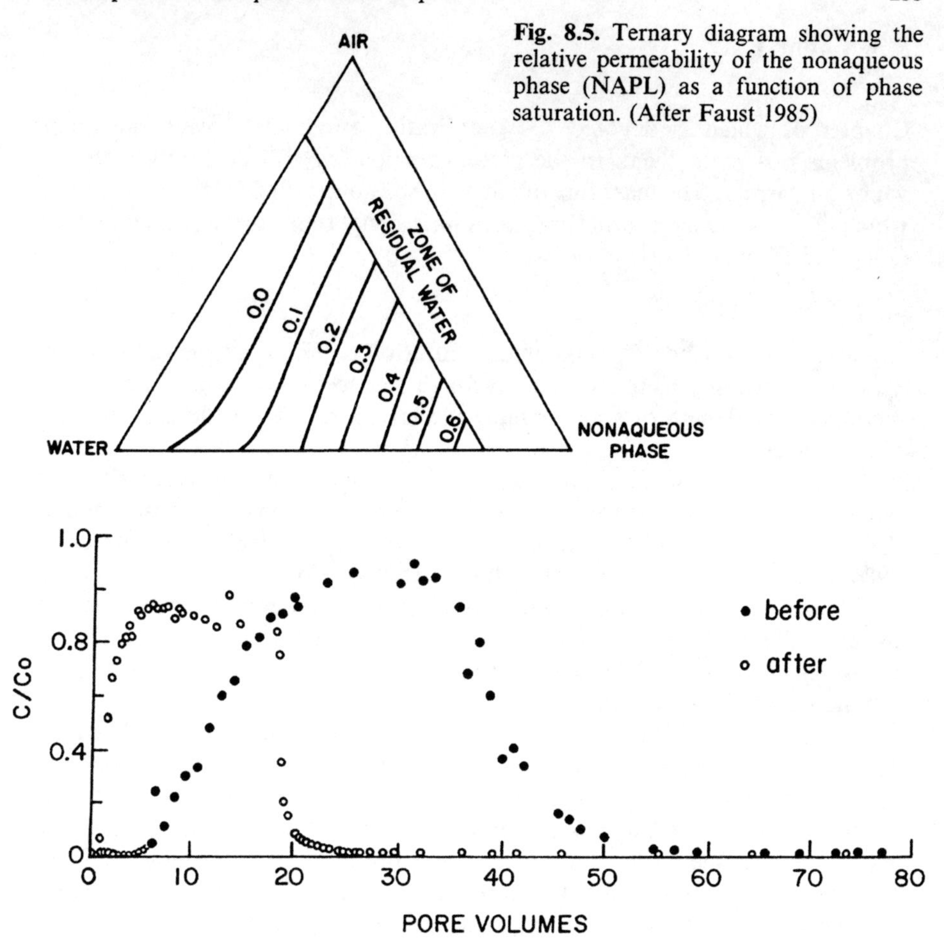

Fig. 8.5. Ternary diagram showing the relative permeability of the nonaqueous phase (NAPL) as a function of phase saturation. (After Faust 1985)

Fig. 8.6. Eluted toluene pulses through saturated zone material contaminated with residual hydrocarbons and through the same material after residual hydrocarbon removal by organic solvent extraction. (After Bourchard et al. 1989)

microdroplets trapped in the pore space by interfacial surface tension. This phase, which is a hydrophobic one, may affect the transport of other neutral organic compounds. Figure 8.6 (after Bourchard et al. 1989) shows the pulses of eluted neutral organic solvent (toluene) through saturated-zone material which was initially contaminated with residual hydrocarbon and through the same material after residual hydrocarbon was removed. It is clear that the toluene pulse was more highly retarded on the material contaminated with the residual hydrocarbons, indicating that these hydrocarbons provided a hydrophobic environment conducive to the sorption of toluene from solution. It may be concluded that, as a result of solvent, a residual hydrophobic film is formed on the porous material, which will affect the consequent transport of other water-immiscible fluids.

8.3 Vapor Flux

Chapter 4, which is devoted to volatilization processes, covers the main problems related to them. In the present section, we will deal only with the vapor flux aspect. The mass flux of vapor in soils obeys Fick's law of diffusion, which takes the general form for a one-dimensional transport in the z direction (Jury and Fluhler 1992):

$$J_g = \xi_g(a) D_g^a \, \partial C_g / \partial z \ , \qquad (8.10)$$

where J_g is the gas flux, D_g^a is the binary diffusion coefficient of the vapor in air, $\xi_g(a)$ is a tortuosity factor to account for the reduced cross-sectional area and increased path length of a vapor molecule in soil, and C_g is the gaseous chemical concentration.

A variety of vapor flux approaches, which include the soil tortuosity and porosity effect, as well as vapor adsorption on the solid phase or dissolution in the liquid phase, are described in the literature (e.g., reviews by Sallam et al. 1984; Collin and Rasmuson 1988; Hutzler et al. 1989).

The mechanisms that are assumed to affect transport in unsaturated systems include advection with air and water, dispersion in air and water, diffusion across immobile water interfaces, mass-transfer resistance at the air-water and mobile-immobile water interfaces, and adsorption.

When a chemical characterized by a high vapor pressure is applied to soil, it can be partitioned among the soil solid phase, the mobile and immobile soil water, and the soil vapor phase. The chemical vapor transport into the soil and from the soil to the atmosphere or to the groundwater occurs both under natural environmental conditions and as a result of an anthropogenic intervention during a remedial vapor-extraction process. Temperature can play a significant role in vapor transport under certain conditions.

The mass transfer of contaminant among the gas phase, the dissolved phase and adsorption may occur at equilibrium or nonequilibrium. The approach of Armstrong et al. (1994) is used to describe the above situation:

1. The advective-dispersive two-dimensional transport equation in the case of a static moisture distribution and an incompressible air phase is:

$$\Theta_a D_{ij} \frac{\partial}{\partial x_i}\left(\frac{\partial C_a}{\partial x_j}\right) - \Theta_a v_i \frac{\partial C_a}{\partial x_i}$$

$$= \Theta_a \frac{\partial C_a}{\partial t} + \Theta_w \frac{\partial C_w}{\partial t} + \rho_b \frac{\partial C_s}{\partial t} \quad i,j = x,z \ , \qquad (8.11)$$

where subscripts a, w, and s designate the gaseous, dissolved, and sorbed phases (also referred to as the air, water, and soil phases) of the contaminant, C_a, C_w (M/L^3) and C_s (dimensionless) are the corresponding concentrations, Θ_a and Θ_w are the volumetric air and water contents (dimensionless), and ρ_b is the soil bulk density (M/l^3).

2. When the contaminant is dissolved in water or adsorbed on soil solid phase, in addition to the gas phase, a mass-transfer process occurs between the phases. Assuming an equilibrium partitioning, Henry's law can be used for defining the equilibrium between gas and water phases and Freundlich isotherm for defining the equilibrium between the solute and the solid phase.

Expressing C_w and C_s, on the right-hand side of Eq. (8.11), in the form of equivalent air phase concentrations, C_a, Eq. (8.11) then becomes

$$\Theta_a D_{ij} \frac{\partial}{\partial x_i}\left(\frac{\partial C_a}{\partial x_j}\right) - \Theta_a V_i \frac{\partial C_a}{\partial x_i} = R\Theta_a \frac{\partial C_a}{\partial t} , \tag{8.12}$$

where R is the retardation factor:

$$R = 1 + \frac{\Theta_w}{\Theta_a H} + \frac{\rho_b K_d}{\Theta_a H} ,$$

where K_d is the distribution coefficient (from Freundlich's isotherm) and H the Henry's law constant.

3. While the equilibrium form is valid for diffusion-dominated conditions, it is not adequate to describe a strongly advective situation. Assuming that diffusive mass transfer between the air phase and the water phase takes place through a boundary layer, Armstrong et al. (1994) expressed the driving force as the concentration gradient between the average concentration in the aqueous phase and the equilibrium concentration at the water/air interface. This idealized diffusive mass transfer process is approximated by a first-order kinetic relationship and represents the aggregate of all physical nonequilibrium processes affecting the migration of contaminant mass between the water and air phases, including diffusion within the soil moisture and dead-end pore effects.

The mass-balance equation for the stationary water phase with mass transfer across the air-water and the water-solid interfaces is:

$$\Theta_w \frac{\partial C_w}{\partial t} = \Theta_a \lambda_{aw}\left(C_a - HC_w\right) - \Theta_w \lambda_{ws}\left(C_w - \frac{C_s}{K_d}\right) , \tag{8.13}$$

where the first term on the right-hand side expresses air-water transfer and the second term describes water-solid transfer, with λ_{aw} and λ_{ws} being the corresponding mass-transfer coefficients; C_w represents the average concentration in the water, while C_a/H and C_s/K_d represent the equilibrium concentrations at the air-water interface and the water-solid interface.

The partitioning between water and solid phases occurs according to a similar first-order mass-transfer process. The kinetically limited desorption between the soil aggregates and interparticle diffusion within the saturated soil particles is considered in the above development. Brusseau (1991) developed a more general approach, assuming both advective and nonadvective domains within the soil and accepting the existence of both equilibrium and nonequilibrium phase transfer in each domain.

8.4 Transport on Suspended Particles

Because of their high degree of sorption on soil colloidal materials, many pollutants, such as radionuclides, heavy metals, or sparingly water-soluble organic compounds, are considered to be immobile in the environment and are, therefore, considered not to be potential contaminants. However, in nature, the transport of the synthetic pollutants involves not only solutes or vapors; it also occurs when they are adsorbed on sediments and colloidal materials dispersed in the aqueous phase. The chemical transport in association with suspended particles occurs either on the ground surface, in eroded sediments, or into the soil profile, in association with organic macromolecules and inorganic colloids. These two aspects will be discussed below.

8.4.1 Transport on Eroded Sediments

Runoff water, which can be generated by either rain or irrigation, is an important means of transport of chemicals from one soil location to another or from the soil to surface water bodies.

What is the difference between the transport of chemicals in runoff water and that in soil water? Usually, in soil water, chemicals are transported only as solutes and the governing factor in their transport is their concentration in solution, whereas in runoff water, they are transported both as solutes and adsorbed on suspended particles. Thus, highly insoluble and adsorbed pollutants may be moved.

In addition to molecular and soil properties, the amount of chemicals found in runoff water is affected by the intensity of the falling water and the slope of the land. The speciation status of the chemical, as well as its method of application to the land surface, may also affect its transport by runoff. Partitioning between the solid and liquid phases will determine the ratio between the amounts of chemicals transported as solute and with particles. The enrichment ratio of an eroded sediment (E_R) in relation to that of the original polluted soil is given by the relation

$$E_R = \frac{\text{g chemical kg}^{-1}\text{ soil} \quad \text{eroded sediment}}{\text{g chemical/kg}^{-1}\text{ soil} \quad \text{original soil}} . \tag{8.14}$$

Menzel (1980) observed that E_R is commonly greater than unity and when this is so, it commonly declines as the amount of soil lost during an erosion event increases. Rose (1993) developed Eq. (8.14) on the basis of experimental results, and expressed the loss of any chemical N (g m^{-2}) by:

$$N = E_R\, M C_N \ , \tag{8.15}$$

where M is the accumulated mass of chemical per unit area of soil during a water erosion event, C_N being the areal concentration of chemical N in the original soil ($g\ m^{-2}$). It is evident from Eq. (8.15) that the erosion-induced transport of any soil-sorbed chemical is given by the product of the amount of sediment lost from the given area of soil and the concentration of the chemical of concern in the eroded layer.

Rose (1993) pointed out that at least two quite different erosion processes result in the enrichment of sorbed chemicals. The first process applies to both water and wind erosion. In both cases, the erosion mechanism appears to be nonselective for all except quite large aggregates and gravel-sized particles. However, once a particle is dislodged into the moving fluid, its rate of falling through the fluid and, hence, the rate of deposition, are strongly size-dependent. The size selectivity of the deposition process, combined with the commonly nonuniform variation of concentration of sorbed chemical with particle size, is one reason for the enrichment of chemicals in the eroded sediment. A second type of process which can lead to chemical enrichment occurs when rainfall impact is an important erosion mechanism and where soil aggregates are stable and do not immediately collapse under such impact. This can occur in some soils whose stable aggregates remain intact for a number of direct hits by raindrops. Such collisions of the drop with soil lead to very fast lateral water flow, and the resultant high stresses strip off the outer layer of the aggregate, producing much finer particles, which are then preferentially eroded. This process is termed raindrop stripping and, especially when depths are shallower than the size of larger stable aggregates, this enrichment process can be dominant. Chemical transport on the land surface can be understood in terms of preferential long-distance transport of the finer sediment fraction (with potentially the highest adsorbed chemical content), in the suspended load, leaving behind the coarser sediment.

8.4.2 Transport in Association with Dispersed Colloidal Material

When colloidal material is present in the aqueous phase, as a result of soil particle disturbance or as colloidal organic material, a three-phase system exists with the chemical sorbed on the soil solid phase, dissolved in the water phase, or adsorbed on dispersed colloidal material. Transport of the colloid-solute species in the environment will depend on the convective transport of the aqueous phase and the degree of colloidal association with stationary soil material. For example, an organic colloid particle dispersed in an aqueous phase will act as a sorbent for a neutral organic molecule and in these conditions, its transport will be greatly facilitated.

The apparently increased solubility of many pollutants in the aqueous phase is enhanced through complexation with naturally occurring humic and fulvic acids and root exudates, or through their binding to colloidal clay metal-oxides or other inorganic colloids.

The apparent solubility of various micropollutants was the object of many studies during the 1980s. The distribution of activated elements in some surface

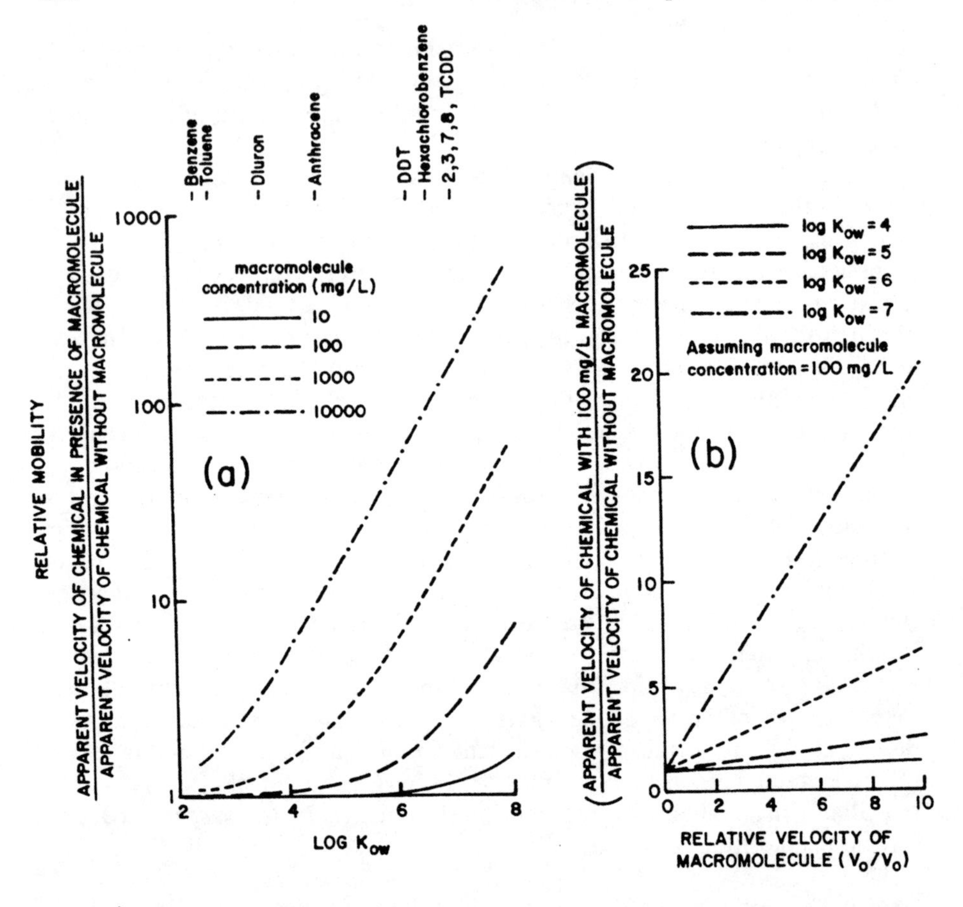

Fig. 8.7a,b. Mobility of a hydrophobic compound relative to the mobility of the same compound without the presence of any COC, **a** as a function of octanol-water partition coefficient and amount of macromolecule (COC) organic C in the mobile phase, and **b** versus the velocity of the COC relative to the velocity of water where the concentration of COC is 100 mg l^{-1}. (After Enfield and Bengisson 1988)

waters, for example, was explained by their binding on colloidal organics (Nelson et al. 1985). It was also observed that inorganic colloids are also important in radionuclide disturbance. The same active elements may associate with colloidal Fe oxides on other inorganic colloids in ground water. Kim et al. (1984), in their review of transport processes involving organo-chemicals and Bourchard et al. (1989) discuss the solute transport in association with dispersed colloidal matter. It is emphasized that in the case of neutral organic compounds in soil (NOC_s), the hydrophobic retention on organic colloids is a much more important solubilizing process than NOC_s associated with inorganic colloids. The sorption of NOC_s to colloidal organics (COC) leads to an increased apparent solubility of NOC_s itself, and this explains why hydrophobic compounds such as organochlorinated pesticides have been observed to be much more mobile in the environment than predicted by models which assume a simple two-phase system. In the same way, it was observed that macromolecule transport in the aqueous phase may facilitate NOC_s transport (Enfield et al. 1982; Enfield 1985). The COC influence on the apparent solubility of NOC_s is most significant for low-soluble, highly sorbed solutes. It is for these hydrophobic NOC_s that the relative increase in solubility due to sorption on colloidal material is the greatest.

Enfield and Bengisson (1988) proposed an approach to describing colloid-facilitated transport which involved dividing a representative elemental volume of the soil system into three phases: a liquid or aqueous phase; a solid phase, the soil; and a mobile organic phase, the macromolecules. The importance of macromolecules to the mobility of hydrophobic compounds is graphically presented in Fig. 8.7a and b. Summarizing the process, Bourchard et al.(1989) showed that COC-facilitated transport of NOC_s will depend primarily on the following: (1) NOC_s sorption to COC, which will be greatest for low-solubility NOC_s and hydrophobic COC; (2) the amount of COC present in the aqueous phase, which, at concentrations normally observed in nature, facilitates the transport of very-low-soluble NOC_s; and (3) COC mobility.

8.5 Factors Affecting Transport Processes

The transport of pollutants from the land surface to the groundwater is affected by a broad range of environmental factors, of which the most important are the spatial variability of field and climate characteristics, and the heterogeneity of the incoming pollutant.

8.5.1 Field and Climate Variability

The fact that variation of each soil property is not completely random in space has always been noticed by pedologists. These properties include those which control the transport of pollutants into soils, such as soil hydraulic properties. The conventional approach describing the spatial variability of soil hydraulic properties treats the observation of a given property as being statistically independent, regardless of the spatial position. The data from the observation of each individual property are used to estimate a probability density function for that particular property (Nielsen et al. 1973). This statistical approach usually neglects the spatially structured arrangement of the natural porous media. Russo and Bresler (1981) incorporated the spatial structure of the properties in their treatment of the field variability of soil hydraulic parameters. In their approach, the spatial structures of the hydraulic properties are considered in terms of the probability density functions, autocorrelation functions, and integral scales. The soil system is regarded as a continuum, the properties of which are continuous functions of the space coordinate. The elementary volume representing the hydraulic properties of the field is large compared with the soil-pore scale but small compared with the scale of heterogeneity in the field, and the hydraulic-conductivity continuum could be represented by a three-dimensional spatial stochastic process governed by probability laws (Bear 1972; Yevjevich 1972). To understand the effect of field spatial variability on the transport of chemicals, we have to consider the above stochastic process as an ensemble with the same statistical properties or as a repeated measurement of a given property in a given time and place under the same conditions. In characterizing the spatial distributions of the soil hydraulic properties, Russo and Bresler (1981) considered the determinations of the sorptivity (S) and five parameters describing the hydraulic conductivity [K(h)] and soil water retentivity [Θ(h)] functions. The parameters are: saturated hydraulic conductivity (K_s); water entry value (h_w); saturated (Θ_s) and residual (Θ_r) water contents; and a constant β characterizing the pore size distribution of the soil. For a given depth, each of these parameters is described as a realization of a stationary two-dimensional isotropic and random process. These stochastic processes are characterized by truncated normal or log-normal probability density functions, independent of the spatial position, and by autocorrelation functions between any two spatial points in the field, which depend solely on the size of the vector separating the two points. The spatial variability of each of the six parameters has a structure that is characterized by a characteristic length – the integral scale, I_s representing the largest distance for which the parameter is correlated with itself. Values of I_s, which are calculated from the autocorrelation functions for each parameter, generally decrease with increasing depth.

Addiscott and Wagenet (1985) suggested a method by which variable soil properties expressed in a distribution may be combined to estimate a variate relationship. The probability distribution (normal, log-normal, etc.) of the input for each property is divided into sections, each related to the same number of observations. The section medians for each property are then combined through the functional relationship involved in all combinations of section and property, to generate a population of values for the required variate, without presupposing any particular type of distribution for it. The method gave estimates of unsaturated hydraulic conductivity, K(Θ), and additionally revealed a marked increase in the variance of K(Θ) as the volumetric moisture content, Θ decreased.

Together with soil variability, climate variability contributes in an unknown manner to the leaching of pollutants from the land surface to the groundwater. Jury and Gruber (1989) coupled the variability of these two factors to produce a probability density distribution of the residual mass fraction remaining in soil after the leaching of the chemicals below the surface degradation zone.

8.5.2 Heterogeneity of the Pollutant

Under land treatment of hazardous organic chemicals or land disposal of wastes, the pollutants reach the soil, not as a single compound with well-defined properties, but as a mixture of compounds with various physico-chemical characteristics. Under such conditions, we are dealing with transport of pollutants from solutions containing mixtures of water and water-miscible organic solvents. The parameters controlling the transport of organic pollutants into soils via the respective processes, such as sorption, aqueous phase solubility, or degradation, will be different, and simple extrapolation of existing data on solute in aqueous solutions to mixed-solvent solutions is not feasible. Rao et al. (1985), considering the transport of hydrophobic organic chemicals (HOC) in aqueous and mixed-solvent systems, assumed that in a soil mixed-solvent system, the HOC sorption is controlled primarily by solubility (hence, activity) in the solvent from which sorption occurs. Solubility in a given solvent mixture decreases exponentially with increasing sorbate hydrocarbonaceous surface area (HSA), leading to an exponential increase in sorption coefficient with increasing HSA. For sorption from binary solvents, for example, an increase in the fraction of organic cosolvents (f_c) results in an exponential increase in solubility and causes an exponential decrease in sorption coefficient. Because sorption and leaching are inversely related, an increase in f_c leads to an enhanced HOC mobility in soils. If during the pollutant transport a degradation process occurs, the extent of that process is also affected by the interactions between the components of the mixtures. Oh et al. (1994) proved that the interactions between hydrocarbons during their biodegradation process can affect the degradation pathway of each hydrocarbon under cross-inhibitory kinetics. As a result, the mobility of each compound of the hydrocarbon mixture could also be affected.

8.6 Examples

In this chapter, we have already discussed the processes controlling the transport of pollutants in the soil medium, the means of transport, and the factors affecting this process. A number of examples were selected to illustrate the main topics previously discussed. It is, however, necessary to underline that space limitations force us to cover only a few examples from the vast quantity of publications in this field.

8.6.1 Diffusion of a Degradable Pesticide

There are two possible ways of treating the diffusion of a pesticide in soil. The first considers the soil to be biologically inert, while the second allows for the presence of bioactive organisms. Gerstl et al. (1979b) reported on an investigation of the diffusion of parathion (O,O-diethyl O-ρ-nitrophenyl phosphoro-thioate) in a biologically inert ("sterile") loamy soil (Gilat, Israel), and on the diffusion of the same pesticide in the same soil before its sterilization.

In the sterilized soil it was found that the apparent diffusion coefficient (D) of parathion over a wide range of moisture contents varied between 0.66×10^{-7} and 3.39×10^{-7} cm^2s^{-1}. The main factors affecting the diffusion of parathion in a sterile soil are the adsorption coefficient of the compound on the soil solids surface (K), the moisture content (Θ), and the impedance factor (f).

Figure 8.8a shows an example of parathion distribution in the Gilat soil after 4 days, when the soil was packed at a bulk density (ρ_b) of 1.4 g cm^{-3}, and at a moisture content of 20% (w/w). The calculated distribution using Eq. (8.16) accurately described the measured parathion distribution. This equation was developed by Crank (1975) for one-dimensional diffusion in a semi infinite medium:

$$C(x,t) = \frac{C_0}{2}\left(\mathrm{erfc}\frac{x}{2\sqrt{Dt}}\right) , \tag{8.16}$$

where C_0 is the concentration of the diffusive species in the source part and erfc is the error function complement. The results fitted the experimental data for all cases.

In the presence of bioactive microorganisms, parathion is degraded during its diffusion in soil. Gerstl et al. (1979a) reported on a complementary experiment where a block of Gilat soil, uniformly mixed with parathion, was

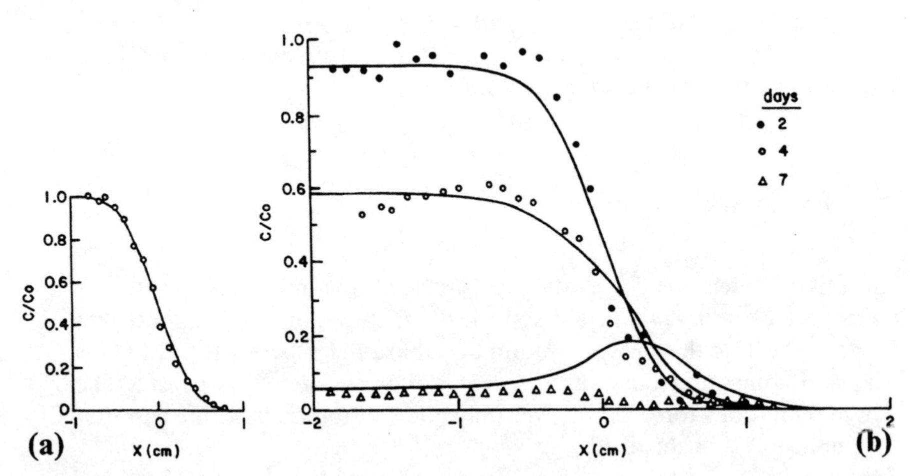

Fig. 8.8. a Distribution of parathion in a biologically inert Gilat soil (20% moisture content) after 3.03 days. The *solid line* was calculated using D = 1.67×10^{-7} cm^2 s^{-1}; the *points* represent experimental measurements. **b** Daily calculated (*solid line*) and measured (*points*) distributions of parathion in a biologically active Gilat soil during the process of diffusion. Volumetric soil moisture content θ = 0.34. (After Gerstl et al. 1979a)

placed in contact with a parathion-free block. The problem was to predict the distribution of parathion in this system, in which the concentration of parathion is affected by both microbial decomposition and diffusion.

The continuity equation for parathion is

$$\partial C/\partial t = D\partial^2 C/\partial x^2 - R_{xt} \ , \tag{8.17}$$

where R_{xt} is the rate of microbial decomposition.

Since parathion is converted quantitatively mole for mole into its decomposition product, the continuity equation for the product is

$$\partial C'/\partial t = D'\partial^2 C'/\partial x^2 + R_{xt} \ , \tag{8.18}$$

when C' is the concentration of the decomposition product. Clearly, the rate of decomposition at any distance and time will depend on the local concentration of parathion and on the microbial activity. These factors will reflect earlier changes not only in concentration, but also in microbial activity at the distance in question.

Figure 8.8b shows the parathion distribution in the nonsterilized Gilat soil after 2, 4, and 7 days, with the soil packed at a bulk density (ρ_b) of 1.4 g cm^{-3} and at a moisture content Θ of 0.34. It was observed that at the very beginning (after 2 days of incubation), when the percentage decomposed was still very low, the distribution of parathion along the diffusion cell was similar to that occurring in the sterile soil. After 4 days of incubation, with the buildup of microbial activity and the increase of decomposition, the distribution pattern was drastically changed. At this time, the parathion distribution in the nonsterile soil differed from that in the sterile soil mainly in the half-cell in which the pesticide was initially added; it differed less in the second half-cell which was initially parathion free. After 7 days of incubation the parathion was almost completely decomposed and its distribution in both half-cells was affected by the decomposition process. This may be explained in terms of the development of microbial activity in the half-cell which was initially free of parathion.

The above example points out the necessity to consider additional processes occurring in soil (e.g., degradation) during pollutant diffusion, and to include their parameters in the prediction of the redistribution of the pollutants in the soil medium.

8.6.2 Miscible Displacement

When a fluid containing a dissolved tracer is displaced from a porous medium by a tracer-free quantity of the same fluid, this miscible displacement results in a tracer concentration distribution which depends upon microscopic flow velocities, tracer diffusion rates, and other chemical and physical processes. This phenomenon will be exemplified by the results of Nielsen and Biggar (1961, 1962), reported by them in a series of three papers which became a benchmark in the understanding of solute transport in soils. The distance a solute travels through a bulk soil is determined by the tortuosity of the total path length it follows. Owing to the differing magnitudes of the convection, diffusion, and chemical processes which occur in different pore sequences, the paths of all ions will not be the same, and the resulting tracer distribution will clearly give a good deal of information about the behavior of the water flowing through various soils.

Miscible displacement was studied by Nielsen and Biggar (1962) in several porous materials under saturated and unsaturated conditions and at several different average flow velocities. Physical differences among porous materials were manifested by changes in the shape and position of breakthrough curves caused by ionic diffusion. The abscissa in Fig. 8.9a gives the volume of effluent divided by the volumetric water capacity of a given porous material, i.e., the number of pore volumes displaced. Thus, if one pore volume of tracer-labeled water completely displaced the tracer-free water without any mixing at the boundary between the two fluids, the breakthrough curve for all materials would be that of the broken vertical line. The displacement of the breakthrough curve to the left of this vertical line or the area between the two lines is a relative measure of the volume of water not displaced but remaining within the sample. The magnitude of the translation would be expected to be associated with soil texture and aggregation, sands having the smallest volume of water not readily displaced, and clays the greatest. Figure 8.9a shows glass beads and Oakley sand to have had the least holdback, while Yolo loam had the most. Because of the large flow velocities used for the four materials, the effects of ionic diffusion on the translation of the breakthrough curves are small. The shapes of the breakthrough curves for the various media were dependent upon microscopic flow velocity distribution and tracer diffusion, but could also be affected by adsorption and ion exchange.

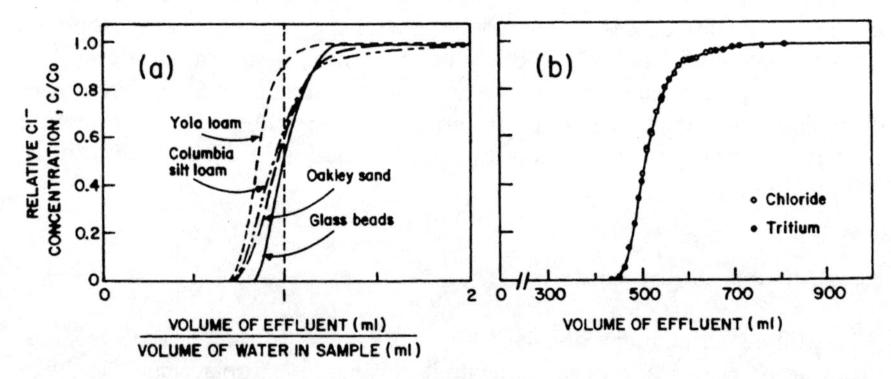

Fig. 8.9. a Breakthrough curves for four saturated porous materials – Yolo loam, Columbia silt loam, Oakley sand and glass beads. The average flow velocities were 1.89, 2.49, 0.30, and 1.77 cm h^{-1}, respectively. **b** Breakthrough curve for tritium and chloride from saturated 200-μm glass beads. The average flow velocity and volumetric capacity of the sample were 2.11 cm h^{-1} and 512 ml, respectively. (After Nielsen and Biggar 1962)

In a subsequent paper, Biggar and Nielsen (1962b) checked the contribution of diffusion to the spreading of solutes by using as tracers chloride and tritium, which are characterized by unequal diffusion coefficients. Since the diffusion coefficient of tritium is greater than that of Cl^- in pure solution, the fact that no separation of tracers could be measured in the case of the glass-bead column indicates that the dispersion which occurred resulted mainly from microscopic flow velocity distribution (Fig. 8.9b).

Based on these results, the authors concluded that the distribution of a tracer at some distance from its source depends upon the geometry of the porous material and the physical and chemical interactions between the tracer solution and the medium during flow. When an incoming fluid is identified by a solute of concentration C_0, the fraction of this solute in the effluent at time t (h) will be C/C_0.

Plots of C/C_0 versus pore volume, commonly called breakthrough curves, are descriptive of the relative times taken for the displacing fluid to flow through the medium or for the solutes in the displacing fluid to come into chemical equilibrium with the soil. C/C_0 is the relative concentration of the incoming liquid found in the effluent. Figure 8.10 (after Nielsen and Biggar 1962) illustrates this for various types of breakthrough curves for miscible displacement. Piston flow would rarely, if ever, occur in soils; for a single capillary tube of constant radius, there is only a narrow range of flow velocities within which this type of flow is approached. Because soils do not have pore sequences of constant radii, the probability that their breakthrough curves

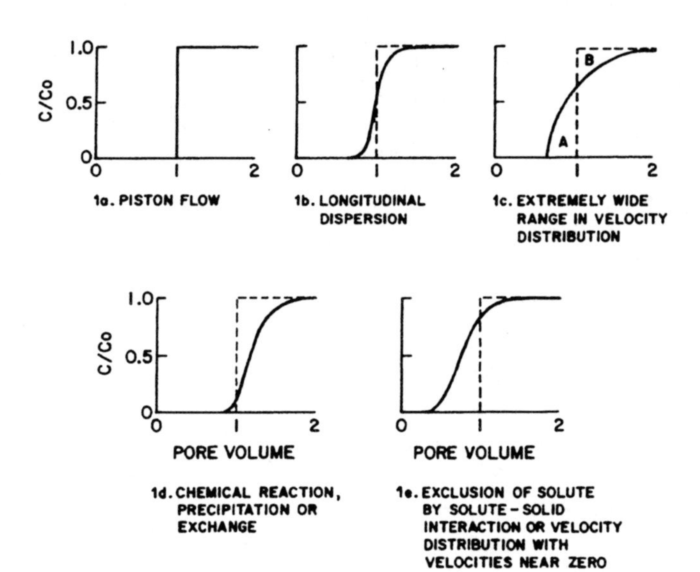

Fig. 8.10,1a–1e. Types of breakthrough curves for miscible displacement. C/C_o is the relative concentration of the invading fluid measured in the effluent. Pore volume is the ratio of the volume of effluent to the volume of fluid in the sample. (After Nielsen and Biggar 1962)

would be piston type is almost nil. The curves presented in Fig. 8.11 exhibit a minimum interaction between the solute, solvent, and solid. This interaction, called holdback, is three to four times greater under unsaturated than under saturated conditions. When displacing fluid or its solutes are retained within the column by any chemical or physical process, the breakthrough curve will be transposed to the right. However, the shape of the curve is not determined by these retaining processes alone, but is affected by the microscopic velocity distribution and other processes (such as anion exclusion), which would transpose the curve to the left. Nielsen and Biggar (1962) give an example of a tritium breakthrough curve at two water contents being translated to the left as a result of transport through unsaturated Yolo loamy soil (Fig. 8.11d); they also present chloride breakthrough curves in Oakley sand and Aiken clay loam.

8.6.3 Transport of Nonaqueous Pollutant Liquids (NAPL)

The flow behavior of immiscible organic liquids (NAPL) entering an unsaturated soil medium follows two well-defined scenarios: the first deals with the infiltration of an NAPL into soil during a short period of time, during which the physical properties of the liquid remain unchanged; in the second, the liquid properties are altered during its transport through the porous medium. Since the dissolution of the organic solvents which control the bulk of

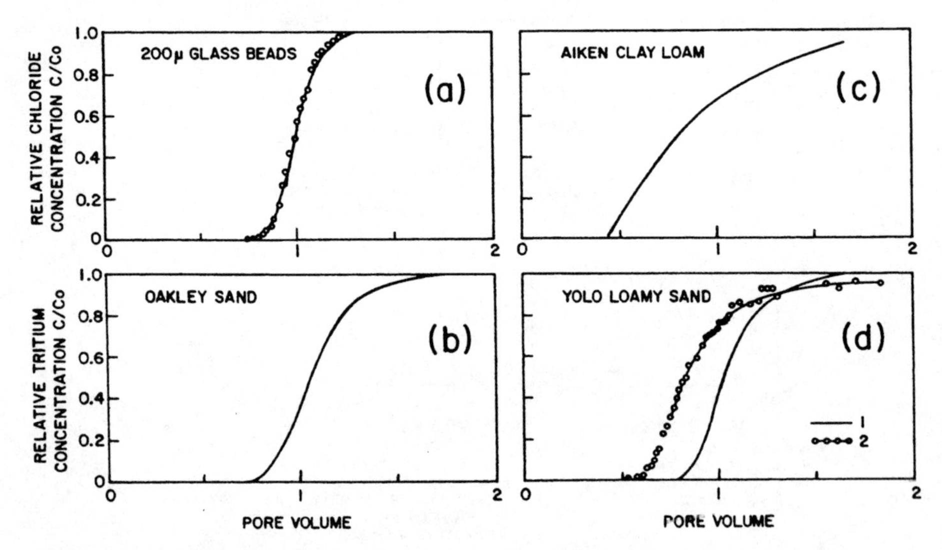

Fig. 8.11a–d. Chloride and tritium breakthrough curves measured for various soils and glass beads. (After Nielsen and Biggar 1962). **a** v = 4.94 cm h^{-1}. **b** v = 0.79 cm h^{-1}. **c** v = 3.06 cm h^{-1}, 1–2 mm aggregates. **d** 1 - v = 0.41 cm h^{-1}, Θ = 0.39 cm^3 cm^{-3}; 2 - v = 0.43 cm h^{-1}, Θ = 0.37 cm^3 cm^{-3}

NAPL in the water phase is negligible, it can be neglected when discussing NAPL transport through soils. The experiments carried out at Richland (WA, USA) and Bet Dagan (Israel), and reported in the papers of Cary et al. (1988a, b,c, 1994), Galin et al. (1990a,b), Gerstl et al. (1994), and Jarsjo et al. (1994) will be used to illustrate this process.

The infiltration and redistribution of two hydrocarbons which have relative viscosities 4.7 (Soltrol) and 77 (mineral oil) times greater than that of water in moist silt loam and loamy sand soils was reported by Cary et al. (1989a, b, c). The distribution of the two hydrocarbons and of the water in the silt loam soil 8 h after adding water and 4 h after adding the hydrocarbons is presented in Fig. 8.12a. As expected, the infiltration rate of hydrocarbons is inversely related to their viscosity. The mineral oil remains in the upper layer of the soil column, soltrol redistribution being similar to that of the water.

For the same hydrocarbons, Cary et al. (1994) simulated the spreading pressure (SP) effects (Fig. 8.12b). Simulations were performed for the case of mineral oil following water in the sand column with a silt loam layer. The spreading pressure was varied between 20 and 100 cm of water. When assaying the amount of oil and water at the end of the experiment, Cary et al. (1994) found that the mineral particles had become largely hydrophobic whenever the transmission oil contained a proprietary additive that contained, in part, alkyl sulfonates that alter interfacial properties. When the water entered these oil-filled pores, it did not wet the mineral particles and vigorously expelled the

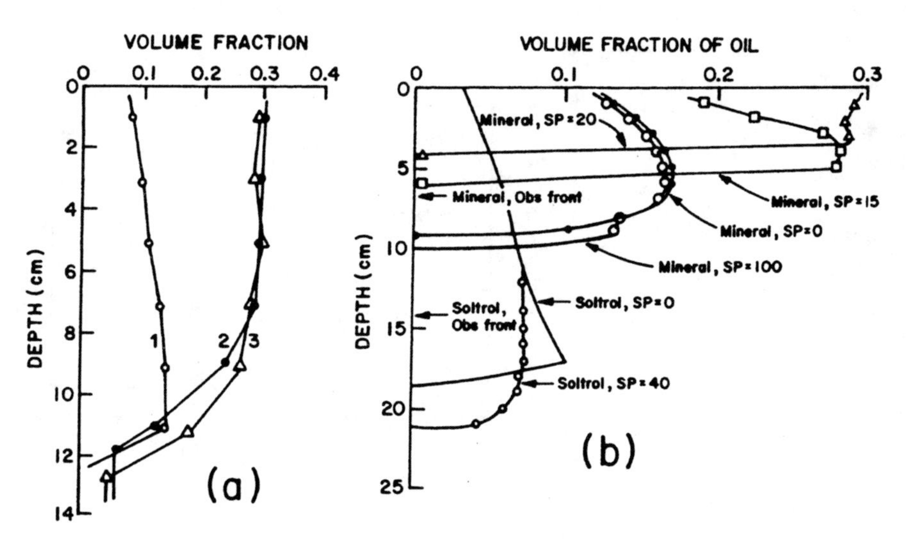

Fig. 8.12a,b. Transport of mineral oil and Soltrol 220. (After Cary et al. 1989a,b, 1994). **a** The distribution of (2) Soltrol-220, (1) mineral oil, and (3) water in a silt loam soil: 8 h after adding water to dry soil and 4 h after adding the hydrocarbons. **b** Distribution of mineral oil and Soltrol-220 as a function of spreading pressure (*SP*) and depth (4 h after infiltration)

transmission oil. The above examples illustrate the core results for a single NAPL compound with a constant viscosity on the infiltration and redistribution process. Galin et al. (1990a,b), Gerstl et al. (1994), and Jarsjo et al. (1994) reported data on the volatilization and temperature dependence of soil conductivity for volatile organic liquid mixtures (VOLM). These experimental results show that volatilization caused a change in the composition of VOLM during its migration in porous media, with an increase in VOLM viscosity as a result, and that changes in the ambient temperature induced a fluctuation in VOLM viscosity. Figure 8.13a shows the effect of viscosity on the conductivity of kerosene – a petroleum product which was used as an example of a VOLM – in three types of soils with different physicochemical properties. The effect of soil moisture content on kerosene conductivity is shown in Fig. 8.13b. The kerosene conductivity of the soil was found to be strongly influenced by the soil texture and initial moisture content, as well as by the kerosene composition. The kerosene conductivity of the sand was two orders of magnitude greater than that of the soils, and was unaffected by initial moisture contents as high as field capacity. The kerosene conductivity of the loam soil was similar in oven-dry and air-dry soils, but increased significantly in soils at 70% and full field capacity. In the clay soil, the kerosene conductivity of the air-dry soil was four times that of the oven-dry soil, and it increased somewhat in the soil at 70% field capacity. No kerosene flow was observed in the oven-dry clay soil at full

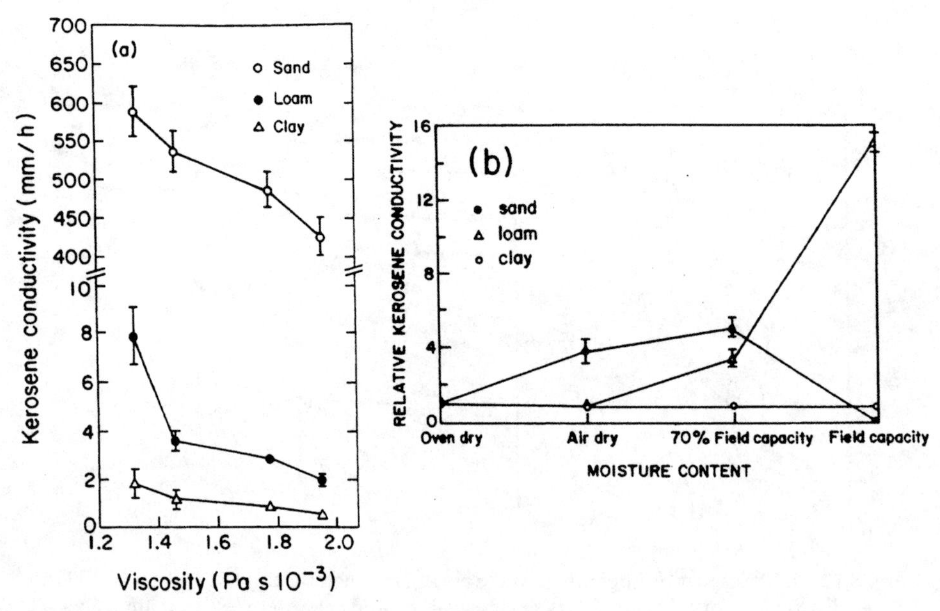

Fig. 8.13a,b. Kerosene conductivity as affected by **a** viscosity, and **b** soil moisture content. (After Gerstl et al. 1994)

field capacity. The differences in kerosene conductivity of these soils and the effect of moisture content were attributed to the different pore-size distributions of the soils. Changes in the composition of the kerosene due to volatilization of the light fractions resulted in increased viscosity of the residual kerosene. This increased viscosity affected the fluid properties of kerosene, which resulted in decreased kerosene conductivity in the sand and in the soils.

In contrast with the change in VOLM viscosity due to selective volatilization of the light fractions, which is a long-term irreversible process, the fluctuation in viscosity due to changes in the ambient temperature is an instantaneous, reversible process. Jarsjo et al. (1996) experimentally studied the changes of porous media conductivity for kerosene in a series of materials with different initial properties, as affected by viscosity changes induced by volatilization and temperature (Fig. 8.14). The kerosene-saturated conductivity K_{sk} in air-dry soil at 24 °C was in the range of 1.8–123 mm h^{-1} for clay soils and 7.9–57.4 mm h^{-1} for loamy soils. In the sandy soil, the KSK was less than 1 order of magnitude greater than in other soils. This conductivity value was used to scale the impact of kerosene viscosity changes on the saturated kerosene conductivity for the soils studied. The temperature effect on kerosene conductivity was tested in coarse sand 2, sandy loam 1 and 2 (see Fig. 8.14), peat, and glacial and post-glacial clay. The ratio between the kerosene viscosities at 24 and 5 °C was 0.71, and is illustrated in Fig. 8.14a (solid line), along with the ratios between the corresponding kerosene conductivities for the various soils (bars). For peat and the sandy soils (coarse sand 2, sandy loam 1, and sandy loam 2, which are characterized by a sand fraction of more than 65%, according to the USDA system of classification), the conductivity ratio K_{sk} (5 °C)/K_{sk} (24 °C) deviates by less than 2% from the viscosity ratio μ (24 °C)/μ (5 °C). For the soils with a

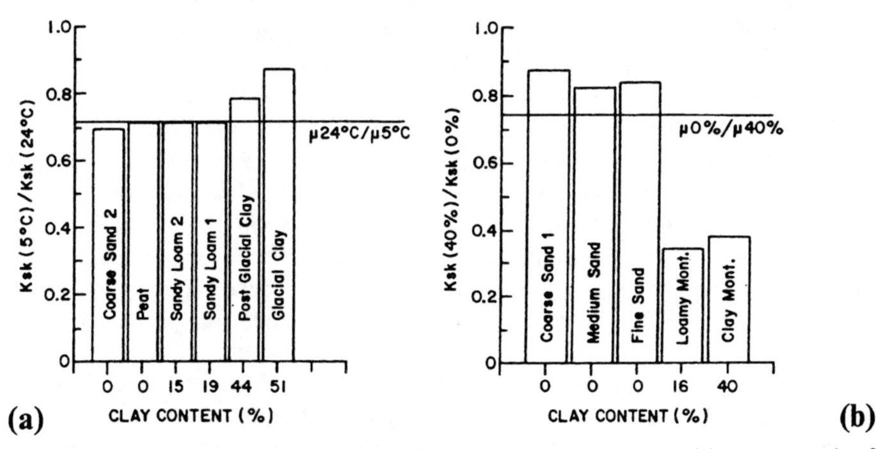

Fig. 8.14a,b. Effect of **a** ambient temperature and **b** kerosene composition as a result of volatilization, on soil kerosene conductivity. (After Jarsjo et al. 1996)

higher clay content, the corresponding deviation reached a maximum of 18%, which is more or less within the range of experimental deviations.

The conductivity ratios for kerosene-0% before volatilization and after volatilization of kerosene-40% in the different soils are illustrated in Fig. 8.14b (solid line), along with the corresponding ratio of kerosene viscosities (solid line). The deviations between the conductivity ratio $K_{sk}(40\%)/K_{sk}(0\%)$ and the viscosity ratio $\mu(0\%)/\mu(40\%)$ for the investigated sands (i.e., the coarse, medium, and fine sands are somewhat greater than the corresponding deviations for different temperatures (Fig. 8.14a). However, they remain within the range of experimental error. In contrast, the clay and loamy montmorillonite soils exhibit large deviations between their conductivity and viscosity ratios (Fig. 8.14b). This deviation may be explained by the fact that volatilization causes changes in both fluid viscosity and chemical composition (Galin et al. 1990a,b; Jarsjo et al. 1994, 1995), and the latter change may lead to new types of interactions with the solid phase. This explanation is supported by the fact that temperature-induced changes, which affect viscosity but not chemical composition, do not lead to the same large deviations between conductivity and viscosity ratios.

8.6.4 Vertical Transport of Pesticides Adsorbed on Colloids

It is generally assumed that organic pollutants with very high K_d values are virtually immobile in the soil, though they may be transported laterally by erosion. The occurrence in groundwater of organic molecules characterized by a very low solubility in the water phase and high K_d values highlights the fact that these molecules are moving through the porous media not as solutes, but together with colloidal materials. Vinten et al. (1983) reported on the vertical transport of DDT and paraquat adsorbed on suspended materials. Figure 8.15A shows the vertical transport of [^{14}C] paraquat adsorbed on Li-montmorillonite suspension through soil columns. When distilled water was the suspension medium, 50% of the pesticide penetrated beyond 12 cm. However, in 1 mM $CaCl_2$ only 5% of the pesticide penetrated deeper than 1 cm. The high [Ca^{2+}] concentration results in rapid immobilization of the clay in the soil through flocculation or through straining by adsorption of the smallest particles on soil surfaces; consequently, little pesticide transport occurs. Under conditions when the clay is dispersed, there is no flocculation and little interaction with the soil solid phase, so the pesticide is readily transported through the soil.

In the second case, a range of behavior of [^{14}C] DDT was observed in three soils (Fig. 8.15B). In the Gilat soil – a silt loam – the flow rate was slow, the pore structure was fine, and little transport of pesticide occurred (only 3% reached 5.4 cm). This is an example of efficient removal of suspended solids by the soil. In the case of the Bet Dagan soil – a sandy loam – there was considerable pesticide transport, as the flow rate was higher and the organic colloids were more mobile; 18% of the pesticide was transported to a depth greater than 9 cm. In the case of the Benei Darom coarse sand, still more transport of pesticide occurred, with 54% penetrating deeper than 5 cm. Again, these differences were due to the differences in mobility of the solids which are directly related to soil type and flow rate.

Having demonstrated the feasibility of transport of strongly adsorbed pesticides on mobile colloids during leaching (Vinten et al. 1983), it is important to indicate under what field conditions such transport might occur and contribute to edge-of-field loss. DDT or paraquat may reach the soil following foliar application and they will be strongly adsorbed close to the soil surface. Other soil-applied herbicides such as the s-triazine group and other organochlorine pesticides have high K_d values and will not be transported much in solution. Hartley and Graham-Bryce (1980) showed the potential hazard in lateral runoff for such pesticides, but considered vertical transport unimportant. However, if soils are leached with rainwater, or with sodic water and subsequently with low-electrolyte-concentration water, dispersion and release of clay can occur (e.g., Shainberg et al. 1981). Under such conditions, when release of soil colloids occurs, pesticides adsorbed in the surface soil may be transported to the drainage water. Pesticides applied as wettable powders may also be transported in suspension form, if leaching conditions occur soon after application.

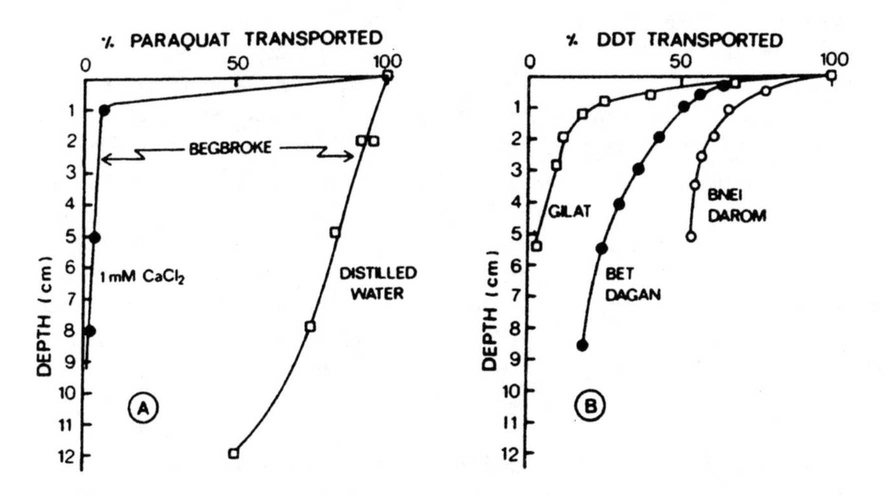

Fig. 8.15A,B. Vertical transport of DDT and paraquat adsorbed on suspended particles (in percentage of total applied). **A** ^{14}C Paraquat adsorbed on Li-montmorillonite. **B** ^{14}C DDT adsorbed on suspended solids in sewage effluent. (After Vinten et al. 1983)

The extent to which this type of transport occurs depends on: the amount of clay or organic matter released by the surface soil on dispersion, the mobility of these colloids in the soil profile, the rate at which soil clogging occurs, the K_d value, and the kinetics of desorption of pesticide from the mobile colloids.

Contaminant transport through unsaturated porous media is an important aspect of subsurface transport phenomena. Colloids generated by weathering, biological activities, and human influence may be much more concentrated in soil water than in groundwater. How these colloids move through the vadose zone and capillary fringe to reach the water table is not well understood. In particular, the role of the gas-water interface, one of the most important interfaces in the vadose zone, has been neglected. Wan and Wilson (1994), in an experiment involving two colloidal materials and three types of porous medium conditions, proved the significance of the gas-water interface for colloid sorption and transport. Three types of common saturation conditions (Fig. 8.16,1) were simulated in packed-sand columns: (1) a completely water-saturated condition, (2) gas bubbles trapped by capillary forces as a nonwetting residual phase (15% gas), and (3) gas present as a continuous phase (46% gas), i.e., in other words, a vadose-zone situation. Different saturations provided different interfacial conditions. Two types of polystyrene latex particles (0.2μm) – hydrophilic and hydrophobic – were used in each of the three saturations. Wan and Wilson (1994) showed that the retention of both hydrophilic and hydrophobic colloids increased with increasing gas content of the porous medium. Colloids preferentially sorbed onto the gas-water interface rather than the matrix surface (Fig. 8.16,2,3). The degree of sorption increased with increasing colloid-surface hydrophobicity. These findings suggest an additional mechanism for filtration and for particulate transport in the subsurface environment, whenever more than one fluid phase is present.

Graber et al. (1995) reported on the enhanced transport of atrazine ($C_8H_{14}ClN_5$) applied to a corn field which was observed after irrigation with secondary effluent. These effluents are characterized by the presence of organic colloids. In the effluent-irrigated cores, atrazine was found to a depth of nearly 4 m. Peaks in atrazine concentration in the effluent-irrigated cores often corresponded to secondary peaks in soil organic carbon, content, suggesting that enhanced transport was affected, at least in part, by complexation with effluent-borne organic matter. In high-quality water-irrigated cores, atrazine was mainly concentrated in the upper 1 m, and soil organic carbon content routinely decreased with increasing depth. Fig. 8.17 shows the distribution of atrazine in the field irrigated with secondary effluent and with a high-quality water.

A different example concerns an experiment designed to study the influence of sewage sludge on atrazine behavior in a drained hydromorphic clay soil cropped with corn (Barriuso et al. 1995). Observations made on drainage water at 0.8 m depth show that concentrations of atrazine were often lower in the presence of sewage sludge (Fig. 8.18b). In this case, the sewage sludge de-

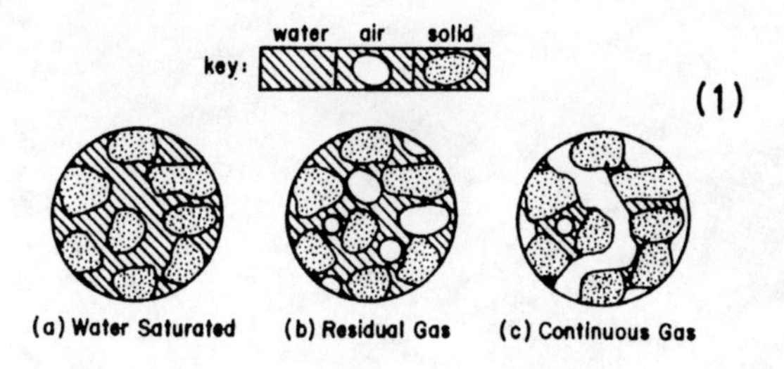

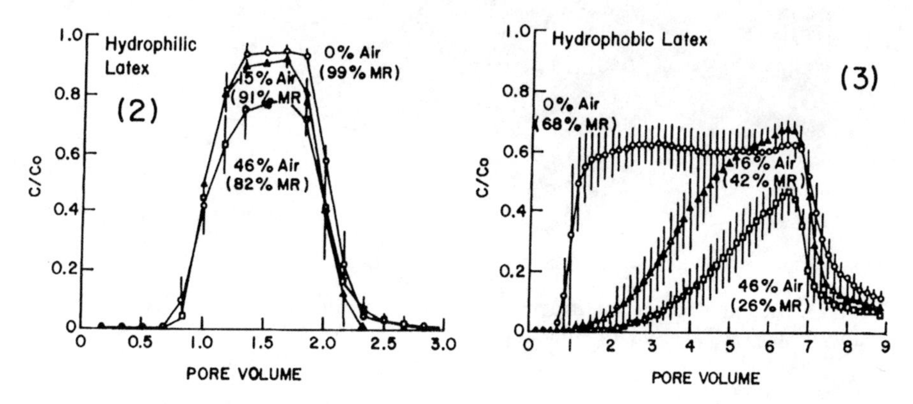

Fig. 8.16,1–3. Transport of colloidal materials as affected by the unsaturated zone air-water content. (After Wan and Wilson 1994). **1** Three types of columns representing three types of groundwater conditions: **a** fully water-saturated; **b** residual gas bubbles; **c** continuous gas and water phases. **2** Breakthrough curves of hydrophilic latex particles from three types of columns: fully saturated; unsaturated with gas bubbles; and unsaturated, with a continuous gas phase. Each *curve* shows an average of five repeated experiments, and the standard deviations are plotted as *error bars*. Percentages show the percentage of pore volume occupied by air (% air) as well as the percentage of total mass recovery (% MR). **3** Breakthrough curves of hydrophobic latex particles from three types of columns: fully saturated; unsaturated, with gas bubbles; and unsaturated, with a continuous gas phase. Each *curve* shows an average of five repeated experiments, and the standard deviations are plotted as *error bars*

creased the transport of the herbicide while it had no apparent effect on its distribution in the profile (Fig. 8.18a).

8.6.5 Preferential Flow

The transport of two herbicides – bromacil and napropamide – with different solubility and adsorption capacity, will be used as an example to illustrate the preferential flow phenomenon.

The leaching of the above pesticides through undisturbed cores of structured Evesham clay soil (Aquic Eutrochrept) under continuous and discontinuous

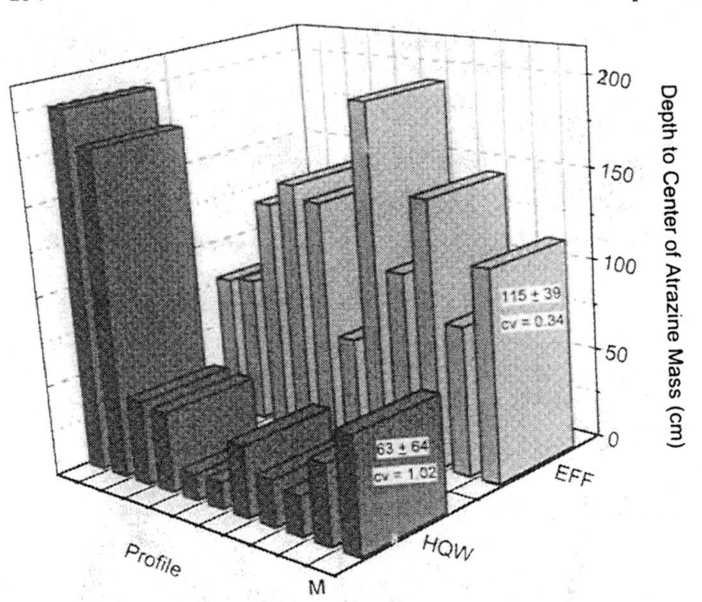

Fig. 8.17. Mean (*M*) center of atrazine mass and individual centers of mass for ten high quality water (*HQW*) and 10 effluent (*EFF*) profiles. (After Graber et al. 1995)

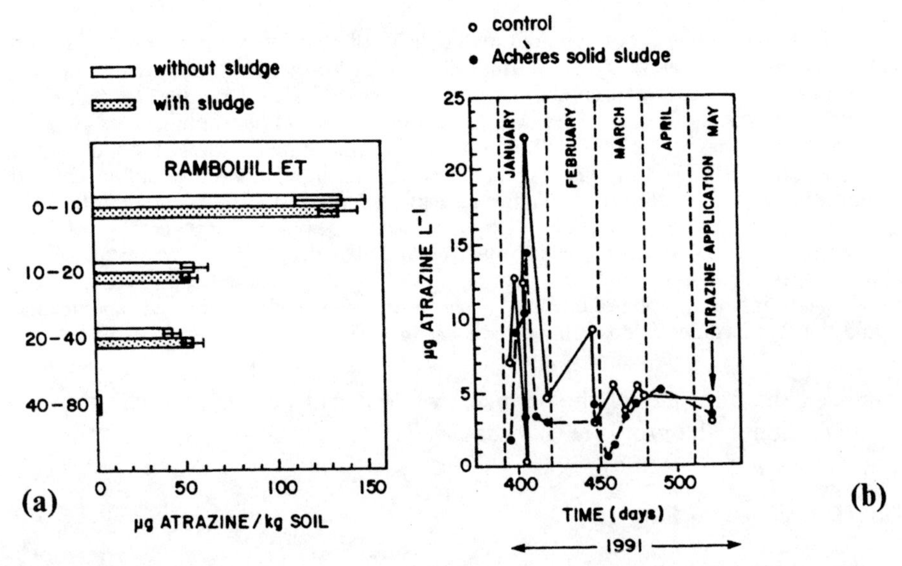

Fig. 8.18. Influence of sewage sludge on atrazine behaviour. This graph shows only a part of observations made at the experimental site of Rambouillet (France); the experiment began at time t = 0 and atrazine was applied at t = 150 d on a soil which was also treated with atrazine one year before. **a** Distribution in the soil profile 12 months after atrazine application (the sampling date is indicated by the arrow). **b** Atrazine concentration in drainage water collected at 0.8 m depth (After Barriuso et al. 1995)

watering regimes and at different initial moisture contents was reported by White et al. (1986). Chloride – a conservative solute – was used as a tracer for water movement in the reported experiments.

The adsorption coefficients for bromacil in the 0–0.02 and 0.02–0.22-m-depth soils were 2.0 and 1.73 l kg^{-1}, respectively; for napropamide the values were 22.7 and 17.7 l kg^{-1}, respectively. The adsorption kinetics for bromacil and napropamide shows that bromacil attains equilibrium almost instantaneously. Napropamide, however, does not reach equilibrium before 2–3 h. These results were obtained for systems in which the herbicide solution was applied to air-dry soil. When napropamide was added to a prewet soil, the system had not attained equilibrium even after 48 h.

Differences in the leaching of bromacil and napropamide through the undisturbed soil were observed. There is a striking effect of initial soil-water content on herbicide transport. Of the strongly adsorbed napropamide, 85% was leached through initially dry soil ($\Theta_i = 0.24$), but only 28% was leached from the wet soil columns. These results were completely consistent with the observed retention of the herbicides in the soil cores, as shown in Fig. 8.19, where the concentration of herbicide (mg kg^{-1} soil) is plotted against soil depth. The prewet soil always retained more herbicide than did the initially dry soil. In the case of bromacil, very little was detected in the dry soil after leaching. Consistently, the highest concentration of herbicide was found in the uppermost soil layer, with an approximately exponential decrease with depth.

During the "discontinuous" leaching period – roughly one-tenth of the total irrigation period – adsorption rather than desorption of herbicide is the dominant process. Napropamide, with a K_d some ten times greater than that of bromacil, should be strongly retained. However, because adsorption equilibrium is not attained instantaneously for napropamide, and the velocity of water flow is so rapid, especially in the dry soil, even napropamide is leached during this period. Discontinuous leaching of the prewet soil permitted more of both herbicides to be retained in the surface layer than continuous leaching, presumably because diffusion of the herbicide into aggregates occurred during the interval between leachings.

Summarizing the results of their experiments, White et al. (1986) showed that with only one pore volume of water, continuously applied at 12 mm h^{-1} to clay soil of this structure, with $\Theta = 0.24$, about 85% of an initial pulse input of the strongly adsorbed napropamide was leached into the effluent. Under the same conditions, nearly 100% of the weakly adsorbed bromacil was leached out of the soil. Clearly, for an undisturbed structured soil in which water flows preferentially down cracks and channels, even strongly adsorbed chemicals are vulnerable to leaching.

Similar behavior in napropamide was observed in a field experiment conducted by Jury et al. (1986) in California. Representative soil-core napropamide concentrations were measured in five of the 19 field cores, together with the field-area-averaged mean value over all 19 cores. Evidence of deep penetration of a fraction of the napropamide pulse was found in virtually all of the

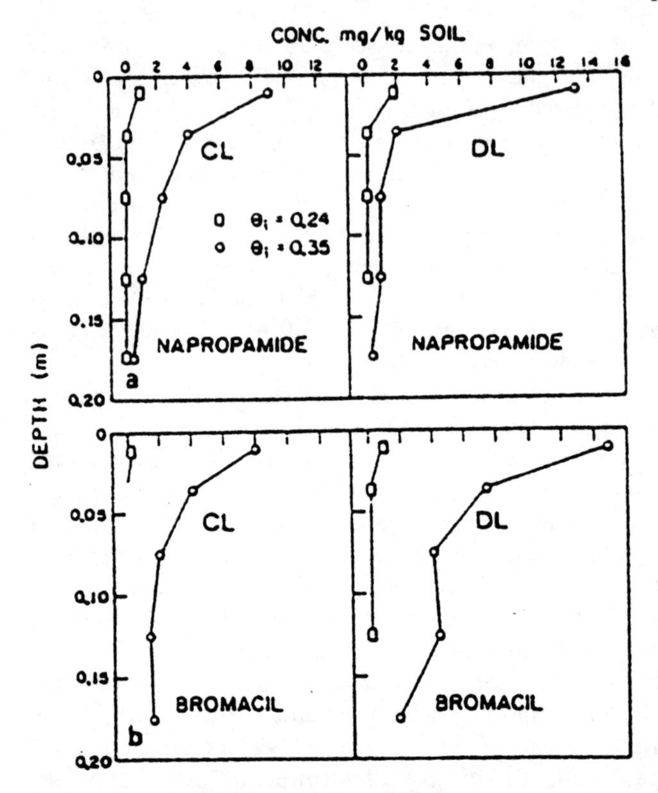

Fig. 8.19. Distribution of napropamide and bromacil in aggregated undisturbed soil column under continuous (*CL*) and discontinuous (*DL*) leaching. (After White et al. 1986)

19 replicates. This deep penetration produced a significant secondary maximum in the mean soil core concentration curve, averaged over all replicates. The average recovery of 75% of the applied pesticide to the field is consistent with the consensus 70-day half-life of the napropamide chemical (Jury et al. 1984). Thus the deep-level concentrations observed in the soil cores were not an artifact. Even though 73% of the recovered pesticide was found in the 0-10-cm layer, over 1% of the chemical penetrated to a depth of 150 cm and in three cores trace concentrations were found at 180 cm.

Jury et al. (1986) pointed out that field studies show evidence of some napropamide movement which is not consistent with the adsorption equilibrium assumption commonly made in transport models and is assumed to apply in characterization of the mobility of an adsorbed chemical. The results do not seem to be attributable exclusively to rate-limited adsorption, but rather are suggestive of the presence of a mobile solution phase. Such behavior could

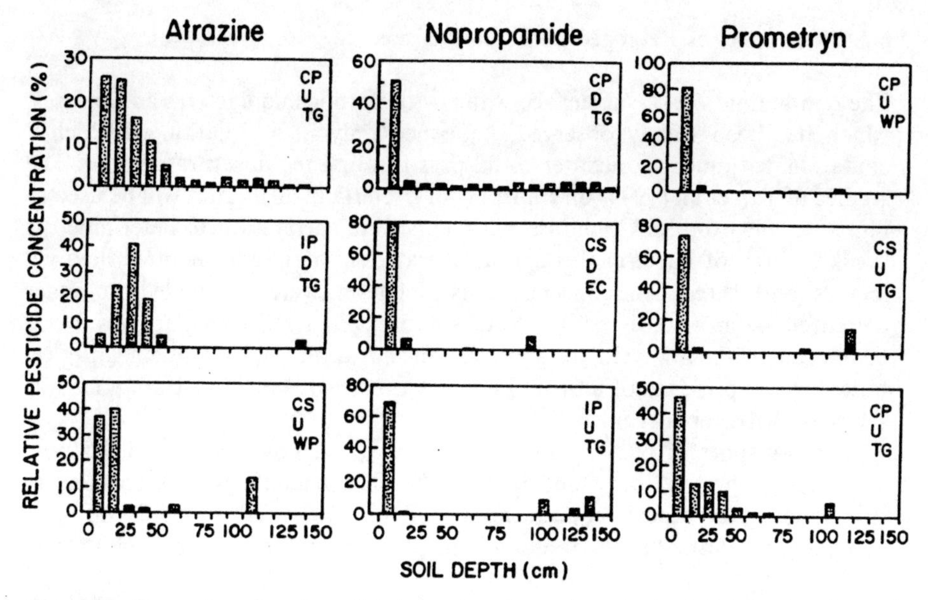

Fig. 8.20. Examples of preferential flow of each pesticide in individual field plots: *C* continuous; *I* intermittent; *P* ponding; *S* sprinkling; *TG* technical grade; *EC* emulsifiable concentrate; *WP* wettable powder; *U* undisturbed; *D* disturbed. (After Ghodrati and Jury 1990)

result from macropore flow or from the presence of a mobile solution of an adsorbed complex of napropamide and dissolved organic matter, or of colloids in suspension.

Examples of preferential flow of pesticides in individual field plots under intermittent, ponding, and sprinkling leaching has been reported by Ghodrati and Jury (1990), and the results obtained are presented in Fig. 8.20.

Ghodrati and Jury (1990) designed a field experiment on an irrigated loamy sand soil. They studied the transport of atrazine, napropamide, and prometryn with two formulations, two soil-surface preparations (undisturbed and repacked) under four flow conditions, obtained by continuous or intermittent sprinkler and flood irrigation. Irrigation was started immediately after the pesticide application and the soil was sampled 6 days later. Figure 8.20 gives partial results from the work of Ghodrati and Jury (1990), showing the deep transport of the three herbicides which was observed whatever the irrigation method and soil surface preparation. During the same experiment, applied chloride has moved to around 30-cm depth. According to their retardation factor values, these three molecules would have migrated to smaller depths. The authors have attributed this discrepancy to preferential flow.

8.6.6 Transport of Pathogenic Microorganisms

The contamination of subsurface water with pathogenic bacteria and viruses, which has been widely observed, happened only in a situation in which a contaminated infiltration water could pass through the unsaturated zone. The studies of Tan et al. (1994) on transport of bacteria in an aquifer will be used to illustrate this process. Column experiments were carried out to determine the breakthrough of bacteria through a saturated aquifer sand at three flow velocities and three cell concentrations. Bacteria were suspended in either deionized water or 0.01 mol l^{-1} NaCl solution. Bacterial transport was found to increase with flow velocity and cell concentration but was significantly retarded in the presence of 0.01 mol l^{-1} NaCl. *Pseudomonas* sp. strain KL2 was selected as the test organism.

The transport of bacteria in a soil column at a flow velocity of 3.5 pore volumes per hour is presented in Fig. 8.21. The bacteria were suspended in deionized water at cell concentrations of 10^7, 10^8, and 10^9 cells ml^{-1}. A companion determination of cloud transport under the same boundary con-

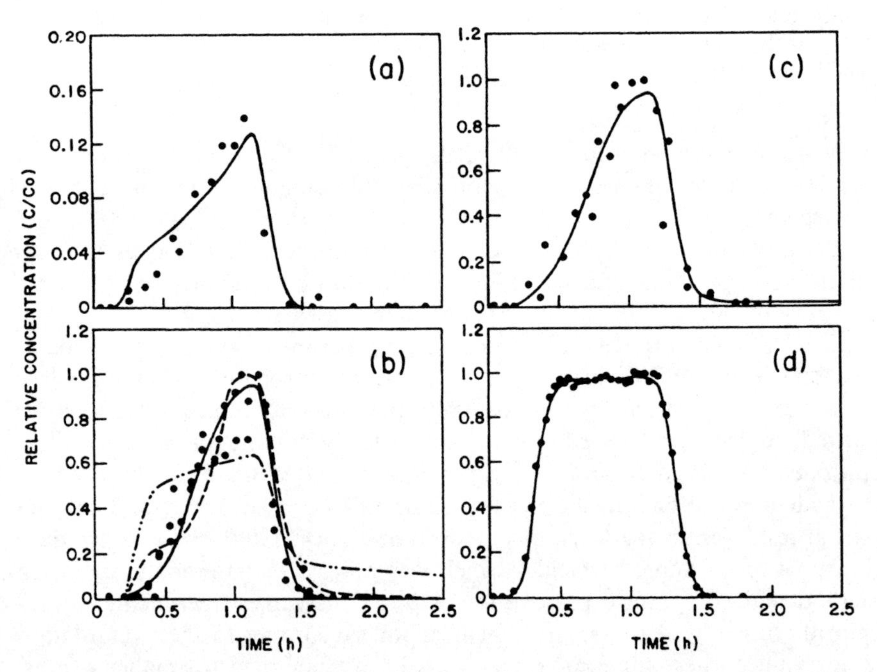

Fig. 8.21a-d. Measured breakthrough curves (BTC_S) of bacteria for experiments in which bacteria were suspended in deionized water with a flow velocity of 0.15 mm s^{-1} and cell concentration of **a** 10^7, **b** 10^8, **c** 10^9 cells ml^{-1}, and **d** of chloride. The aquifer used in the experiment contained 90.3% sand (0.5–2 mm), 3.1% medium sand (0.25–0.5 mm), 4.6% fine sand (50 μm to 0.25 mm), 1.9% silt (2–50 μm), and 0.1% clay (< 2 μm). (After Tan et al. 1994)

ditions was also performed. As expected, bacterial breakthrough was retarded compared with that of chloride. The relative concentration of chloride reached unity much faster than that of bacteria. The C_0 values reached unity only when the cell concentration was 10^8 cell ml^{-1} or higher, the flow velocity was 0.1 m s^{-1} or greater, and the suspending medium solution was deionized water. The increase of C/C_0 for bacteria was gradual, and this resulted in asymmetrical BTCs for bacteria.

The experimental results of Tan et al. (1994) show that bacteria can move through an aquifer sand readily, but that such transport is affected by water flow velocity, cell concentration, and ionic concentration. The experimental data support the possible existence of a threshold retention capacity in the absence of pore plugging; i.e., a finite number of retention sites needs to be filled before significant bacterial breakthrough can occur. The existence of a threshold retention capacity is determined by factors such as the properties of porous materials, cell concentration, and the ionic concentration of the bacterial suspension.

Bacterial transport was enhanced at higher cell concentrations. The explanation may be that once the retention capacity was reached, bacteria would be able to move freely or to replace the retained bacteria. Higher cell concentration would saturate the finite retention sites much earlier, and lead to enhanced bacterial breakthrough. Higher cell concentrations also resulted in the maximum retention capacity S_{max} being higher. This was probably because S_{max} reflects the capacity for both adsorption and straining. The absolute straining capacity may well increase with cell concentration of the suspension because the same pore volume for straining (straining sites) would retain a larger number of bacteria when the cell concentration was higher.

References

Abriola LM, Pinder GF (1985) A multiphase approach to the modelling of porous media contamination by organic compounds. Water Resour Res 21: 11–18

Addiscott TM, Wagenet RJ (1985) A simple method for combining soil properties that show variability. Soil Sci Soc Am J 49: 1365–1369

Armstrong YE, Frind OD, McClellen RD (1994) Nonequilibrium mass between the vapour, aqueous and solid phase in unsaturated soils during vapour extraction. Water Resour Res 30: 355–367

Barriuso E, Calvet R, Houot S (1995) Field study of the effect of sewage sludge application in atrazine behavior in soil. Intern J Environ Anal Chem 59: 107–121

Baveye P, Sposito G (1985) Macroscopic balance equations in soils and aquifers: the case of space- and time-dependent instrumental response. Water Resour Res 21: 116–120

Bear J (1972) Dynamics of fluids in porous media. Elsevier, New York, 19 pp

Beven K, Germann P (1982) Macropores and water flow in soils. Water Resour Res 18: 1311–1325

Biggar JW, Nielsen DR (1962a) Miscible displacement and leaching phenomenon. Agron Monog 111: 254–274

Biggar JW, Nielsen DR (1962b) Miscible displacement: II. Behavior of tracers. Soil Sci Soc Am Proc 26: 125–128
Bode LE, Day CL, Gebhardt MR, Goring CE (1973a) Mechanism of trifluralin diffusion in silt loam soil. Weed Sci 21: 480–484
Bode LE, Day CL, Gebhardt MR, Goring CE (1973b) Prediction of trifluralin diffusion coefficients. Weed Sci 21: 485–489
Bourchard DC, Enfield CG, Piwoni MD (1989) Transport processes involving organic chemicals in soils. In: Shawhney BL, Brown K (eds) Reactions and movement of organic chemicals in soils. SSSA Spec Publ 22. Soil Science Society of America, Madison, pp 349–371
Brusseau ML (1991) Transport of organic chemicals by gas advection in structured or heterogeneous porous media: development of a model and application to column experiments. Water Resour Res 17: 3189–3199
Calvet R (1984) Behavior of pesticides in unsaturated zone. Adsorption and transport phenomenon. In: Yaron B, Dagan G, Goldschmid J (eds) Pollutants in porous media. Springer, Berlin Heidelberg New York, pp 143–151
Cary JW, McBride JF, Simmons CS (1989a) Observation of water and oil infiltration into soil: some simulation challenges. Water Resour Res 25: 73–80
Cary JW, Simmons CS, McBride JF (1989b) Oil infiltration and redistribution in unsaturated soils. Soil Sci Soc Am J 53: 335–342
Cary JW, Simmons CS, McBride JF (1989c) Permeability of air and immiscible organic liquids in porous media. Water Resour Bull 25: 1205–1216
Cary JW, Simmons CS, McBride JF (1994) Infiltration and redistribution of organic liquids in layered porous media. Soil Sci Soc Am J 58: 704–711
Collin M, Rasmuson A (1988) Comparison of gas diffusivity models for unsaturated porous media. Soil Sci Soc Am J 52: 1559–1565
Crank J (1975) Mathematics of diffusion. 2nd edn Clarendon Press, Oxford, 414 pp
Ehlers W, Letey J, Spencer WF, Farmer WJ (1969) Lindane diffusion in soils. I. Theoretical considerations and mechanisms of movement. Soil Sci Soc Am Proc 35: 501–508
Enfield CG (1985) Chemical transport facilitated by multiphase flow systems. Water Sci Technol 17: 1–12
Enfield CG, Bengisson G (1988) Macromolecule transport of hydrophobic contaminants in aqueous environments. Ground Water 26: 64–70
Enfield CG, Carsel RF, Cohen SZ, Pitan T, Walters DM (1982) Approximating pollutant transport to ground water. Ground Water 20: 711–722
Faust CR (1985) Transport of immiscible fluids within and below unsaturated zone. A numerical model. Water Resour Res 21: 587–596
Galin TS, Gerstl Z, Yaron B (1990a) Soil pollution by kerosene products: III. Kerosene stability in soil columns as affected by volatilization. J Contam Hydrol 5: 375–385
Galin TS, McDowell C, Yaron B (1990b) Effect of volatilization on mass flow of a nonaqueous pollutant liquid mixture in an inert porous media: experiments with "Kerosene." J Soil Sci 41: 631–641
Gerstl Z, Nye PH, Yaron B (1979a) Diffusion of biodegradable pesticide as affected by microbial decomposition. Soil Sci Soc Am J 43: 843–848
Gerstl Z, Yaron B, Nye PH (1979b) Diffusion of biodegradable pesticides. I. In a biologically inactive soil. Soil Sci Soc Am J 43: 839–842
Gerstl Z, Galin TS, Yaron B (1994) Mass flow of a volatile organic liquid mixture in soils. J Environ Qual 23: 487–493
Ghodrati M, Jury WA (1990) A field study using dyes to characterize preferential flow of water. Soil Sci Soc Am J 54: 1558–1563
Graber ER, Gerstl Z, Fischer E, Mingelgrin U (1995) Enhanced transport of atrazine under irrigation with effluent. Soil Sci Soc Am J 59: 1513–1519

Graham-Bryce IJ (1969) Diffusion of organophosphorus insecticides in soils. J Sci Food Agric 20: 489–494

Hartley GS, Graham-Bryce IJ (1980) Physical principles of pesticide behavior, vol I, chap. 5. Academic Press, London

Hill DE, Parlange JY (1972) Wetting front instability in layered soils. Soil Sci Soc Am Proc 36 (5): 697–702

Hillel D, Baker RS (1988) A descriptive theory of fingering during infiltration into layered soils. Soil Sci 146: 51–56

Hutzler NJ, Gierke YS, Krause LC (1989) In: Shawhney BL, Brown K (eds) Reactions and movement of organic chemicals in soils. SSSA Spec Publ 22. Soil Science Society of America, Madison, pp 373–403

Jarsjo J, Destouni G, Yaron B (1994) Retention and volatilization of kerosene laboratory experiments on glacial and post glacial soils. J Contam Hydrol 17: 167–185

Jarsjo J, Destouni G, Yaron B (1996) On the relation between viscosity and hydraulic conductivity values of volatile organic liquid mixtures in soils (in preparation)

Jury WA (1983) Chemical transport modeling: current aproaches and unresolved problems. In: Shawhney BL, Brown K (eds) Chemical mobility and reactivity in soil systems. Soil Sci Soc Am Spec Publ No 11. Soil Science Society of America, Madison, pp 49–64

Jury WA, Fluhler H (1992) Transport of chemicals through soil: mechanisms, models and field applications. Adv Agron 47: 142–202

Jury WA, Gruber J (1989) A stochastic analysis of the influence of soil and climatic variability on the estimate of pesticide ground water pollution potential. Water Resour Res 25: 2465–2474

Jury WA, Spencer WF, Farmer WJ (1984) Behavior assessment model for trace organics in soil. III. Application of screening model. J Environ Qual 13: 573–579

Jury WA, Elabad H, Reskelo M (1986) Field study of napropamide movement through unsaturated soil. Water Resour Res 22: 749–755

Jury WA, Dyson JS, Butters GL (1990) A transfer function model of field scale solute transport under transient water flow. Soil Sci Soc Am J 54: 327–332

Kim JI, Buckau G, Baumgartner F, Moon HC, Lux D (1984) Colloid generation and the actinide migration in Gorleben groundwater. In: Scientific basis for nuclear waste management vol 7. Elsevier, New York, pp 31–40

Lenhard RJ, Parker JC (1987) Measurement and prediction of saturation pressure relationships in three phase porous media system. J Contam Hydrol 1: 407–425

Letey J, Farmer WJ (1974) Movement of pesticides in soil. In: Guenzi WD (ed) Pesticides in soil and water. Soil Science Society of America, Madison, pp 67–94

Mackay DM, Roberts PV, Cherry JA (1985) Transport of organic contaminants in ground water. Environ Sci Technol 19: 384–393

Menzel RG (1980) Enrichment ratios for water quality modelling. In: Knisel WG (ed) Cream S – a field scale model for chemicals runoff and erosion from agricultural management systems. USDA Cons Res 26, US Department of Agriculture, Washington, DC

Nelson DM, Penrose WR, Kattunen JO, Mehlhaff P (1985) Effects of dissolved organic carbon on the adsorption properties of plutonium in natural waters. Environ Sci Technol 19: 127–131

Nielsen DR, Biggar JW (1961) Miscible displacement in soils: I. Experimental information. Soil Sci Soc Am Proc 25: 1–5

Nielsen DR, Biggar JW (1962) Miscible displacement in soils. III. Theoretical considerations. Soil Sci Soc Am Proc 26: 216–221

Nielsen DR, Biggar JW, Erh KT (1973) Spatial variability of field measured soil-water properties. Hilgardia 42: 215–259

Nye PH (1979) Diffusion of ions and uncharged solutes in soils and soil clays. Adv Agron 31: 225–272

Oh YS, Shareefdeen Z, Baltzis BC, Bartha R (1994) Interactions between benzene, toluene and p-xylene (BTX) during their biodegradation. Biotechnol Bioeng 44: 533–538
Olsen SR, Kemper WD (1968) Movement of nutrients to plant roots. Adv Agron 20: 91–151
Parker JC, Lenhard RJ, Kuppsamy T (1987) A parametric model for constitutive properties governing multiphase flow in porous media. Water Resour Res 23: 618–624
Pinder GF, Abriola LM (1986) On the simulation of nonaqueous phase organic compounds in the subsurface. Water Resour Res 22: 1095–1195
Quinsberry VL, Phillips RE (1976) Percolation of surface-applied water in the field. Soil Sci Soc Am J 40: 484–489
Quinsberry VL, Phillips RE, Zelenik JM (1994) Spatial distribution of water and chloride macropore flow in a well saturated soil. Soil Sci Soc Am J 58:1294–1300
Rao PSC, Jessup RE (1983) Sorption and movement of pesticides and other toxic organic substances in soils. In: Chemical mobility and reactivity in soil systems. Soil Sci Soc Am Spec Publ 11. Soil Science Society of America, Madison, pp 183–201
Rao PSC, Hornsby AG, Kilcrease DP, Nkedi-Kizza P (1985) Sorption and transport of toxic organic substances in aqueous and mixed solvent systems. J Environ Qual 14: 376–383
Rose CW (1993) The transport of sorbed chemicals IV eroded sediments In: Russo D, Dagan G (eds) Water and solute flow in soils. Springer, Berlin Heidelberg New York
Russo D, Bresler E (1981) Soil hydraulic properties for stochastic processes. I. An analysis of the field spatial variability. Soil Sci Soc Am J 45: 682–687
Sallam A, Jury WA, Letey J Jr (1984) Measurement of gas diffusion coefficient under relatively low air-filled porosity. Soil Sci Soc Am J 48: 3–6
Schwille F (1984) Migration of organic fluids immiscible with water in the unsaturated zone. In: Yaron B, Dagan G, Goldschmid J (eds) Pollutants in porous media. Springer, Berlin Heidelberg New York pp 27–47
Scott HD, Paetzold RF (1978) Effect of soil moisture on diffusion coefficients and activation energies of tritiated water, chloride and metribuzin. Soil Sci Soc Am Proc 42: 23–27
Scott HD, Philips RE (1972) Diffusion of selected herbicides in soil. Soil Sci Soc Am Proc 36: 714–719
Shainberg I, Rhoades JD, Prather RJ (1981) Effect of low electrolyte concentration on clay dispersion and hydraulic conductivity of a sodic soil. Soil Sci Soc Am J 45: 273–277
Steenhuis TS, Boll J, Shalit G, Selker JS, Merwin IA (1994) A simple equation for predicting preferential flow solute concentrations. J Environ Qual 23: 1058–1064
Stone HL (1973) Estimation of three phase relative permeability and residual oil saturation. J Can Petrol Technol 12: 53–61
Tan Y, Gannon JT, Baveye P, Alexander M (1994) Transport of bacteria in an aquifer sand: experiments and model simulations. Water Resour Res 30: 3243–3252
van J Dam (1967) The migration of hydrocarbon in a water bearing stratum. In: Hepple P (ed) The joint problems of the oil and water. Inst Petrol, London, pp 55–96
van Genuchten MT, Wierenga PJ (1976) Mass transfer studies in sorbing porous media. I. Analytical solutions. Soil Sci Soc Am Proc 40: 473–480
Vinten AJ, Yaron B, Nye PH (1983) Vertical transport of pesticides into soil when adsorbed on suspended particles. J Agric Food Chem 31: 61–664
Wan J, Wilson JL (1994) Colloid transport in unsaturated porous media. Water Resour Res 30: 857–864
White RE, Dyson JS, Gerstl Z, Yaron B (1986) Leaching of herbicides through undisturbed cores of a structured clay soil. Soil Sci Soc Am J 50: 277–282
Yevjevich V (1972) Stochastic processes in hydrology. Water Resources Publications, Fort Collins

Part IV
Prediction and Remediation

CHAPTER 9

Modeling the Fate of Pollutants in the Soil

9.1 Overview of Models

Many activities lead to the introduction of various chemicals into natural media: industrial operations, waste disposal, pesticide and fertilizer applications, irrigation with secondary effluents, etc. To these must be added accidents such as fires, tank leakages, and spills, which locally add large amounts of chemicals. The cumulative result is a growing threat to soils and waters which makes the prediction of the fate of pollutants necessary. Modeling may be a rather efficient tool for making this prediction, so that a very great number of models can be found in the literature. In a survey recently published, Johnson (1994) listed more than 1000 environmental software products. However, models do not constitute a universal or perfect way to achieve this prediction, because they are only more or less simplified representations of the real world, and their application encounters many difficulties which need to be well known by users. Nevertheless, the modelling approach has been shown to be very useful in many instances: research, management, regulatory purposes, and teaching.

Modeling of solute transport in soils has been studied for a long time. Mathematical models have generally been based on the principle of miscible displacement of soluble substances. The governing equation is the convection-dispersion equation which can be coupled with sorption and decay of the chemicals. A lot of papers have been published on this subject, examples of which are those of Lapidus and Amundson (1952), Day and Forsythe (1957), Nielsen and Biggar (1962), van Genuchten and Wierenga (1976), Bresler (1980), Wagenet and Rao (1990), and Jury and Fluhler (1992). The first models were developed for one-dimensional transport, homogeneous media, and simple initial and boundary conditions. With the improvement of knowledge about solute behavior in soil, on the one hand, and of computer performance, on the other, models have become more and more complex. At least in principle, they enable us to take into account equilibrium and nonequilibrium sorption phenomena, and the kinetics of chemical tranformations and degradation under various initial and boundary conditions. Progress has also been achieved in the description of two- and three-dimensional transport and with the introduction of preferential flow in heterogeneous media.

Consequently, a variety of models has been published during the last 10 years. These models vary widely in their conceptual approach and their degree of complexity and, therefore, in their data requirements. Models can be classified according to the purpose for which they have been developed. The following categories are usually considered:

- Research models: these give a detailed and comprehensive description of phenomena; they may be useful for testing various hypotheses in relation to mechanisms of retention, transformation and transport.
- Management models: these generally differ from research models in using less precise descriptions; they are able to estimate the integrated effects of various processes that determine, for example, the fate of a pesticide under a given set of practices. They may be used in management decision procedures; for example, in agriculture for defining the characteristics of plant protection treatments, in industry for designing a waste disposal system.
- Screening models: these are simpler than the preceding ones in both the number of phenomena accounted for and their description. They essentially allow the classification of molecules for given pedo-climatic situations. They must be simple enough to give a rough basis for regulatory decisions.
- Teaching models: these are also simpler than research models but they must emphasize the main aspects of chemical behavior, while being easily handled for the training of students.

The complexity of the models, the number of input data, and the processing time increase as we progress from teaching models to research models.

There are several modeling approaches, which have been clearly described by Addiscott and Wagenet (1985). First, it is important to distinguish between deterministic and stochastic models. With the second category, uncertainties of input parameter values can be taken into account, which is not possible with deterministic models. A second distinction is between mechanistic and functional models. The former, which incorporate the most fundamental mechanisms, according to current knowledge, are more complex than functional ones, for which simplified descriptions are used. This distinction corresponds approximately to that between rate and capacity models. Finally, models are also classified according to the mathematical procedure applied in solving the set of equations; this leads to the distinction between analytical and numerical models.

In dealing with the fate of a chemical in the soil, all models have the same general structure, which can be described from three points of view:

1. Phenomena: the central part of the model is constituted by the water and solute transport equations. These equations are coupled with other equations which represent sink and source processes. The number of equations and their mathematical expressions depend on the model category and vary accordingly. Table 9.1 indicates the main phenomena involved in the fate of various possible pollutants.

Table 9.1. Phenomena implied in the fate of various pollutants

<table>
<tr><td colspan="6">Inorganic pollutants</td><td colspan="2">Organic pollutants</td></tr>
<tr><td>**Unionized molecules**</td><td colspan="5">**Ions**</td><td>**Small molecules**</td><td>**Polymers**</td></tr>
<tr><td></td><td colspan="3">Anions</td><td colspan="2">Cations</td><td colspan="2"></td></tr>
<tr><td>N, S molecules</td><td>Br^-, Cl^-</td><td>NO_3^-</td><td>PO_4H^{2-} $PO_4H_2^-$</td><td>NH_4^+</td><td>Heavy metals</td><td>Pesticides, organochlorines, solvents</td><td>Hydrocarbons</td></tr>
<tr><td colspan="3">*Transport in solution*</td><td colspan="5">*Transport in solution and sorbed on solid particles*</td></tr>
<tr><td colspan="7">*Miscible displacement*</td><td>*Immiscible displ.*</td></tr>
<tr><td>*Possible gas transport*</td><td colspan="5">*No gas transport*</td><td colspan="2">*Possible gas transport*</td></tr>
<tr><td>*Transformations*</td><td>*No*</td><td colspan="3">*Transformations*</td><td>*No*</td><td colspan="2">*Transformations*</td></tr>
<tr><td colspan="8">*Sorption*</td></tr>
</table>

2. Compartmentalization of the soil profile: the soil profile is divided into several layers in order to take into account the characteristics of the different horizons. The layers are, in turn, frequently divided into segments for calculation purposes. The numbers of layers and segments vary among models, depending on their complexity.
3. Outputs: in general, models simulate the amounts of chemicals which are transferred to the atmosphere, to surface water, and to groundwater. They often give the distribution of chemicals in the soil profile.

Attention must be drawn to the fact that some models, used for environmental purposes, do not have these general characteristics. This is the case for models which specifically describe the chemical speciation in solution (e.g., GEOCHEM, MINTEQ, STEADYQL, ref. cited in Ghadiri and Rose 1992), or biotic or abiotic transformations of organic compounds (e.g., the pesticide degradation model of Soulas and Lagacherie 1990; model NCSoil of Molina et al. 1983).

9.2 Description of Models

Modeling the fate of soil pollutants is based on the description of solute transport coupled with sink/source phenomena. Several more or less complex approaches have been proposed to reach this goal. It is outside the scope of this book to review all published models; rather, the aim is to give the characteristics of those which seem to be more interesting either because of their underlying hypotheses, or because of their potential applications. For details of the mathematical techniques which are used, the reader is referred to model developers.

Three scales are usually distinguished: (1) the microscopic scale (at the pore level), where elementary laws of fluid mechanics apply, (2) the macroscopic scale (classically, the laboratory column), for which an equivalence between the real dispersed medium and a fictitious continuous medium is assumed, (3) the megascopic scale (the field), where spatial variability of soil properties must be taken into account through a stochastic approach. Models for the first two scales are deterministic. They may be used locally in the field, but generally they cannot be extrapolated.

It is worth noting that water movements may lead to pollutant transport through the transfer either of dissolved molecules or of molecules sorbed on solid particles. In the latter case, the transport takes place essentially on the soil surface during runoff/erosion processes. It may also take place to a lesser degree within the soil profile in association with colloidal materials and hydrosoluble humic substances (see Chap. 8), which can bind pesticide molecules and make them mobile and readily transportable by water movements (Ballard 1971; Vinten et al. 1983). The question of transport mechanisms merits more attention, particularly in relation to the soil organic matter transformations. The present chapter is limited to transport in the soil of pollutants dissolved in the aqueous phase. For modeling of runoff and erosion, the reader is referred to the book of Ghadiri and Rose (1992) and for gas transport to Chapter 8 and to reviews such as that of Taylor and Spencer (1990). This chapter will deal only with transport in the liquid phase.

9.2.1 Deterministic Models

Numerous examples can be found in the literature. Most of them are based on the well-known convection-dispersion (CD) equation. Others rely on quite different approaches, based on, for example, the residence time distribution, the chemical fugacity, transfer functions, or the reservoir analogy.

9.2.1.1 Models Based on the CD Equation

From a general point of view, solute transport is the result of three processes: diffusion in the aqueous phase, diffusion in the gas phase, and convection combined with hydrodynamic dispersion. Although transport in the gas and the liquid phases is simultaneously introduced in some models (the BAM model, for example), they are often modeled separately.

The CD equation has been used for a long time to describe solute transport in porous media. Various initial and boundary conditions have been applied, which lead to several solution methods, either analytical or numerical. Approximate solutions have been proposed by De Smedt and Wierenga (1978) for transport during infiltration and redistribution. Later, van Genuchten and Alves (1982) listed several analytical solutions for the one-dimensional CD equation.

For one-dimensional vertical transport, the CD equation is:

$$\underset{(a)}{\frac{\partial}{\partial t}(\rho_b \mathrm{Si})} + \underset{(b)}{\frac{\partial}{\partial t}(\Theta \mathrm{Ci})} = \underset{(c)}{\frac{\partial}{\partial z}\left[\Theta D(\Theta, q)\frac{\partial \mathrm{Ci}}{\partial z}\right]} - \underset{(d)}{\frac{\partial}{\partial z}(q\mathrm{Ci})} \pm \underset{(e)}{\Sigma\Phi} \, , \tag{9.1}$$

where i denotes a solute, Si is the concentration of sorbed solute (expressed on a mass basis ($M\ M^{-1}$), Ci is the concentration of the solute in the liquid phase ($M\ L^{-3}$), Θ is the volumetric water content ($L^3\ L^{-3}$), ρ_b is the bulk density ($M\ L^{-3}$), q is the soil macroscopic water flux ($L\ T^{-1}$), Φ stands for any sink/source phenomenon for the solute ($M\ L^{-3}T^{-1}$), $D(\Theta,\ q)$ is the hydrodynamic dispersion coefficient ($L\ T^{-2}$), which incorporates the effect of mechanical (induced flow) dispersion and those of molecular diffusion according to the following expression:

$$D(\Theta, q) = \frac{Dp(\Theta)}{\Theta} + Dm(q) \, . \tag{9.2}$$

Dm may be estimated from Ogata (1970, in Wagenet and Rao 1990):

$$Dm(q) = \lambda \cdot v \tag{9.3}$$

$v = q/\Theta$ is the pore water velocity and λ (L) is the dispersivity, which depends on the nature of the medium and on the scale of observation. A first acceptable approximation for λ may be given by (1/10) th the scale of observation (Gelhar et al. 1985, in Wagenet and Rao, 1990). $Dp(\Theta)$ is the effective diffusion coefficient, which may be evaluated from Kemper and van Schaik (1966) in Wagenet and Rao (1990):

$$Dp(\Theta) = \mathrm{Doaexp}(b\Theta) \, , \tag{9.4}$$

Do being the molecular or ionic diffusion in solution, a and b are two empirical constants.

The several terms in Eq. (9.1) relate to:

- term (a) : time variation of the sorbed solute concentration,
- term (b) : time variation of the solute concentration in the liquid phase,
- term (c) : transfer due to hydrodynamic dispersion,
- term (d) : convective transfer,
- term (e) : sink/source phenomena.

Details on the derivation of this equation can be found in several papers (e.g., Wagenet and Rao 1990).

Two situations must be considered when solving Eq. (9.1): steady-state water flow and transient water flow. For steady state, Θ and q are constant, and Eq. (9.1) becomes:

$$\rho_b \frac{\partial Si}{\partial t} + \Theta \frac{\partial Ci}{\partial t} = D \frac{\partial^2 C_i}{\partial z^2} - q \frac{\partial Ci}{\partial z} \pm \Phi \ . \tag{9.5}$$

This situation is simpler and allows the relative weights of the various phenomena to be determined quite easily. It may correspond to either saturated or unsaturated media. An example of a model based on this formulation is the BAM model (see Sect. 9.2.3).

For transient water flow, Θ and q must be known as functions of time and depth. When the movement of water is assumed to take place predominantly in the soil matrix and not in macropores or in any kind of bypass, this may be achieved through a mechanistic description of water flow based on an equation derived by combining Darcy's law and the equation of continuity. For one-dimensional transient vertical flow this equation is:

$$\frac{\partial \Theta}{\partial t} = \frac{\partial}{\partial z}\left[K(\Theta)\frac{\partial H(\Theta, z)}{\partial z}\right] - A(z, t) \ , \tag{9.6}$$

where $H(\Theta, z)$ (L) is the hydraulic head which is the sum of matric $h(\Theta)$ and gravitational $g(z)$ potential energy, neglecting other potential components. $K(\Theta)$ ($L\ T^{-1}$) is the hydraulic conductivity, a function of the soil water content Θ. $A(z, t)$ (T^{-1}) represents the uptake of water by the plant.

Equation (9.6) may be solved for given initial boundary conditions and soil properties, to give the function $H(z, t)$. This allows us to obtain $h(z, t)$, which in turn gives $\Theta(z, t)$, provided $\Theta(h)$ is known. Then, the water flux, q, may be calculated from:

$$q(z, t) = \int_{z_1}^{z_2} \left(\frac{\partial \Theta}{\partial t}\right) \partial z \ , \tag{9.7}$$

where z_1 and z_2 refer to two given depths of the soil profile.

To use Eqs. (9.6) and (9.7) demands knowledge of the functions $K(\Theta)$ and $H(\Theta)$ as precisely as possible, and this is a serious limitation. A better way to proceed is to use measurements of hydraulic conductivity and of matric potential for the site where the model is to be applied. This is time-consuming, and may be very expensive when many data are required, so that measured values are replaced with estimates. These are obtained with empirical equations, the parameters of which are correlated with soil characteristics (clay and silt contents, bulk density). Examples of such equations have been reviewed and discussed by Hutson and Cass (1987) and by Tietje and Tapkenhinrichs (1993). They may be particularly useful for using soil survey databases, as demonstrated by Petach et al. (1991). Nevertheless, it should be kept in mind that hydraulic conductivity and matric potential values obtained in this way are only approximate and are sometimes poor estimates. Solutions of Eq. (9.6) are obtained for given sets of initial and boundary conditions. Depending on these conditions, these solutions can be derived either analytically or numerically. Equation (9.6) does not apply directly to water transport in nonhomogeneous soils where there are nonmatrix flows, earthworm holes, and structural cracks. Modeling for heterogeneous soils will be discussed later.

Transport of pollutants in the gas phase is described in models based on Henry's law and on molecular diffusion. Exchange at the soil-atmosphere interface is described in terms of diffusion in a boundary layer (see the model LEACHP for example).

9.2.1.2 Other Modeling Approaches

Other approaches are not based on the CD equation. Three selected approaches are interesting to mention, because of their successful use.

Models based on chemical engineering theories. The theory of residence time distribution was developed for chemical engineering purposes in order to describe the transport of reactive solutes in chemical reactors (Villermaux 1982). It gives a theoretical framework which allows the modeling of simultaneously occurring phenomena: convection, hydrodynamic dispersion, mass transfer between a mobile and a stationary phase, multisite reversible adsorption, and first-order chemical reactions. Basically, this theory gives the distribution of residence times of a solute in a porous medium as a function of the characteristic times of elementary processes. This distribution is calculated from the time variation of input and output concentrations. This approach has been applied to metal transport in soil columns and also in the field (Jauzein 1988; Tevissen 1993). Another transport theory has also been applied for modeling solute transport: that of networks of mixing cells (Villermaux 1982).

Models using a compartment description for calculating water flow. Owing to the difficulties encountered when solving Eq. (9.1), simpler descriptions of water transport, often called capacity models, have been proposed for modeling purposes. In these descriptions, the soil is assumed to be equivalent to a series of reservoirs which can correspond either to the horizons or to any other layers. Each reservoir is assumed to be able to hold a maximum volume of water which is generally taken as equal to that retained at field capacity. When the amount of water entering a reservoir exceeds this maximum volume, the excess water is allowed to flow to the next reservoir. The transfer of water along the soil profile is described in this way and stops at the layer where the water content is below the field capacity. The amount of infiltrating water in the first layer is calculated as the difference between the applied water (rain and/or irrigation) and evapotranspiration. No kinetic equations are introduced in this description; the time interval associated with the transfer is generally set at 1 day, but other values may be chosen, depending on available data. These models have been improved to account for two liquid phases, corresponding to immobile and mobile water, by introducing a water content limit, below which no water movement is possible. Various values have been given to this limit but they are arbitrary. As an example, in the model VARLEACH, mobile water is assumed to be that retained between matric potential values of –10 and –200 kPa. Another improvement has been introduced to describe the effect of a low-permeability layer on water transfer (model PRZM, see Sect. 9.3). In this case, the volume of water held in a soil layer can be greater than that corresponding to the field capacity, and water is progressively drained according to an exponential function of the amount of excess water.

All reservoir-based models describe the downward movement of water but some are also able to represent the upward flow when evapotranspiration is greater than the water application rate.

When water movements are described with reservoir-based models, only convective transport is considered. If piston flow is assumed, the solute concentration remains constant during the transport. This description is a rather approximate one, so that improvement may be obtained by allowing partial or total mixing between the solution entering a layer and the solution already present in that layer. One of the first modeling approaches of this kind was proposed by Burns (1974).

Fugacity-based models. Fugacity models were developed for describing pollutant distribution in aquatic systems (Mackay 1979; Mackay and Paterson 1983). Instead of concentration C (mol m^{-3}) as controlling variable, the models use the fugacity f (Pa).

$$C = fZ \;, \tag{9.8}$$

where Z is the fugacity, which is characteristic of the chemical, the medium and the temperature. For example, if a compound is distributed among the solid, liquid, and gas phases, the following relation holds at equilibrium:

$$f = \frac{Cg}{Zg} = \frac{Cl}{Zl} = \frac{Cs}{Zs} \;, \tag{9.9}$$

where Cg and Zg, Cl and Zl, Cs and Zs are the concentration and the fugacity capacity of the gas, liquid, and solid phases, respectively.

It can be shown that fugacity capacity factors are simply related to partition coefficients, so that Eq. (9.9) can be written:

$$f = Cg(RT) = Cl\frac{Po}{Co} = Cs\frac{Po}{\rho K_d Co} \;, \tag{9.10}$$

where R is the gas constant, T the thermodynamic temperature, Po the saturated vapor pressure at T, Co the water solubility at T, ρ the solid phase density, and K_d, the solid/liquid distribution coefficient for the chemical under consideration.

Similar relations can be established for chemical distribution among all the compartments of the environment, including living organisms. Furthermore, the authors have shown that the rates of various processes may be described using the product of a fugacity and a transport or transformation coefficient. This model, which is essentially a global model, was developed for simulating the transfer of pollutants in large and complex aquatic systems such as lakes.

The fugacity model was further modified by Mackay and Diamond (1989) in order to use it for nonvolatile chemicals. In this case, the fugacity can no longer be used as an equilibrium criterion because there is no measurable vapor-phase concentration. To overcome this difficulty, the authors suggested the use of a new variable, based only on liquid concentration and named "equivalent concentration". This was applied to simulation of PCBs and Pb distribution in Lake Ontario, and it has proved to be an interesting tool.

9.2.1.3 Transport in Heterogeneous Media

Soils are often nonuniformly textured and are structured so that their pore space is characterized by complex pore shapes and size distributions, which are time- and space-variable. This has a profound influence on water and solute transport, because one fraction of the liquid phase is immobile while another is greatly mobile, which leads to preferential flow. One of the first models describing the role played by the immobile phase was proposed by Coats and Smith (1964). Immobile water could comprise stagnant water around solid particles and/or water held in intraaggregate pores (Nkedi-Kizza et al. 1983).

Preferential flow seems to be observed in various soils, often in structured clay soils, but sometimes also in sandy soils. Numerous published field experimental results are often poorly or not at all simulated by models which describe the fate of soil-applied chemicals. Some weakly adsorbed, but also some strongly adsorbed compounds are transported to great depth, as if they were subject to an accelerated movement (Ghodrati and Jury 1992). Two examples are given in this chapter to illustrate the observations. Although flow mechanisms are not yet completely understood, three kinds of preferential flow have been suggested (Luxmoore and Ferrand 1993).

- channeling in pores:
  - in macropores ($\phi > 1$ mm), often comprising biopores,
  - in mesopores ($\phi < 1$ mm) comprising a network of cracks and biological channels,
- fingering flow, due to wetting front instability; and
- funnel flow, caused by redirection of flow by heterogeneities in the profile.

When such flow patterns occur, solute transport may greatly deviate from a near piston-like displacement. Attempts have been made to model such a phenomenon, mainly by distinguishing mobile and immobile water, the former being partially or totally responsible for the rapid transport. Several models have been proposed to account for preferential flow in laboratory columns (e.g., Coats and Smith 1964; van Genuchten and Wierenga 1976; Gaudet and Vauclin 1990) and in the field (e.g., Addiscott et al. 1986; Corwin et al. 1991; Hutson and Wagenet 1993; Jarvis 1994).

Pore heterogeneity and its influence on water and solute transport have been introduced into models via two approaches. In the first, an attempt is made to use the pore-size distribution either by solving the CDE for several pore classes (Lindstrom and Boersma 1971) or by defining probability density functions of the solute travel times (Destouni 1993). The second approach more simply involves dividing the pore space into two fractions, corresponding to mobile and immobile water. This division may be based on hydraulic and water-retention porperties of the soil as in the model MACRO proposed by Jarvis (1994). It may also be roughly defined by considering that immobile water is held below a given water potential, for example, the field capacity or an assumed fraction of it. These fractions are introduced to describe the water flow in either capacity models (Addiscott et al. 1986; Corwin et al. 1991), CDE-based models (van Genuchten and Wierenga 1976; Gaudet and Vauclin 1990, or semiquantitative descriptions (Calvet et al. 1978).

Another aspect of modeling transport in heterogeneous media is the introduction of solute exchange between mobile and immobile water, which is assumed to be diffusion-controlled and described either by a first-order rate function (van Genuchten and Wierenga 1976) or by Fick's law if the geometry of the media is known (van Genuchten and Dalton 1986). As also suggested by Luxmoore and Ferrand (1993), mixing of solute between the two water phases may also result from disconnection/reconnection of water flow paths during drying and wetting cycles.

Besides the mathematical difficulty due to complex flow patterns, two questions still remain without a satisfactory answer: how must soil heterogeneities be characterized, and how can the partitioning between mobile and immobile water be defined in order to feed the models with pertinent input data? This is probably a challenge for future researchers.

9.2.1.4 Sink/Source Phenomena

Soil pollutants are often affected by several phenomena which provide sinks or sources for solutes in the fluid phase. Transformation, degradation, retention, and precipitation processes are sink phenomena; desorption, solubilization, and, in some cases, transformations (e.g., nitrate production) are considered as sources. Modeling these phenomena can be done with various degrees of approximation, for both the physical description and their dependence on soil water composition and temperature.

Several criteria may be used to classify reactions corresponding to sink-source phenomena. According to Rubin (1983), reactions may be classed into the following categories:

- fast and slow reactions, the former corresponding to local equilibrium,
- homogeneous and heterogeneous reactions, the latter category being obviously relevant for the soil,
- surface (sorption) and classical reactions (oxidation, reduction, precipitation, dissolution, and complex formation).

Pollutants are not equally affected by all transformation reactions, as schematically indicated in Table 9.1.

Retention Phenomena. Retention is the transfer of chemical species form the liquid or the gas phase to a solid phase because of reversible (sorption) and/or irreversible phenomena. Sorption is always included in models, even in the simpler ones, but irreversible retention is practically never considered. *Nonsingularity* of retention was only recently introduced into a few models (e.g., Piver and Lindstrom 1990). This is not a satisfactory situation, because an increasing body of observations is adding increasing interest to this phenomenon (Calvet 1994). There are many unanswered questions about molecules retained by irreversible retention (bound residues), as regards their nature, their properties, and their mobility (see Chap. 6). Modeling the fate of pesticides in soil should incorporate such a phenomenon. Nevertheless, whether the questions are fully answered or not, pesticide-leaching models would be improved by a better description of sorption. Two aspects have to be considered. The first, probably connected with bound residue formation, is the time dependence of the sorption coefficient. The estimation of sorption coefficients for modeling is not a simple matter (Green and Karickoff 1990; Calvet 1994) and it is further complicated by its time dependence. It appears that the apparent sorption coefficient increases with time, as suggested by several observations (Walker 1987; Boesten et al. 1989; Lehman et al. 1990; Barriuso et al. 1992) but, unfortunately, experimental and theoretical procedures are not yet available to evaluate the sorption-time relationship.

Sorption kinetics and equilibrium characteristics must be taken into account when introducing sorption phenomena into Eq. (9.1). The simplest situation corresponds to transfer with local equilibrium, in which case only sorption isotherms are necessary. This applies to media in which all active surfaces are equally and readily accessible, such as homogeneous slightly swelling soils, poorly aggregated soils, and soils with low organic matter content. For other situations, the assumption of local equilibrium is an approximation which may be rather irrelevant in highly heterogeneous soils and soils rich in organic matter. So the most frequent situation is that of nonequilibrium which involves the introduction of sorption kinetics.

Brusseau et al. (1991) grouped the rate-limiting processes into two categories: transport-related and sorption-related nonequilibrium. The former results from the existence of several flow domains, related to macroscopic heterogeneities (aggregates, macropores, cracks). Sorption-related nonequilibrium may result from chemical nonequilibrium due to rate-limited interaction between sorbates and sorbents. This may be relatively unimportant for hydrophobic molecules, but not for polar molecules. Sorption-related nonequilibrium may also result from diffusive mass transfer which involves three processes: (1) film diffusion (probably negligible), (2) retarded intraparticle diffusion (diffusion into the micropores of aggregates), (3) intrasorbent diffusion. This process corresponds to the diffusion of hydrophobic molecules into the organic matter particles and of polar molecules into clay particles. Several models have been developed in order to take sorption kinetics into account (Pignatello 1989). Equilibrium characteristics are provided by sorption isotherms which can be linear or nonlinear whatever the solute species, ionized or nonionized molecules. However, linear sorption isotherms have often been observed or used for the past 50 years to explain chromatographic separation (Martin and Synge 1941). It is obviously more easy to deal with linear isotherms when modeling solute transport. For nonionized molecules, the distribution coefficient K_d or the related coefficients K_{oc} and K_{om} are used (see Chap. 5). For ion exchange reactions, equilibrium constants must be known, which implies the knowledge of ion selectivity coefficients. Many experiments have beeen conducted to study the relationship between sorption and transport, and a complete description of these has been published by Schweich and Sardin (1981).

Models for heavy metals deal with precipitation/solubilization phenomena through speciation calculations and solubility relationships (e.g., Ghadiri and Rose 1992).

Chemical and biochemical transformations. Solutes may be transformed during transport, because of biotic or abiotic reactions, which results in the disappearance of more or less chemicals usually from the liquid phase, but sometimes also from the solid phase.

Soulas (1995) has recently reviewed the models used to describe the biodegradation of pesticides. Experimental data show that biodegradation conforms to either first- and second-order kinetics or to hyperbolic (Michaelis-Menten) kinetics, depending on the molecule and the soil. However, these are only global descriptions, so that attempts to model biotransformations more precisely have been undertaken. This was done by introducing microbial processes in order to take into account both the degradation and the maintenance and growth of microbial populations. Transformation of nitrogen compounds has also been extensively studied and has lead to several models, particularly for nitrate and ammonium ions. Examples of such models are those of van Veen and Paul (1981), Molina et al. (1983), and Jenkinson et al. (1987).

For transport modeling, it is generally assumed that transformations may be described by global first-order kinetics, whatever the number of reactions involved. For organic pollutants, abiotic and biotic transformations and degradation are generally not distinguished; they are lumped into a global equation. When first-order rate kinetics are assumed to be valid, it is necessary to know only one parameter, the first-order rate constant, and its relationships with the soil water content and temperature. The basic equation is:

$$\frac{\partial Ci}{\partial t} = -\mu Ci, \tag{9.11}$$

where μ is a first-order rate constant.

This is only approximately true, but it is an acceptable approximation for most modeling purposes. Nevertheless, in all cases, the main difficulty is due to variation of μ with environmental factors such as soil temperature and humidity. Some simple relationships have been proposed for pesticides.

An important point is the coupling of sorption with transport and transformation-degradation processes. Sorption which is rate-limited by transport is related to the presence of heterogeneous flow domains and, particularly, to the more rapid flow in macropores which prevents pesticide molecules from reaching adsorbing sites in micropores. Some modeling approaches have been proposed to describe this coupling. Sorption may modify biotic transformations and degradation in two ways. The first is the reduction of the amount of degradable pesticide, which happens because sorbed molecules are not mobile and are thus inaccessible to microorganisms. The second way involves a kinetic effect related to the diffusion of molecules out of the micropores, which is necessary if they are to be accessible. Models to account for this kinetic effect have recently been developed by Scow and Hutson (1992) and by Duffy et al. (1993). Introduction of such a coupling, at least in research models, would enable us to simulate the fate of pesticides more precisely.

9.2.2 Stochastic Models

The uncertainty of model parameter values resulting from lateral and vertical soil variability explains why deterministic models fail to describe water and solute transport in the field correctly. As stated by Jury (1982), from an environmental point of view, it is generally the minimum residence time which should be estimated, particularly for prediction purposes. Several approaches may be followed to achieve this goal, which also take the spatial variability into account.

Scaling theory had been applied by several modelers to describe water and solute transport in terms of the deterministic flux and mass balance theory applied to parallel and non-interacting regions where vertical flow is taking place. Each region is characterized by a set of hydraulic and retention properties and is assumed to possess similar structure characteristics (Dagon and Bresler 1979). Another approach is to consider soil properties as random space functions and to run models with values derived from these functions (Russo 1993). As a consequence, the resulting flow equations are of a stochastic nature and dependent variables are also random space values. This approach may use Monte Carlo calculations or mean solutions, determined analytically or numerically (ref. cited in Villeneuve et al. 1990). This procedure involves the determination of the spatial distribution of soil properties, which may often be impractical or, at least, very time-consuming and expensive.

An alternative is obtained by considering the field soil as a transfer medium, characterized by transfer functions, without defining any transport mechanism. According to Jury (1982), the transfer of solute from the soil surface to any given depth in the soil is considered to be a stochastic function of time, or of the net amount of water applied to the soil. Modeling expresses the solute concentration at a given time and at a given depth, as a function of both the probability density distribution of water application rate, and of soil travel times, which depend on the variability of soil properties. This transfer function is a concept similar to that used in chemical reaction engineering, but is defined in a stochastic instead of a deterministic way. A generalization of transfer function modeling to solutes undergoing physical, chemical, and biological transformations was proposed by Jury et al. (1986b). They introduced a lifetime density function to represent the net influence of soil processes and modes of solute input on the solute lifetime. This modeling approach has been shown to be able to simulate bromide and chloride transfers during unsteady flow in an unsaturated well-structured clay soil (White et al. 1986). The transfer function of the soil medium has also been described in terms of a continuous time variation (Knitton and Wagenet 1987). The calculation of the probability of movement of solute molecules between adjacent soil layers enabled the solute concentration to be calculated. The authors have shown that this model is able to describe the transport of bromide and nitrate in layered soil columns.

9.3 Examples of Models Describing the Fate of Pesticides in Soils

The purpose of this section is to give an idea of the great variety of models and to illustrate how some of them are constructed. The reader's attention must be drawn to two points. The first is that the following presentation is not exhaustive and does not claim to be complete; a more detailed review was recently published by Ghadiri and Rose (1992). The second is that, although these models were developed for pesticides, they may be used for other organic chemicals as well. They may also be used for metals if degradation processes are removed and if routines describing precipitation processes and speciation are introduced. Models describing the fate of nitrogen in soils present the same features as regards water and solute transport; however, their biodegradation part is different and applies specifically to mineral and organic nitrogen transformations. Examples of models which are discussed below are presented in Table 9.2.

Relations describing water and solute transport and the water balance in the soil constitute the master set of equations. These equations are coupled with other equations representing sorption/desorption, transformation/degradation, plant uptake, and volatilization. These phenomena are described below, in various degrees of detail.

Table 9.2. Pesticide leaching models

Acronym	Model	Category	Reference
LEACHMP3	Leaching estimation and chemistry model	Research	Hutson and Wagenet (1992[a])
X	Mathematical model for describing transport in the unsaturated zone of soils	Research	Piver and Lindstrom (1990)
Y	Modeling the influence of sorption and transformation on pesticide leaching and persistence	Research	Boesten and van der Linden (1991)
PESTFADE	Pesticide fate and transport model	Research	Clemente et al. (1993)
VARLEACH		Research/ management	Walker (1987)
MACRO	Model of water movement and solute transport in macroporous soils	Research/ management	Jarvis (1994)
PRZM2	Pesticide root zone model	Management	Mullins et al. (1992)
VULPEST	Vulnerability to pesticides	Management	Villeneuve et al. (1990)
GLEAMS	Groundwater loading effects of agricultural management systems	Management	Leonard et al. (1987)
BAM	Behavior assessment model	Screening	Jury et al. (1983)
CMLS	Chemical movement in layer soils	Educational	Nofziger and Hornsby (1986)

[a]Date of the last published version.

9.3.1 Transport Phenomena

Table 9.3 gives an overview of the transport phenomena included in the various models, and Table 9.4 indicates how water transport is described in the models. When the Richards equation is resolved for transient flow, hydraulic conductivity-water content and matric potential-water content relations are needed, and this is often a serious limitation to model application. In constrast, the PRZM2, VARLEACH, and CMLS models demand only some water content data such as the field water capacity, the wilting point, and the water content, at some specified matric potentials (e.g., –10 and –200 kPa).

In some models, the effect of a water table can be accounted for by simulating upward flow. This is done by modeling capillary rise in a model based on the water transport Eq. (9.6).

Table 9.3. Transport phenomena included in the models

Model	Evapotranspiration	Solute transport: Transport in the soil		Solute transport: Volatilization	Heat transport in the soil	Runoff
		Convection	Convection dispersion			
LEACHMP3	X		X	X	X	
X	X		X	X	X	
Y	X		X			
PESTFADE	X		X			
VARLEACH	X	X				
MACRO	X	X	X		X	
PRZM2	X	X	X	X	X	X
VULPEST	X	X				
GLEAMS	X	X				X
BAM		X		X		
CMLS	X	X				

Solute transport may be obtained by the solution of the dispersion/convection equation (LEACHMP, X, Y, PESTFADE), by multiplying a water flux by a solution concentration (VULPEST, BAM), by piston flow (CMLS), or by displacing given volumes of water and allowing complete mixing of solutions (PRZM, VARLEACH). Volatilization and losses from the soil to the atmosphere are obtained by coupling the liquid/gas partition of the pesticide (Henry's law) with diffusion in the gas phase of the soil and in the atmosphere just above the soil surface.

All the models cited above simulate the water and solute transport in the unsaturated zone for given sets of boundary conditions. These boundary conditions (Table 9.5) cover several field situations but necessitate the numerical solution of the water-transport and convection-dispersion equations.

9.3.2 Sink/Source Phenomena

Several sink/source phenomena are introduced in the various models described above. Characteristics which are used are summarized in Table 9.6.

It is interesting to note some additional information:

1. Concerning *sorption*
   - in the X model, the linear sorption coefficient is expressed as a function of the granulometric composition and of the organic matter content. A first-order rate-irreversible sorption is also introduced.
   - in LEACHMP2, two-site sorption kinetics can be used; one fraction of the sites exhibits a local chemical equilibrium, and a second fraction is characterized by kinetically controlled sorption and desorption which are described by linear isotherms.

Table 9.4. Modeling water transport

Model	Modeling water transport	Comments
LEACHMP3	Generalized Richards equation	Possible choice between steady and transient flow
X	Generalized Richards equation	Equation can also be solved for transport in the vapor phase
Y	Generalized Richards equation	
PESTFADE	Generalized Richards equation	
VARLEACH	Mass balance applied to water in soil layer, flow drainage is calculated as the excess of water compared with field water capacity	Distinction is made between mobile and immobile water for calculating the flow of the soil solution
MACRO	Generalized empirical Richards equation for flow in micropores drainage rule in macropores	Distinction is made between macro- and micropores to account for preferential flow
PRZM2	Mass balance applied to water in soil layer, flow drainage is calculated as the excess of water compared with field water capacity	A fraction of water held in each layer may be allowed to drain
VULPEST	No calculation of water transport	Amount of water which infiltrates is an average monthly value of the difference between rainfall and evapotranspiration
GLEAMS	Mass balance applied to water in soil layer; flow drainage is calculated as the excess of water compared with field water capacity	
BAM	The Richards equation is solved for steady flow	The soil profile is assumed to be at a uniform water content
CMLS	Mass balance applied to water in soil layer, flow drainage is calculated as the excess of water compared with field water capacity	

- in, PESTFADE a two-sites approach is also developed; the two types of sites are characterized by first-order rate kinetics but have different localizations, one being available in micropores, the other in macropores. A sorption kinetic model formulated by Gamble and Khan (in Clemente et al. 1993) may be incorporated.

2. Concerning *transformation and degradation*
   - in BAM, the degradation rate constant may vary with depth according to an empirical reduction factor.
   - in the Y model, the degradation rate constant may also vary according to an empirical numerical function of depth proposed by the authors.

Table 9.5. Various boundary conditions (BC) used in the models. There may be a choice among several BCs for some models

Model	Water transport		Solute transport	
	Upper BC	Lower BC	Upper BC	Lower BC
LEACHMP3	Ponded, non-ponded infiltration, water evaporation, zero flux at the soil/atmosphere interface	Permanent water table, free drainage, zero flux, zero flux + constant potential fluctuating water table	Volatilization, diffusion through a surface soil film and a stagnant atmospheric film; nonvolatile chemicals, zero flux	Water table, unit gradient drainage
X	As for LEACHMP3	Fixed water table	As for LEACHMP3	Fixed water table
Y	Variable flux given by rainfall-evapotranspiration difference	Fixed water table at 1 m depth	Zero flux	Only convection flow is allowed out of the system at 3 m depth
PESTFADE			Zero concentration at the soil surface	Zero flux at a given depth
VARLEACH	X[a]			
MACRO	Variable flux given by rainfall-evapotranspiration difference	Free drainage	Zero flux for a soil-applied pesticide, complete mixing between incoming rainfall and water in shallow surface	Only convection flow at a given depth
PRZM2	X[a]	Free drainage	Diffusion through a limited film at the soil/atmosphere interface	Free drainage
VULPEST	Variable flux given by rainfall-evapotranspiration difference	Not specified	Constant concentration	Zero concentration at infinity
GLEAMS	Variable flux given by rainfall-evapotranspiration difference	Free drainage	Zero flux	Not specified
BAM		Semiinfinite		Semiinfinite
CMLS	X[a]	Free drainage		Free drainage

[a]Comment given for the model MACRO applies also to models VARLEACH, PRZM-2 and CMLS.

Table 9.6. Sink/source phenomena

Model	PU	Sorption			Biotic transformation and degradation					Abiotic transformations	
		Equil.		N.E	Liquid				Gas	Liq.	Sol.
Model		L	NL	NE	1st.	Θ	T	z			
LEACHMP	X	X	X	X	X	X	X	X	?	X	
X		X		X	X		X	X		X	X
Y	X	X			X	X	X	X			
PESTFADE		X		X	X	X	X			X	
VARLEACH		X			X	X	X	X			
MACRO	X	X			X						
PRZM2	X	X			X						
VULPEST		X			X						
GLEAMS	X	X			X						
BAM		X			X			X			
CMLS		X			X						

PU: plant uptake; Θ: water content-dependent rate constant; T: temperature-dependent rate constant; z: depth-dependent rate constant; liq.: abiotic transformation in the liquid phase; sol.: abiotic transformation at the soil constituent surfaces; Equil.: sorption equilibrium; L: linear isotherm; NL: nonlinear isotherm; 1st.: overall first-order reaction; MACRO: the rate constant of degradation can be different in micropores and in macropores.

- in many models, a different value of degradation rate can be introduced for each soil horizon as input data.
- LEACHMP3 can simultaneously simulate the fate of several chemicals (up to 10) undergoing biotic and abiotic reactions, all being of the first order. The several rate constants are allowed to vary with depth, soil water content, and temperature. Transformation in the gas phase is also introduced.

9.4 Some Problems Connected with Modeling

Using models is not a simple matter, even when the software can be easily handled. Answers must be given to several questions, three of the most important being: At what scale is modeling to be performed? What parameter values must be used in relation to model sensitivity? When is a model valid?

9.4.1 The Scale of Modeling

Spatial scale. As is well known, soil is a porous medium which presents a high spatial variability from the molecular to the regional scale. As a result, it is important to specify the scale at which modeling is to be performed and what models are to be used. Wagenet (1993) has recently proposed a hierarchical organization of spatial scales, together with the possible corresponding modeling approaches (Fig. 9.1). He emphasized the fact that mechanistic deterministic models cannot describe a variable field at any particular scale in a single run of the model. These models must be used with a deterministic/stochastic approach which appears to be possible only when simplified descriptions of transport are used, as shown by Hutson and Wagenet (1993). In particular,

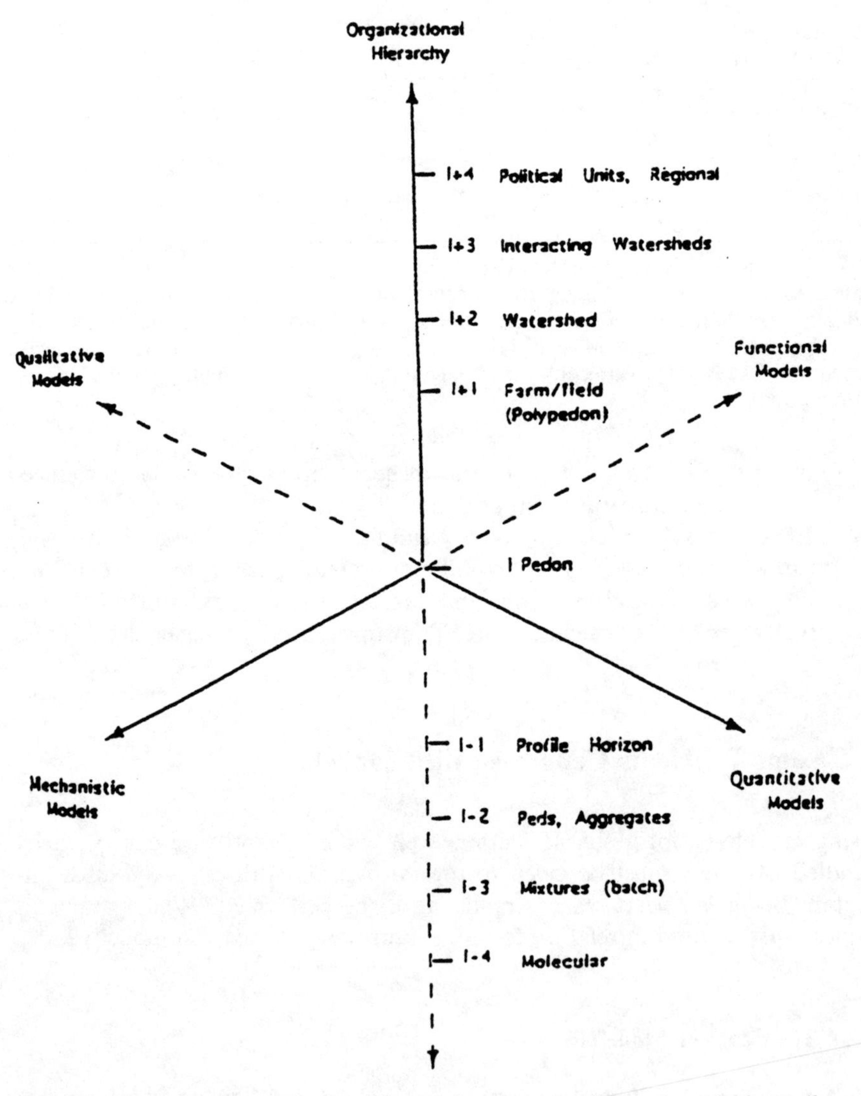

Fig. 9.1. Hierarchical organization of spatial scales related to soil pesticide studies and modeling approaches. (After Wagenet 1993)

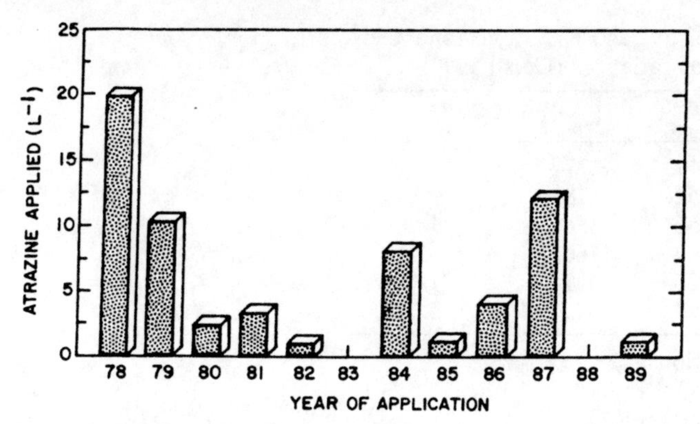

Fig. 9.2. Predicted leaching of atrazine in a soil column for different years. (After Walker and Hollis 1993)

multiple runs, in which various combinations of climate, soil, crop and chemical applications are used, enable us to deal with the (i + 3) and (i + 4) scale levels (Fig. 9.1).

Temporal variability. Temporal variability may characterize soil properties, climate, and the coupling between the two. It seems from the literature that only climatic variations have been taken into account. Figure 9.2 gives an example of the effect of climatic variation on the simulation of atrazine leaching in soil columns under several climatic conditions (Walker and Hollis 1993). It clearly appears that temporal extrapolation is not possible from a single simulation but necessitates a frequential analysis.

9.4.2 Model Sensitivity

Model sensitivity to variations of input parameters is key information for the correct application of pesticide leaching models. This is so because input parameters are highly variable, whether they are measured or estimated.

The variability of measured values has two components, one due to uncertainties associated with protocols and analytical methods, the other being the result of spatial variability of the soil properties. This spatial variability has been described by Rao et al. (1986) in terms of factors linked to the site pedogenesis (intrinsic factors) and to cultivation and pesticide application practices (extrinsic factors). As a result, the soil properties display a wide in-situ variability, as was emphasized above. Field variations of about 30% in the sorption coefficient and half-life are currently observed (e.g., Allen and Walker 1987).

It is also important to be aware of the variability of estimated parameters, which are frequently used for predictive purposes. Some important parameters such as the normalized sorption coefficient, K_{oc} and the degradation half-life, $T_{1/2}$ have to be estimated when measured values are lacking. The only way to

Table 9.7. Example of average K_{oc} values and their associated coefficient of variation. (Gerstl 1990)

COMPOUND	n	K_{oc}	CV
Atrazine	217	227	158
Carbofuran	52	78	229
Diuron	156	384	74
Lindane	94	1160	176
Napropamide	36	487	71
Trifluralin	22	11 035	72

n: number of observations.
K_{oc}: average value (l kg $^{-1}$).
CV: coefficient of variation (%).

Table 9.8. Examples of half-life range values reported by Mullins et al. (1992)

Pesticide	Degradation rate constant days $^{-1}$
Aldicarb	0.1486–0.0023
Dicamba	0.2140–0.0197
2, 4-D	0.0693–0.0197
Malathion	2.9100–0.4152
Parathion	0.2961–0.0046
Picloram	0.0354–0.0019
Trifluralin	0.0956–0.0026

do this is to refer to published values, which are generally widely variable. Table 9.7 gives a series of K_{oc} values compiled by Gerstl (1990) and their associated coefficients of variation. The same situation is also observed for the first-order rate constant, as indicated in Table 9.8.

Thus, it is important to know the consequences of such variability for simulations. This knowledge is useful for defining the necessary precision of input parameter measurements and also for interpreting simulated results. These consequences are indicated by the sensitivity analysis which is based on the ratio of the output changes to input changes over the full range of likely parameter values. Furthermore, as stated by Addiscott (1993), sensitivity must be analyzed for changes in parameter values and for changes in variance of parameter values. The sensitivity of pesticide leaching models to variations of sorption and degradation parameters is high (e.g., Villeneuve et al. 1988; Boesten 1991). Studying the simulation of aldicarb leaching by means of the PRZM model, Villeneuve et al. (1988) showed that an uncertainty of 15% in the degradation rate and of 24% in the sorption coefficient generate a possible modification of 100% in the predicted cumulative quantity of pesticide reaching the water table after 3 years. The papers of Boesten (1991), Jarvis (1994), and Booltink and Bouma (1993) give examples of detailed sensitivity analysis. Two examples of sensitivity analysis results are given in Fig. 9.3. From the high sensitivity of the models to sorption and degradation parameters, it can be concluded that these parameters must be known as precisely as possible, including their variation along the soil profile.

From a general point of view, sensitivity assessment must be discussed carefully, taking into account the nature of the parameters and the period corresponding to the simulation. It is also important to be aware that the use of different criteria such as the distribution of the chemical in the soil profile, the amount remaining in a given soil layer, the amount leached and the amount transported beyond a given depth, may lead to different conclusions. Thus, comparison between models should use the same criteria.

9.4.3 Model Validation

As emphasized by Loague and Green (1991), "a model should be assumed suspect until it is proven correct." Model validation is obviously an important step in the modeling process. Models must be validated before they can be used with confidence in decision-making and risk-assessment processes. There are relatively few published papers reporting validation works based on objective validity criteria and, generally, they concern only one site and one year of experiments. Thus, it is not easy to know how precisely the various models ar able to describe the fate of pollutant and how they can be used for application purposes. Validation encounters two main difficulties: the first concerns the acquisition of data and the second relates to validation procedures.

To validate a model properly, it is necessary to obtain many data. Input parameter values can be obtained by appropriate measurements, by estimates based on some correlations between parameters and some soil properties (e.g., pedotransfer functions), by calibration, and from published values. The best way is to measure the required parameters, but this is not always possible and estimates are often used. Nevertheless, whatever the acquisition procedure, it is necessary to estimate the associated uncertainty. This may seem obvious, but published complete parameter data sets are not very numerous. A second set of data concerns those related to the experiment. This raises the problem of the soil-sampling procedure, which has been discussed, among others, by Smith et al. (1987). Solute concentrations at a given location and at a given time must be known, together with their associated uncertainities. Thus, the number of soil samples must be compatible with the desired accuracy for a given level of confidence, and it is important to make sure that sampling points are randomly distributed and independent. The great number of measurements needed makes experiments very expensive and explains why many published studies are not based on detailed data sets. Nevertheless, any modeling work should be accompanied by well-designed field experiments, which enable both observed results and simulation uncertainties to be correctly estimated. Under these conditions, the validation procedure can be based on sound comparisons be-

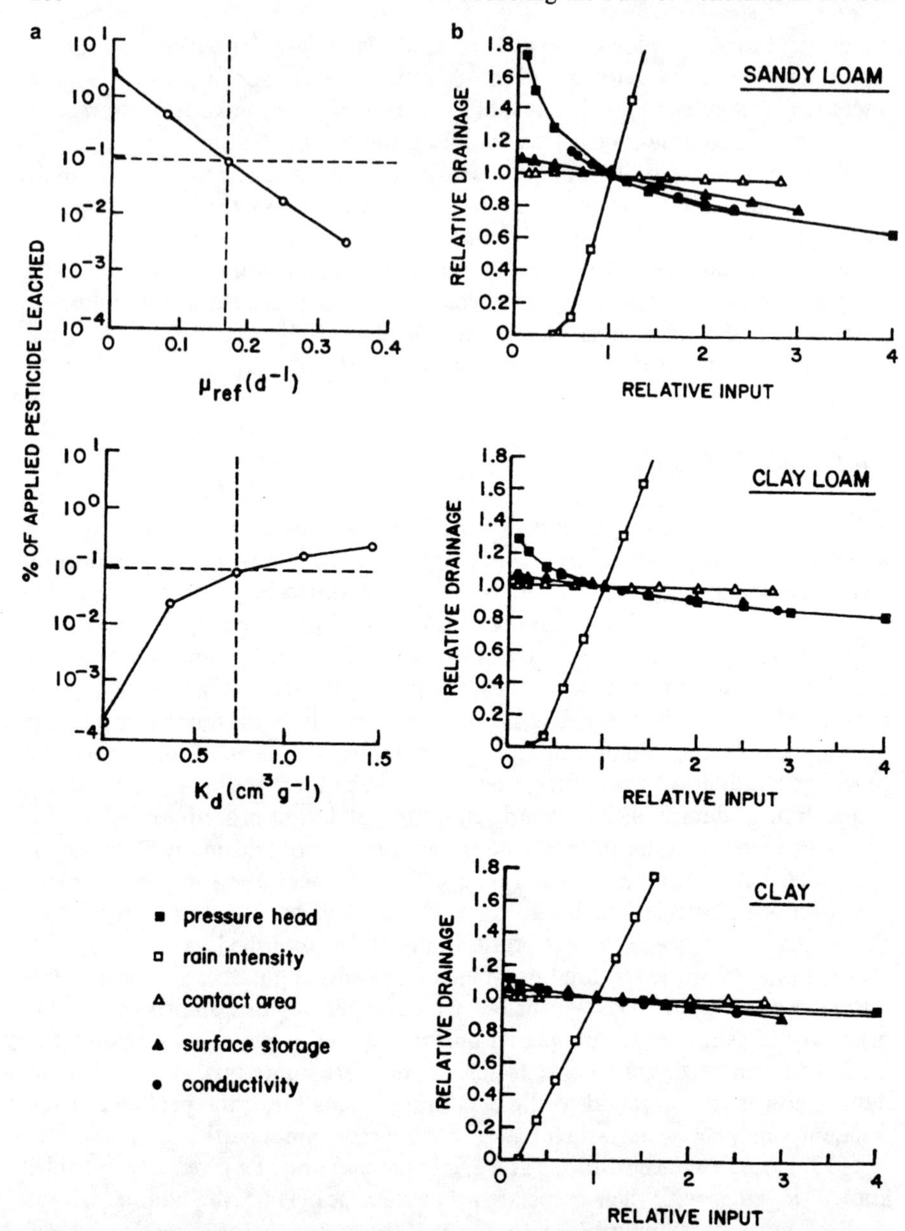

Fig. 9.3. Examples of sensitivity analysis. **a** Model MACRO (after Jarvis 1991). Variation of % of applied pesticide leached with the degradation rate (μ_{ref}) and with the sorption coefficient (K_d). **b** Model of Booltink and Bouma (1993). Variation of the relative drainage as a function of the relative input values of pressure head, rain intensity, contact area, surface storage, and conductivity.

tween observed and simulated data. Attention must be drawn to the fact that calibrating a model against a given field data set is not a validation. This only shows that the model outputs are able to represent this data set numerically. The second difficulty stems from the validation procedure to be applied. This question has been discussed by several authors (e.g., Loague and Green 1972; Parrish and Smith 1990). The validation is based on the camparison between experimental data and simulated data. Various experimental data are currently used: total mass of solute, position of the center of mass, peak concentration, time necessary for a critical concentration to reach a given depth, depth of the leaching front, mass flux through a given surface, etc. Several statistical criteria may be used: maximum error, root mean square error, coefficient of determination, modeling efficiency, and coefficient of residual mass (see Loague and Green (1972) for definitions). However, another question remains: what is the maximum acceptable difference between observed and simulated data for predictive purposes? A complete and definite answer probably does not exist today, and this is a fundamental point. Parrish and Smith (1990) have discussed model performance testing and they have stressed the fact that, due to their approximate nature, models are incapable of calculating the true values. They can only be expected, at best, to come close to them. According to these authors, a model may be considered valid if it gives values nearly equal to or sufficiently close to true values that the criterion of model validity lies in the definition of what is meant by "sufficiently close". Finally, it is for model users to decide what differences are acceptable, in the light of the nature of the problem under consideration, the type of model, the available data, and the objectives assigned to the modeling procedure.

References

Addiscott TM (1993) Simulation modelling and soil behaviour. Geoderma 60: 15–40

Addiscott TM, Wagenet RJ (1985) Concepts of solute leaching in soils: a review of modelling approaches. J Soil Sci 36: 411–424

Addiscott TM, Heys PJ, Whitmore AP (1986) Application of simple leaching models in heterogeneous soils. Geoderma 38: 185–194

Allen R, Walker A (1987) The influence of soil properties on the rates of degradation of metamitron, metazachlor and metribuzin. Pestic Sci 18: 95–111

Ballard TM (1971) Role of humic carrier substances in DDT movement through forest soils. Soil Sci Soc Am Proc 35: 145–147

Barriuso E, Koskinen W, Soerenson B (1992) Modification of atrazine desorption during field incubation experiments. Sci Total Environ 123/124: 333–344

Boesten JJTI (1991) Sensitivity analysis of a mathematical model for pesticide leaching to groundwater. Pestic Sci 31: 375–388

Boesten JJTI, van der Linden AMA (1991) Modeling the influence of sorption and transformation on pesticide leaching and persistence. J Environ Qual 20: 425–435

Boesten JTI, van der Pas LJT, Smelt JH (1989) Field test of a mathematical model for non-equilibrium transport of pesticides in soil. Pestic Sci 25: 187–203

Booltink HWG, Bouma J (1993) Sensitivity analysis on processes affecting bypass flow. Hydrol Processes 7: 33–43

Bresler E (1980) Models for predicting distribution of chemicals in soil during irrigation. CR Colloque Franco Israelien, INRA, Versailles, France pp 27–47

Brusseau ML, Jessup RE, Rao PSC (1991) Modeling the transport of solutes influenced by multiprocess nonequilibrium. Water Resour Res 25, 9: 1971–1988

Burns IG (1974) A model for predicting the redistribution of salts applied to fallow soils after excess rainfall or evaporation. J Soil Sci 25, 2: 165–178

Burns LA (1985) Validation methods for chemical exposure and hazard assessment models. Workshop 11–13 November 1985, pp 148–172

Calvet R (1993) Comments on the characterization of pesticide sorption is soils. In: Del Re AM, Capri E, Evans SP, Natali P, Trevisan M (eds), IX Symp Pesticide Chemistry, Piacenza, 11–13 October 1993

Calvet R (1994) Modeling pesticides leaching in soils: main aspects and main difficulties. In: Guipponi C, Marani A, Morari F (eds) Modelling the fate of agrochemicals and fertilizers in the environment. European Society for Agronomy, Padova, Italy

Calvet R, Le Renard J, Tournier C, Hubert A (1978) Hydrodynamic dispersion in columns of pumice particles. J Soil Sci 29: 463–474

Carsel RF, Mulkey LA, Lorber MN, Baskin LB (1985) The pesticide root zone model (PRZM): a procedure for evaluating pesticide leaching threats to groundwater. Ecol Model 30: 49–69

Carsel RF, Jones RL, Hansen JH, Lamb RL, Anderson MP (1988a) A simulation procedure for groundwater quality assessment of pesticides. J Contam Hydrol 2: 125–138

Carsel RF, Smith CN, Mulkey LA, Dean JD, Jowise PP (1988b) User's manual for the pesticide root zone model (PRZM): release 1. USEPA EP-600/3–84–109. US Government Printing Office, Washington, DC

Clemente RS, Prasher SO, Barrington SF (1993) Pestfade, a new pesticide fate and transport model: model development and verification. Am Soc Agric Eng 36, (2): 357–367

Coats KH, Smith BD (1964) Dead-end pore volume and dispersion in porous media. J Soc Petrol Eng 36, 4: 73–84

Corwin DL, Waggoner BL, Rhoades JD (1991) A functional model of solute transport that accounts for bypass. J Environ Qual 20: 647–658

Dagan G, Bresler G (1979) Solute dispersion in unsaturated heterogeneous soil at field scale. I: Theory. Soil Sci Soc Am J 43: 461–467

Day PR, Forsythe WM (1957) Hydrodynamic dispersion of solutes in the soil moisture stream. Soil Sci Soc Proc 21: 477–480

De Smedt F, Wierenga PJ (1978) Solute transport through soil with nonuniform water content. Soil Sci Soc Am J 42: 7–10

Destouni G (1993) Field-scale solute flux through macroporous soils. In: Russo D, Dagan G (eds) Water flow and solute transport in soils. Springer, Berlin Heidelberg New York

Duffy MJ, Carski TH, Hanafey MK (1993) Conceptually and experimentally coupling sulfonylurea herbicide sorption and degradation in soil. In: Del Re AM, Capri E, Evans SP, Natali P, Trevisan M (eds) IX Symp Pesticide Chemistry, Piacenza, 11–13 October 1993

Gaudet JP, Vauclin M (1990) Transport de l'eau et de substances chimiques dans les sols partiellement saturés: la colonne et le milieu naturel. In: Sardin M, Schweich D (eds) Impact of physico-chemistry on the study, design and optimization of processes in natural porous media. Int Conf, Nancy, 1990, pp 253–274

Gerstl Zev (1990) Estimation of organic chemical sorption by soils. J Contam Hydrol 6: 357–375

Ghadiri H, Rose CW (1992) Modeling chemical transport in soils. Lewis, London
Ghodrati M, Jury WA (1992) A field study of the effects of soil structure and irrigation method on preferential flow of pesticides in unsatured soil. J Contam Hydrol 11: 101–125
Green RE, Karickoff SW (1990) Sorption estimates for modeling. SSSA Book Ser 2: 79–101
Hutson JL, Cass A (1987) A retentivity function for use in soil-water simulation models. J Soil Sci 38: 105–113
Hutson JL, Wagenet RJ (1992) Leaching estimation and chemistry model Vers.3. Dept Soil, Crop and Atmospheric Sciences, Research Ser 92–3, Cornell Univeristy, New York
Hutson JL, Wagenet RJ (1993) A pragmatic field-scale approach for modeling pesticides. J Environ Qual 22: 494–499
Jarvis N (1994) Macro – a model of water movement and solute transport in macroporous soils. Report monograph no.19. Department of Soil Science, Uppsala
Jauzein M (1988) Methodologie d'étude du transport transitoire de solutés dans les milieux poreux. Thèse de Doctorat de l'Institut Nationale Polytechnique de Lorraine, Nancy
Jenkinson DS, Hart PBS, Rayner JH, Parry LC (1987) Modeling the turnover of organic matter in long term experiments at Rothamsted. Intercol Bull 15: 1–8
Johnson SM (1994) Environmental software survey. Air Waste 44: 79–111
Jury WA (1982) Simulation of solute transport using a transfer function model. Water Resour Res 18, 2: 363–368
Jury WA, Fluhler H (1992) Transport of chemicals through soil: mechanisms, models, and field applications Adv Agron 27: 141–201
Jury WA, Spencer WF, Farmer WJ (1983) Behavior assessment model for trace organics in soil. I. Model description. J Environ Qual 12: 558–564
Jury WA, Farmer WJ, Spencer WF (1984a) Behavior assessment model for trace organics in soil: II. Chemical classification and parameter sensitivity. J Environ Qual 13, 4: 567–572
Jury WA, Spencer WF, Farmer WJ (1984b) Behavior assessment model for trace organics in soil: III. Application of screening model. J Environ Qual 13, 4: 573–579
Jury WA, Elabd H, Clendening LD, Resketo M (1986a) Evaluation of pesticide transport screening models under field conditions. American Chemical Society, Washington DC, pp 385–395
Jury WA, Elabd H, Resketo M (1986b) Field study of napropamide movement through unsaturated soil. Water Res 22: 749–755
Jury WA, Sposito G, White RE (1986c) A transfer function model of solute transport through soil. 1. Fundamental concepts. Water Resour Res 22 (2): 243–247
Knitton RE, Wagenet RJ (1987) Simulation of solute transport using a continuous time Markov process. 1. Theory and steady state application. Water Resour Res 23, 10: 1911–1916
Lapidus L, Amundson NR (1952) Mathematics of adsorption in beds. IV. The effect of longitudinal diffusion in ion exchange chromatographic columns. J Phys Chem 56: 984–988
Lehmann RG, Miller JR, Laskowski DA (1990). Fate of fluroxypyr in soil: II. Desorption as a function of incubation time. Weed Res 30: 383–388
Leonard RA, Knisel WG, Still DA (1987) Gleams: Groundwater loading effects of agricultural management systems. Trans ASAE 30 (5): 1403–1418
Lindstrom FT, Boersma L (1971) A theory of mass transport of previously distributed chemicals in a water-saturated sorbing porous medium. Soil Sci 111 (3): 192–199
Loague K, Green RE (1991) Statistical and graphical methods for evaluating solute transport model: overview and application. J Environ Qual 22: 155–161

Luxmoore RJ, Ferrand LA (1993) Towards pore-scale analysis of preferential flow and chemical transport. In: Russo D, Dagan G (eds) Water flow and solute transport in soils. Springer, Berlin Heidelberg New York

Mackay D (1979) Finding fugacity feasible. Environ Sci Technol 3: 1218–1223

Mackay D, Diamond M (1989) Application of the QWASI (Quantitative water air sediment interaction) fugacity model to the dynamics of organic and inorganic chemicals in lakes. Chemosphere 18, 7–8: 1343–1365

Mackay D, Paterson S (1981) Calculating fugacity. Environ Sci Technol 15: 1006–1014

Mackay D, Joy M, Paterson S (1983) A quantitative water, air, sediment interaction (QWASI) fugacity model for describing the fate of chemicals in lakes. Chemosphere 12: 981–997

Martin MJP, Synge RLM (1941) A new form of chromatogram employing two liquid phases, I. A theory of chromatography. Biochem J 35: 1359–1370

Molina JAE, Clapp CE, Shaffer MJ, Chichester FW, Larson WE (1983) NCSOIL, a model of nitrogen and carbon transformation in soil: description, calibration and behavior. Soil Sci Soc Am J 47: 85–91

Mullins JA, Carsel RF, Scarborough JE, Ivery AM (1992) PRZM-2, a model for predicting the fate in the crop root and unsaturated soil zones. User manual for release 2.0, US-EPA, Athens, Georgia

Nielsen DR, Biggar JW (1962) Miscible displacement: III. Theoretical considerations. Soil Sci Soc Au Proc 26: 216–221

Nkedi-Kizza P, Biggar JW, van Genuchten Th., Wierenga PJ, Selim HM, Davidson JM, Nielsen DR (1983) Modeling tritium and chloride 36 transport through an aggregated oxisol. Water Resour Res 19, 3: 691–700

Nofziger DL, Hornsby AG (1986) A microcomputer-based management tool for chemical movement in soil. Appl Agric Res 1: 50–56

Parrish RS, Smith CN (1990) A method for testing whether model predictions fall within a prescribed factor of true values, with an application to pesticide leaching. Ecol Model 51: 59–72

Petach MC, Wagenet RJ, DeGloria SD (1991) Regional water flow and pesticide leaching using simulations with spatially-distributed data. Geoderma 48: 245–270

Pignatello JJ (1989) Sorption dynamics of organic compounds in soils and sediments. SSSA Spec Publ 22: 45–80

Piver WT, Lindstrom FT (1990) Mathematical models for describing transport in the unsaturated zone of soils. In: Hutzinger O (ed) The handbook of environmental chemistry, vol 5, part A. Springer, Berlin Heidelberg New York, pp 125–259

Rao PSC, Edwardson KSC, Ou LT, Jessup RE, Nkeddi-Kizza P, Hornby AG (1986) Spatial variability of pesticide sorption and degradation parameters. In: Gardner WY, Honeycutt RC, Nigg HN (eds), Evaluation of pesticides in ground water. ACS Symp Ser 315, American Chemical Society

Rubin J (1983) Transport of reacting solutes in porous media: relation between mathematical nature of problem formulation and chemical nature of reactions. Water Resour Res 9, 5: 1231–1252

Russo D (1993) Analysis of solute transport in partially saturated heterogeneous soils. In: Russo D, Dagan G (eds) Water flow and solute transport in soils. Springer Berlin Heidelberg New York

Schweich D, Sardin M (1981) Adsorption, partition, ion exchange and chemical reaction in batch reactors or in columns – a review. J Hydrol 50: 1–33

Scow KM, Hutson J (1992) Effect of diffusion and sorption on the kinetics of biodegradation: theoretical considerations. Soil Sci Soc Am J 56: 119–127

Smith CN, Parrish RS, Carsel RF (1987) Estimating sample requirements for field evaluations of pesticide leaching. Environ Toxicol Chem 6: 347–357

Soulas G (1995) Modeling of biodegradation of pesticides in the soil. In: Bitton G, Tarradellas J, Rossel D (eds) Soil ecotoxicology. CRC Press, Lewis publishers, New York (in press)

Soulas G, Lagacherie B (1990) Modelling of microbial processes that govern degradation of organic substrates in soil, with special reference to pesticides. Philos Trans R Soc Lond B 329: 369–373

Taylor AW, Spencer WF (1990) Volatilization and vapor transport processes. Soil Sci Soc Am 2: 213–269

Tevissen E (1993) Methodologie d'étude et modélisation du transport de solutés en milieux poreux naturels. Thése de Doctorat de l'Institut Nationale Polytechnique de Lorraine, Nancy

Tietje O, Tapkenhinrichs M (1993) Evaluation of pedo-transfer functions. Soil Sci Soc Am J 57: 1088–1095

van Genuchten M Th, Alves WJ (1982) Analytical solutions of the one-dimensional convective-dispersive solute transport equation. US Dept Agric Tech Bull 1661: 1–151

van Genuchten MT, Wierenga PJ (1976) Mass transfer studies in sorbing porous media I. Analytical solutions. SSSA 40, 4: 473–480

van Genuchten Th, Dalton FN (1986) Models for simulating salt movement in aggregated field soils. Geoderma 38: 165–183

van Veen JA, Paul EA (1981) Organic carbon dynamics in grassland soils. I. Background information and computer simulation. Can J Soil Sci 61: 185–201

Villeneuve JP, Lafrance P, Banton O, Frechette P, Robert C (1988) A sensitivity analysis of adsorption and degradation parameters in the modeling of pesticide transport in soils. J Contam Hydrol 3: 77–96

Villeneuve JP, Banton O, Lafrance P (1990) A probabilistic approach for the groundwater vulnerability to contamination by pesticides: the vulpest model. Ecol Model 51: 47–58

Villermaux J (1982) Genie de la réaction chimique. Conception et fonctionnement des réacteurs. Technique et Documentation Lavoisier Paris, 401 pp

Vinten AJA, Mingelgrim U, Yaron B (1983) The effect of suspended solids in wastewater on soil hydraulic conductivity. Soil Sci Soc Am J 47: 402–412

Wagenet FJ (1993) A review of pesticide leaching models and their application to field and laboratory data. In: Del Re AM, Capri E, Evans SP, Natali P, Trevisan M (eds) IX Symp Pesticide Chemistry, Piacenza, 11–13 October 1993

Wagenet RJ, Hutson JL (1986) Predicting the fate of nonvolatile pesticides in the unsaturated zone. J Environ Qual 15, 4: 315–322

Wagenet FJ, Rao PSC (1990) Modeling pesticide fate in soils. In: Cheng HH (ed) Pesticides in the soil. Environment: processes, impacts and modeling. SSSA Book Series 2, Madison, Wisconsin, pp 351–400

Walker A (1987) Evaluation of a simulation model for prediction of herbicide movement and persistence in soil. Weed Res 27: 143–152

Walker A, Hollis JM (1993) Prediction of pesticide mobility in soils and their potential to contaminate surface and groundwater. British Crop Protection Council Monogr 59, Brighton

Wania F, Mackay D, Paterson S, di Guardo A, Mackay N (1993) Compartmental models in environmental science. In: Del Re AM, Capri E, Evans SP, Natali P, Trevisan M (eds) IX Symp Pesticide Chemistry. Piacenza, 11–13 October 1993

White RE, Dyson JS, Haigh RA (1986) A transfer function model of solute transport through soil. 2. Illustrative applications. Water Resour Res 22, 2: 248–254

CHAPTER 10

Risks and Remedies

Soil remediation has to be considered when potentially toxic elements or substances present in soils may have adverse effects on biodiversity or human health. The choice of the most suitable treatment is based mainly on the assessment of the risk to the environment imposed by the presence of toxic substances in the soil.

The risk assessment includes diagnosis and prognosis of pollution. The diagnosis is the characterization of the polluted land at the present time (nature and state of pollutants; spatial extent of the pollution; hydrologic, pedologic, and geologic characteristics of the site; etc.). The prognosis is a prediction of the evolution of the pollution with time as a result of a change in some soil parameter(s) [pH, redox potential (Eh), etc.]. The risk assessment is based partly on the scientific data discussed in the preceding chapters, which described the physicochemical state, the location, the transformation, the mobility, and the bioavailability of pollutants; it also considers data characterizing the soil. The methodology developed for the risk assessment can be used to monitor the pollution after remediation.

The choice of treatment takes into account the risk assessment before, during, and after remediation, but must also consider its cost and the designated use of the land.

This chapter is dedicated to risk assessment, treatments, impacts of the treatments, pollution control, and cost of remediation.

10.1 Risk Assessment

The risk for the environment of potentially toxic elements or compounds present in soils is strongly related, on the one hand, to the state, mobility, and bioavailability of these substances and, on the other, to the geologic and pedologic characteristics and the physicochemical and biological properties of the soil. The risk assessment comprises two parts. The first, diagnosis, is the characterization of the pollution at the present time including a description of the soil properties. The second part, prognosis, is a description of changes which may occur in the course of time or in case of changes in the physical or biological parameters of the soil, e.g., pH or Eh.

10.1.1 Diagnosis

Contaminated land results from human activities. Among contaminants we may mention: metals, such as Cd, Pb, Cu; metalloids, such as As, Se; organics, such as pesticides; and mineral compounds, such as chlorides and nitrates. Pollution may be localized or dispersed. Localized pollution can result from an accident during transport or storage (oil, etc.), or from fixed activities (industrial plants, smelters, etc.), as a consequence of which, quite high concentrations of potentially toxic materials may be found on small areas. Dispersed pollution occurs when pollutants are spread at low concentration over large areas. In both cases, several determinations have to be made.

Identification of the pollutants and characterization of the site can be easy in the case of a transport accident, when the spilled product is known, or of a specific industrial activity (lead-battery plant, crystal factory, etc.) They could be more difficult if several different activities succeeded one another at the same site or if toxic pollution results from leaching from a waste disposal site or from crop protection practices. In all cases, a geologic, pedologic, hydrologic, and geographic study has to be conducted, to obtain information on the polluted site environment. The impacts of this environment on the mobility, bioavailability, and transformation of the pollutants and, hence, on their impact on the wider environment, are strongly dependent on these characteristics.

Spatial extent of the pollution has to be determined by sampling, which must be done by a method which takes into account the spatial heterogeneity of the surface and the pedologic and geologic horizons: samples have to be collected in homogeneous domains on the surface and in each horizon of the soil profile. We must avoid sampling in depth simply as a function of the distance from the surface, since this has no significance. Sampling should include water from the water table if there is one. The spatial spreading of the pollution with time is a function of the pollutant mobility.

Plants, lichens, microfauna, etc. can be used as bioindicators, giving indications of toxicity and biodiversity. The chemical analysis of plants may yield information about the bioavailability of toxics. More generally, the impacts on the food chain (water, vegetables, etc.) and on people living in the vicinity also have to be considered.

Physicochemical state and location of pollutants. The chemical analysis of collected samples may yield more information on the location of toxics, on the scale of particular domains of the surface, or of horizons in the soil profiles. It is also important to know the location of toxics on the soil-constituent scale (one element may be preferentially associated with one constituent: Cu with organic matter, etc.) but also on the molecular scale (elements can be complexed by organic matter or fixed by iron, aluminum, or manganese oxihydroxides, or adsorbed on clay surfaces, etc.) This may help us to understand the mechanisms involved in the retention of pollutants by soil constituents which were discussed in the previous chapters.

The mobility (how elements or substances move from one compartment of the soil to another), bioavailability (the capacity of elements or substances to be absorbed by living organisms), retention, and transformation of toxic materials are strongly related to their physicochemical state and their location in the soil solid phase. For example, metals naturally included in the structure of minerals have no adverse effect on the environment (biodiversity, plant toxicity, etc.), because the mobility and bioavailability of such trapped elements is very low. This generally applies to elements naturally present in soils (geochemical background). These potentially toxic elements, whose mobility and bioavailability are very low, present no risk to the environment. Thus, it is necessary to determine precisely not only the total amounts of toxic materials, but also their physicochemical states and their locations. The relationships between the physicochemical state of pollutants (e.g., speciation) and the locations of potentially toxic elements and substances in soils are largely described in the previous chapters.

10.1.2 Prognosis

This comprises the prediction of the evolution of the pollution as a function of time, and of the consequent physicochemical and biological changes. The prognosis has an impact on the way to manage the contaminated land. If the spatial spreading of the pollution as time goes on were expected to be negligible, it could be better to leave the site in its existing state; on the other hand, if the spatial extent of the pollution would increase rapidly, remediation has to be undertaken. Prognosis is an important part of the risk assessment, which, therefore, includes both diagnosis and prognosis.

As a consequence, the risk cannot be assessed only on the basis of the total amount of potentially toxic elements or substances in the soil. The state of the soil and contaminants at the present time, and also its evolution as a function of time and as a result of changes in the physicochemical or biological parameters of both pollutants and site, have to be considered. Risk assessment is a wider concept than the characterization of pollution; it includes several parameters whose sum cannot necessarily be expressed numerically. Thus, it is more complicated to introduce risk assessment into regulations.

10.2 Selection of Remedies

The objective of treatments is to reduce the risk related to the presence of toxics in soils to an environmentally acceptable level. The treatment does not necessarily mean that pollutants have to be removed. Treatments which have a positive effect on each factor considered in the risk assessment (mobility, bioavailability, transformation, etc.) also have to be considered. Thus, the choice of the suitable treatments will be based on the risk assessment itself and on the effects of the treatments on the parameters involved in the risk assessment. The following methods of selection of remedies should be considered, under the various indicated conditions.

10.2.1 There Is No Pollution Risk in the Land Under Consideration

If the risk to the environment is very low, with no impact of the potentially toxic elements or substances on the health of living organisms, including humans, through water or the food chain, nothing need be done, even if the content of these toxic materials is higher than the threshold values defined in some regulation. It is important to keep in mind that in some places the natural content of potentially toxic elements may be much larger than these threshold limits, with absolutely no risk to the environment. Such a situation exists in the eastern part of France, where the natural content of Ni in the various horizons (120–200 ppm) is larger than the threshold value (50 ppm) given in the regulation governing the spreading of sludges produced by the treatment of waste water (Buatier et al. 1994). In this region, the Ni is included in the structure of some mineral constituents and is not mobile or bioavailable. The weathering of these minerals as a function of the local physicochemical or biological conditions does not introduce amounts of Ni into the soil solution and from that into the food chain of living organisms, which could have a detectable effect on the environment. In fact, the very presence of these Ni-containing minerals in these soils is an indication that weathering is not efficient under the prevailing conditions.

In the same way, if materials spread on the soil as a result of industrial or agricultural activities contain potentially toxic elements or substances, but those materials present no risk to the environment, in the sense developed above involving diagnosis and prognosis, nothing need be done. It follows from this that in-situ remediation may be possible. In fact, if some in-situ treatment with specific amendments leads to the fixation of pollutants in such a way that they no longer form a risk to the environment, the remediation technique is successful even if it does not change the total amount of toxic materials in the soil.

10.2.2 There Is No Risk Around the Contaminated Site

In the case of localized pollution due to accidents, industrial activities, storage of wastes, etc., the risk assessment, including both diagnosis and prognosis, may lead to the conclusion that the natural or artificial fixing of pollutants is good enough to prevent any spatial spreading of the pollution in the atmosphere, or horizontally, or vertically in the soil. Again, no remediation is necessary, but, as a precaution, a monitoring device can be installed to be sure that nothing is changing. The concept of waste disposal implicitly includes such a practice.

Thus, remediation of locally contaminated land can include the confinement of the pollutants in such a way as to eliminate the risk to the environment around the contaminated site. Existing confinement techniques involve the use of ion-exchange materials, generally bentonite, which may sometimes present problems in relation to particular pollutants, which have to be confined. This applies particularly in the case of cations which can be exchanged for the compensating cations of bentonite, which is a very good cation exchanger. However, these technologies can be improved to provide the best barrier for the relevant pollutants.

10.2.3 There Is a Risk to the Environment

The diagnosis and prognosis of the pollution indicate that pollutants may have an adverse effect on the environment. In such a case, remediation has to be applied. Two cases could be considered.

10.2.3.1 Diffused Pollution

Pollutants are at very low concentrations in soils, having accumulated in the course of years because of soil amendments, fertilizers, or phytosanitary treatments, or from dust, vapors, etc., and be entrapped by some soil constituents. Such is the case with pesticide molecules, which accumulate in humic substances; but if some physicochemical or biological changes occur, these xenobiotics could be released, and they could then destroy crops. In agronomy, this phenomenon is called the persistence effect. Some such ions, for example NO_3^-, simply move to the water table. If this occurs in chalky soils, NO_3^- cannot be transformed because the size of the pores is smaller than that of the microorganisms involved in denitrification and, consequently, it will progressively contaminate the water table. This forms a typical environmental time bomb.

There are possibilities of overcoming such problems. For example, research could be focused on the development of pesticide molecules which cannot be trapped by humic substances. Alternatively, in the case of chalky soil contaminated by NO_3^-, farmers could manage the land in such a way as to have evapotranspiration close to the rainfall.

Techniques developed in agronomy can be used to decrease the mobility and bioavailability of the pollutants. For example, pH or Eh may be controlled by liming, drainage, etc., to maximize the fixation of heavy metals to soil constituents. The remediation techniques applicable in case of diffused pollution demand a very good knowledge of soil processes, which is why most of them come, or will come, from agronomic practices.

Liming As was shown in the preceding chapters, pH has a definitive impact on the physicochemical state of potentially toxic elements or organic substances and on their fate in the environment (transformation, degradation, etc.) Liming to raise the pH is the most efficient means of increasing cadmium retention in the soil, since a medium with high pH immobilizes much more cadmium than one with low pH. Between pH 4 and 8, the cadmium fixation capacity increases threefold for each pH unit (Impens and Avril 1994).

Water control The redox potential (Eh) depends on the availability of oxygen in soils, waters, and sediments, and upon biochemical reactions by which microorganisms extract oxygen for respiration (Salomons 1993). Redox conditions influence the mobility of metals in two different ways. Firstly, the valence of certain metals changes. For example, under reducing conditions, Fe^{3+} is transformed to Fe^{2+} and, similarly, the valence of manganese and arsenic is subject to direct changes. Since the reduced ions are more soluble, increased concentrations of these metals have been observed in reducing environments such as groundwaters and sediment solutions. Under reducing conditions, sulfate reduction will take place: for example, in sediments, lead sulfide with a low solubility is formed. On the other hand, an increase in the redox potential will cause lead sulfide to become unstable, with a subsequent rise in dissolved lead concentrations. This occurs when polluted sediments are dredged and dumped on land. In such oxygenated environments, lead sulfide is not stable and will be oxidized. More generally, storage of dredged contaminated soil on land causes an increase in the mobility of heavy metals in the surface layers of the dump site. These processes, which occur in sludges or sediments, may also occur during agricultural practices (drainage, irrigation, etc.)

10.2.3.2 Localized Pollution

Localized pollution is generally the result of accidents or of industrial or domestic activities, and the quantity of pollution could be high while its spatial extent is small. Generally, the contamination spreads from its source (waste dump, factory chimney, sewage outfall, etc.), and the concentration of pollutants decreases with distance from the source. The upper surface of industrial waste land is generally highly disturbed and heterogeneous, which makes sampling very difficult. If the contaminated site is, for example, an old factory, several activities may have occurred, generating different pollutants. Such contaminated sites may cover hundreds of hectares, including highly and weakly polluted zones.

In such contaminated areas, risk assessment includes some special measures, such as historic examination of the activities on the site, identification and location of pollutants within the site, etc. Special techniques are used to remediate such contaminated sites.

10.3 Remediation Technologies

Technologies developed to remediate localized polluted sites can be divided into two groups: hard and soft techniques.

10.3.1 Hard Techniques with Excavation (Surgical Operation)

If a small area is highly contaminated, so that the risk to the environment – especially to humans, through the food chain or the atmosphere – is very high, the soil should be cleaned. The simplest way is to remove contaminated materials and to put them into a waste disposal site, but this simply displaces the problem. In other respects, excavation may introduce other problems related to the reactivation of pollutants.

The excavated material may also be treated on the site, in a mobile plant, by means of equipment which can: remove the clay fraction in which pollutants are generally located, wash the soil material with water or specific solutions, heat the soil material to burn organic substances, etc.

Soil washing. The contaminated soil is mixed with washwater to which chemicals may be added to disperse clay and to chelate metals (Opatken 1993). The soil and washwater are then separated and the soil is rinsed with clean water and removed from the process as a product. Soil particles can easily be separated because of their high settling velocity. Coarse soil particles are generally separated with a trommel. The separated small particles carry higher levels of pollutants than the average of the original soil and, therefore, should be targeted for either further treatment or secure disposal. Water used in the soil-washing process is treated to remove pollutants and to enable it to be recycled for further use. Of course, contaminated soils can also be washed with specific solvents to remove organics. Depending on the residue content and the contaminants, 10–40 ton h^{-1} can be treated in such plants.

Thermal treatment. Volatilization and oxidation of organic pollutants depend on temperature, oxygen partial pressure, and residence time. Operating temperatures range from 400 to 700 °C. Kiln off-gasses are passed through an afterburner (a few seconds at 800–1200 °C), where complete oxidation of volatilized organics is achieved. Afterburner off-gases are cooled in a heat exchanger and passed through a particulate and sulfur dioxide removal system (Honders and Meeder 1993).

Thermal treatment is suitable for all types of soil with organic pollutants. However, due to off-gas emission, there are restrictions on the input concentration of halogenated organics and heavy metals. Depending on the type of soil (sand/clay) and moisture content, 15–40 ton h^{-1} can be treated. These techniques produce a large amount of residues, which have to be transported to a waste dump. The cleaned materials are then returned to the excavated place. Because of the treatment to which they had been subjected, these cleaned materials have properties different from those of the original soil.

Biological treatments may also be applied, particularly when soils are contaminated with organic compounds. The basic idea is to create the best conditions for microorganisms to develop in the polluted materials and to use the pollutants as part of their diet. The collection of contaminated materials has to be done in such a way as to provide the microorganisms with a good environment (porosity, water, food, etc.) in which to grow. This means that these materials have to be prepared, as in the case of a field before sowing, using agricultural practices. A complementary feeding system for microorganisms could be necessary. The control of temperature may have an important impact on the yield of the xenobiotic transformation process. If the ambient temperature is low, the contaminated materials can be put in a "greenhouse" to increase the temperature.

There are several techniques involving the role of microorganisms for the remediation of excavated soils contaminated with organic compounds (biotertre, land farming, etc.). In biotertre, contaminated soils are put in heaps, several meters high, on a geomembrane, which serves to collect the leaching solution. The heap is covered with a plastic sheet to retain the humidity. Plastic tubes are placed inside the heap to control the atmosphere. Nutrients can be introduced to obtain the best yield from the microorganisms which already exist in the soil or which can be introduced to deal with specific pollutants. After a few months, a biotertre 1.8 m high, containing 2300 m^3 of oil-contaminated soil, can transform a large amount of hydrocarbons, reducing their concentration from 2800 to 800 ppm.

Land farming is another way to treat excavated soil materials or organic wastes, especially those from the oil industry. These contaminated materials are spread, 20 cm thick, on a surface made or chosen to prevent the spread of contamination, both in depth and laterally. This 20-cm layer is then tilled in the same way as a field, using agricultural practices. The process for molecules whose transformation is difficult (polychlorophenol, HAP, etc.) is much slower than biotertre, but it does work. A 15-cm layer containing 1500 m^3 of a soil contaminated with oil, covering 1.5 ha, was examined after 100 days, and the hydrocarbon concentration was found to have fallen from 2800 to 38 ppm.

The difficult techniques reported here as examples require excavation. This limits their application to the worst sites. Indeed, if a contaminated site of 1 ha, polluted to a 1-m depth, is considered, the volume of contaminated materials is 10 000 m^3. The excavation, treatment, and storage of residues of such amounts of material make this form of remediation expensive, which is why other technologies have to be developed.

10.3.2 Soft Techniques Without Excavation (Chemotherapy)

The pollution of contaminated land is generally heterogeneous. There are highly polluted areas, and also less polluted areas which can be contaminated by the former.

Even if excavation is necessary (surgical operation), for the former, in situ treatments (chemotherapy) may be used for the others.

The main objectives of chemotherapy are to reduce the transfer of pollutants among the several parts of the environment – soil, air, water, plants, etc. – and to increase their transformation. The parameters involved are the same as those used for risk assessment. Those related to the content, i.e., to the pollutants, are the physicochemical state (speciation), location, mobility, bioavailability, and degradation. Those related to the container, i.e., to the soil, are pH, Eh, and agronomic conditions for microorganisms, plants, etc. to grow. The two sets of parameters interact.

In situ immobilization. As in the case of transformations, mobility is strongly related to the physicochemical state and the location of pollutants. If elements or organic compounds (pesticides) become trapped within the structure of minerals or humic substances, they are no longer mobile or bioavailable and, particularly in the case of organics, they are physically protected and not accessible to microorganisms which might be able to transform them. All in-situ treatments which lead to a long-term fixation of pollutants may have a positive effect. The characterization of fixation according to a time scale is questionable. Time scale is a fundamental parameter with respect to processes involved in contaminated soils. The order of magnitude of the "long term" involved in the retention of pollutants after remediation is probably close to the time scale involved in agronomic and pedologic processes. However, the fixation of heavy metals by microorganisms, plants, or soil organic matter occurs on a short time scale. Indeed, after the death of the microorganisms or the transformation of organic matter, toxic elements will again be mobile and bioavailable. On the other hand, if heavy metals are incorporated in mineral structures, a long-term fixation may be expected.

Thus, the techniques considered to modify the physicochemical state and the location of toxic materials in soil should take this parameter into account, and only those which give long-term fixation should be used.

Several soil conditioners or fertilizers may be used to fix toxics or to change the physicochemical properties of the soil (pH, Eh, organic matter, clay, etc.), to improve the fixation of pollutants by the solid phase of the soil or by chemicals. Changes in the physicochemical properties of the soil have already been discussed in Section 10.2.3.1, so here we will deal with the fixation of pollutants by soil conditioners or fertilizers. Data published by Mench et al. (1994) provide a good example of the possibilities and efficiency of these soft treatments.

The experiment of Mench et al. (1994) was conducted on a silty leached soil whose characteristics are given in Table 10.1. Slags used as fertilizers (TBS), beringite a modified aluminosilicate (Vangronsveld et al. 1990), steel shot (ST), and hydrous manganese oxide (HMO) were applied as immobilizing agents (Didier et al. 1992). Chemicals were added to soil on a 1% w/w basis for ST and HMO and 5% w/w basis for beringite. TBS was added until the soil pH was increased to 8.5.

Table 10.1. Main physicochemical characteristics and total metal content of the soil used in the experiments[a]. (After Mench et al. 1994)

Sand	($g\ kg^{-1}$)	236
Silt	($g\ kg^{-1}$)	555
Clay	($g\ kg^{-1}$)	209
Organic matter	($g\ kg^{-1}$)	31
Carbonates	($g\ kg^{-1}$)	8
C/N		11
pH (water)		7.8
CEC ($cmol_c\ kg^{-1}$ of soil DW)		22
Soil total metal content ($mg\ kg^{-1}$ of soil DW; aqua regia extract)		
Cd		18
Cu		43
Cr		29
Fe		19 272
Mn		439
Ni		17
Pb		1112
Zn		1434

[a]0–20-cm top layer.

The mobility of metals was determined by extraction with distilled water, 0.1 N $Ca(NO_3)_2$ solution and 0.05 M EDTA solution. Bioavailability was determined by growing dwarf beans, ryegrass, and tobacco by means of the pot technique.

Results presented in Table 10.2 show that EDTA is the most efficient reagent to extract Cd, Pb, or Zn. In untreated soil, the amounts of Cd, Zn, and Pb extracted by $Ca(NO_3)_2$ represented 6.6, 1, and 0.1% of the respective total contents. Application of ST, HMO, and beringite to the soil reduced the extractibility of Cd and Zn by water and by $Ca(NO_3)_2$. To a lesser degree, TBS had a similar effect. Ryegrass and tobacco absorbed much less Zn than beans did in treated pots.

Table 10.3 shows changes in metal concentration in the aerial parts of plants as a function of the treatment. HMO, ST, and beringite significantly reduced the Cd level in the aerial parts of each species.

The decrease in mobility and plant uptake of Pb and Cd after soil treatment with soil conditioners or fertilizers is related to the affinity of these elements to HMO (Manceau et al. 1992). In the case of steel shot, iron oxide is formed on the surface, where lepidocrocite and maghemite were identified (Didier et al. 1992). Cd is probably trapped in these iron compounds.

Among the tested additives, steel shot, oxihydroxides of manganese, and beringite exhibited the greatest potential for reducing Cd plant uptake (Mench et al. 1994).

Other in-situ techniques are being developed. To increase the efficacy of these treatments, it could be necessary, at the same time, to lime or drain the soil to control pH and Eh, in order to provide the best conditions for fixing toxics.

Table 10.2. Extractable fractions of the metals (mg kg^{-1} of soil DW), and linear correlation coefficients between the metals extracted from either untreated or treated soils and the metal uptake by ryegrass and tobacco.[a] (After Mench et al. 1994)

Soil treatment	Water			$Ca(NO_3)_2$			EDTA			Soil pH
	Cd	Pb	Zn	Cd	Pb	Zn	Cd	Pb	Zn	
Untreated	0.15	0.15	1.8	1.2	0.8	14.7	11.4	599	267	7.8
TBS	0.06	0.16	0.7	1.0	0.5	7.4	11.0	576	212	8.5
HMO (%)	0.03	0.94	0.7	0.0	0.1	2.2	10.6	500	268	7.8
ST (1%)	0.02	0.08	0.4	0.4	0.4	3.4	7.6	536	185	7.8
Beringite	–	–	–	0.3	0.2	1.7	–	–	–	7.3
Correlation coefficient (r^{-2})										
Ryegrass	0.64[a]	0.50	0.58	0.94[b]	0.96[b]	0.64[a]	0.09	0.82[a]	0.29	
Tobacco	0.79[a]	0.40	0.49	0.95[b]	0.36	0.14	0.07	0.60	0.95[b]	

[a]$p < 0.05$; [b]$p < 0.01$

Table 10.3. Metal contents (μg^{-1} DW) in the aerial parts of plants grown on untreated and treated soils. (After Mench et al. 1994)

Soil treatment	Ryegrass (shoots)			Tobacco (shoots)			Bean (primary leaves)			
	Cd	Pb	Zn	Cd	Pb	Zn	Cd	Pb	Zn	Mn
Untreated	4.5	14.9	182	40.5	36.6	120	1.8	5.4	29.2	18.9
TBS	4.0	8.9	177	30.2	16.5	86	–	–	–	–
HMO (%)	1.3	3.2	117	14.8	54.9	108	0.4	4.1	35.0	83.6
ST (1%)	2.2	6.7	75	24.5	50.9	90	0.8	4.0	36.5	14.1
Beringite	–	–	–	–	–	–	1.1	6.9	39.5	21.2

Transfer and transport limitation. One way to decrease mobility is to act at the molecular level to change the physicochemical state of pollutants. Other techniques act at a macroscopic level to control the spatial spreading of the pollution, i.e., contamination; the following example, reported by Clijsters and Vangronsveld (1995), gives an idea of what can be done.

The zinc smelter involved, in Belgium, was closed about 20 years ago after decades of activity. The high content of heavy metals (Cd, Zn, etc.) had a bad effect on vegetation: only resistant plants grew, others disappeared or were in a bad shape, leaving an almost bare surface. As a consequence, erosion became an important factor in the spatial spreading of the pollution, through the transport of small contaminated particles by the wind. At the same time, the poor vegetation in the area concerned caused reduced evapotranspiration, which increased the speed of advance of the pollutants to the water table. One of the objectives of the remediation in this case was to create conditions for plants to grow, in order to decrease the spatial spreading of the contamination, both on the surface and in depth. The experimental data reported by Clijsters and Vangronsveld (1995) show how the vegetation was effective.

As a result of industrial activities, about 150 ha had become a bare sandy area without any vegetation, and metal contamination of the site soil was heterogeneous. Therefore, a rectangular 300 × 100-m plot was chosen in the most polluted area (1500–16 000 ppm zinc; 700–1800 ppm lead; 480–1650 ppm copper; 10–70 ppm cadmium). Soil phytotoxicity was principally due to the highly assimilable zinc.

Compost (100 tons ha^{-1}) and 120 tons ha^{-1} of a modified aluminosilicate (beringite) were mixed with the upper 35-cm soil layer. After 1 year, a drastic decrease in Zn and Cd contents in the grass was observed (Table 10.4) and the 3-ha plot was completely covered with vegetation. After 4 years, the metal–immobilizing capacity of the treated soil persists and the vegetation has further developed and diversified. The landscape was highly improved and further dispersion of the historical contamination by wind erosion and water percolation were drastically reduced. Thus, soil immobilization of heavy metals, combined with revegetation, represents a valuable sanitation technique for industrial sites which are strongly contaminated with heavy metals.

Leaching and venting. The preceding example illustrates the possibility of reducing the mobility of pollutants and, at the same time, limiting their spatial spreading. The mobility of some other pollutants may be used to remove them from the soil. Such pollutants include, for example, water-soluble compounds and substances with a high vapor pressure.

Sometimes, in a semiarid region, irrigation is applied without drainage, which very often leads to salinization of the soil. The resulting amount of NaCl can be so high that the fields become infertile. To recover the fertility of the land, NaCl has to be removed, which can be done by irrigation combined with drainage. Field experiments have been performed which show that it is possible to remove (it is a function of the irrigation level) NaCl from the soil rapidly and to restore soil fertility. To avoid the loss of the swelling clay fraction, it would be better to irrigate with water enriched in Ca.

Although this form of remediation is not always easy, it is a practicable way to remove some pollutants.

The well-known technique of venting is generally applied in the case of substances which have a high vapor tension under field conditions (oil, chloride

Table 10.4. Metal content (mg kg^{-1} DW) of aerial organs of 15-month-old grasses. (After Clijsters and Vangronsveld 1995)

	Zinc	Cadmium
Uncontaminated control	150	<0.5
Untreated contaminated soil	1138	279
Treated soil samples*1	267	0.49
4	73	0.11
12	142	0.36

* replications

compounds, etc). Pumping of the soil atmosphere increases the rate of vaporization and facilitates the collection of toxic vapors.

Phytoremediation. Microorganisms and plants take up not only those elements they need for growth but also others. Agronomists have shown that the amounts of elements living organisms are able to take up are smaller than the total amounts present in the soil and larger than the amounts which are exchangeable by $CaCl_2$; N, for example. In fact, living organisms produce specific molecules which can make some fixed elements available, which makes the assessment of the amount of element actually bioavailable very difficult. Nevertheless, it has been demonstrated that one way to control bioavailability is to act on the physicochemical state and on the location of elements or organic compounds. The field experiment using a modified aluminosilicate reported by Clijsters and Vangronsveld (1995) illustrates how reduction of the availability of heavy metals for plants allows revegetation and decreases the spatial spreading of pollution.

Contrariwise, the availability of toxics for some plants may be used to remove them from the soil. Some plants are able to take up a large amounts of toxic elements without any visible toxicity effect; therefore, the possibility is being considered of using these plants to remediate soils contaminated with toxic elements (phytoremediation). This technique appears to be an elegant one, but the yield still remains poor, even with these specific plants, and the harvested plant material has to be treated.

In situ transformation. As described in Chapter 7, transformation involves both physicochemical and biological processes. Xenobiotics (pesticides) may be transformed by physicochemical processes catalyzed by clay constituents, or may simply migrate to the atmosphere by evaporation. Evaporation is also important for the light fractions of oil. In the same way, soil parameters (pH, Eh) may have an impact on the physicochemical state of elements, particularly on their valency and on the mechanisms involved in their fixation. For organics, transformation means degradation and, finally, the disappearance of the toxic materials. For elements, transformation means change in their physicochemical state.

With respect to remediation, the impact of the physicochemical state of elements has been discussed already; therefore, we shall focus now on the degradation of organics. Some aspects of bioremediation were discussed in the preceding paragraph. Two conditions should be satisfied: the accessibility of organics to microorganisms and the growth of the microorganisms. In agricultural soils, the second condition is generally satisfied but the first may not be, because some pesticides may be trapped in humic substances. In such a case, one solution is to use pesticides whose molecules have no affinity for humic substances. In industrial waste land highly contaminated with organics, one way to remediate is to create the best conditions for living organisms to grow, which can be done through the biotertre and land farming techniques described above.

10.3.3 Control and Cost

There are at least two features which make the soft techniques of remediation (chemotherapy) interesting: the cost of rehabilitation and the quality of the soil after treatement.

Soft remediation chemotherapy techniques may leave pollutants in the soil so, as a precaution, it is necessary periodically to make a risk assessment to determine if the risk related to potentially toxic elements is decreasing, increasing, or remaining constant. This is a supplementary requirement in relation to the hard techniques, where the surgical operation has removed the source of contaminants.

Little information is available in the literature about the cost of remediation. Honders and Meeder (1993) reported the data presented in Table 10.5. Performance and cost are given for two different treatments, including residue

Table 10.5. Performance and cost data. (After Honders and Meeder 1993)

Entry		1990–1991	1992
Thermal treatment			
Market share		70%	53%
Treatment efficiency	PAH	98%	99%
	Min. oil	92%	96%
	CN	92%	99%
Achieved A-value[a]		28%	51%
Treatment costs	D Fl/ton		95–170
	US $/ton		55–95
Extraction-classification			
Market share		30%	47%
Treatment efficiency	PAH	87%	91%
	Min. oil	78%	95%
	CN	88%	98%
	Cd	83%	96%
	Zn	78%	89%
	Pb	74%	81%
	Cu	n.a.	90%
	Ni	n.a.	94%
	Cr	n.a.	92%
Achieved A-value[a]		11%	45%
Treatment costs[b]	D Fl/ton		45–230
	US $/ton		80–130
Total			
Achieved A-value[a]		23%	48%
Treatment costs[b]	D Fl/ton	180	150
	US $/ton	100	83

[a] A-value not always requested.
[b] Including residue disposal. This adds DF125–30/ton input (ca. 15 US $/ton) to the bare treatment costs.

disposal. Coming back to the preceding example of 1 ha of land contaminated to 1 m depth, the remediation cost, according to these data, would be $10\,000 \times 1.5 \times 100 = 1\,500\,000$ US dollars.

Of course, remediation can be carried out at a lower cost by means of soft techniques. This could be acceptable if, when account is taken of the planned use of the land (parking, storage area, etc.), there is no remaining risk to the environment. The full significance of the concept of risk assessment is indicated here, in comparison with that of toxic material contents, as applied in regulations by definition of threshold values. Because of the tremendous impact of pollution on the price of industrial waste sites, which are generally located in downtown city areas, it is commercially interesting to develop soft techniques of remediation, which are much less expensive than the hard ones discussed above, but which can be as safe.

References

Buatier C, Gomez A, Feix I, Wiart J, Ducaroir J, Rambert P (1994) Caracterisation et analyse de la mobilité et de la biodisponibilité du nickel dans les sols agricoles. Le cas du Pays de Gex (01). Rapport INRA, ADEME Chambre d'Agriculture de l'Ain, Versailles, 180 pp

Clijsters H, Vangronsveld J (1995) Soil remediation by in situ immobilization of metals and metalloids. In: Du Bouiguier HC, Hofman M (eds) Proc Symp "L'assainissement des sols pollués". Ecotop 95, Bruxelles, pp 55–59

Didier V, Gomez A, Mench M, Mason P, Lambrot C (1992) Contrôle de la mobilité et de la biodisponibilité des metaux de sols contaminés: efficacité de différents composés du fer et d'autres composés mineraux. Rapport final Convention de Recherche. 90231 INRA, Ministère de l'Environnement

Honders A, Meeder TA (1993) Ex-situ soil decontamination in the Netherlands. In: Neretnieks I (ed) Proc Workshop on Contaminated Soils - Risks and Remedies, Stockholm, October 26–28, 1993. Royal Institute of Technology, Thomas Agren, Kemakta Konsult AB, Stockholm, pp 83–94

Impens R, Avril C (1994) A strategy for soil remediation. In: Avril C, Impens R (eds) Proc symp on soil remediation, Brussels, May 27, 1994. Academie Royale des Sciences, des Lettres et des Beaux Arts de Belgique, pp 1–10

Manceau A, Gorshkov AI, Drists VA (1992) Structural chemistry of Mn, Fe, Co and Ni in Mn hydrous oxides. II. Information from EXAFS spectroscopy, electron and X-ray diffraction. Ann Mineral 77: 1144–1157

Mench M, Vangronsveld J, Didier V, Clijsters H (1994) Evaluation of metal mobility, plant availability and immobilization by chemical agents in a limed-silty soil. Environ Pollut 86: 279–286

Opatken E (1993) Ex-situ remediation technologies as practiced in USA. In: Neretnieks I (ed) Proc Workshop on Contaminated Soils - Risks and Remedies, Stockholm, October 26–28, 1993. Royal Institute of Technology, Thomas Agren, Kemakta Konsult AB, Stockholm, pp 103–127

Salomons W (1993) Time delayed responses of chemicals in soils. In: Neretnieks I (ed) Proc Workshop on Contaminated Soils - Risks and Remedies, Stockholm, October 26–28, 1993. Royal Institute of Technology, Thomas Agren, Kemakta Konsult AB, Stockholm, pp 1–17

Vangronsveld J, van Assche F, Clijsters N (1990) Immobilization of heavy metals in polluted soils by application of a modified alumino-silicate: biological evaluation. In: Barcels J (ed) Environmental contamination. Proc 4th Int Conf Barcelona, CEP Consultants, Edinburgh, pp 283–288

Subject Index

Springer-Verlag and the Environment

We at Springer-Verlag firmly believe that an international science publisher has a special obligation to the environment, and our corporate policies consistently reflect this conviction.

We also expect our business partners – paper mills, printers, packaging manufacturers, etc. – to commit themselves to using environmentally friendly materials and production processes.

The paper in this book is made from low- or no-chlorine pulp and is acid free, in conformance with international standards for paper permanency.